GENERATIVE ARTIFICIAL INTELLIGENCE FOR NEUROIMAGING

GENERATIVE ARTIFICIAL INTELLIGENCE FOR NEUROIMAGING

Methods and Applications

Edited by

DEEPIKA KOUNDAL

University of Eastern Finland, A.I. Virtanen Institute for Molecular Sciences, Kuopio, Finland

UPES, Dehradun, Uttarakhand, India

D. JUDE HEMANTH

Department of ECE, Karunya University, Coimbatore, Tamil Nadu, India

ELSEVIER

ACADEMIC PRESS

An imprint of Elsevier

Academic Press is an imprint of Elsevier
125 London Wall, London EC2Y 5AS, United Kingdom
50 Hampshire Street, 5th Floor, Cambridge, MA 02139, United States

ISBN: 978-0-443-45010-5

For information on all Academic Press publications visit our website at https://www.elsevier.com/books-and-journals

Publisher: Mara Conner
Acquisitions Editor: Saniya Puri
Editorial Project Manager: Pratishtha Gupta
Production Project Manager: Fizza Fathima
Cover Designer: Mark Rogers

Typeset by TNQ Tech

Contents

PART III Applications in neuroscience

Contributors

Amira S. Ashour
Tanta University Faculty of Engineering, Electronics and Electrical Communications Engineering Department, Tanta, Egypt

Abraham Olufemi Asuku
Bioresources Development Centre, National Biotechnology Research and Development Agency, Department of Medical Biotechnology, Ogbomoso, Nigeria

Dheerja Babbar
Sarvodaya Hospital, Department of Physiotherapy, Faridabad, Haryana, India

Prateek Bhadauria
MITS-DU Gwalior, Department of Electronics Engineering, Gwalior, Madhya Pradesh, India

Kanika Bhalla
Washington University School of Medicine, St. Louis, MO, United States

Shiveta Bhat
Department of Electronics and Communication Engineering, Model Institute of Engineering and Technology, Jammu, Jammu and Kashmir, India

Rosepreet Kaur Bhogal
Lovely Professional University, Phagwara, Punjab, India

Ankur Chaudhary
Department of Information Technology, Noida Institute of Engineering & Technology, Greater Noida, Uttar Pradesh, India

Yanhui Guo
University of Illinois, Department of Computer Science, Springfield, IL, United States

P. Hemalatha
Rajalakshmi Engineering College, Department of Computer Science and Business Systems, Chennai, Tamil Nadu, India

Akshay Juneja
COER University, Computer Science and Engineering, College of Smart Computing, Roorkee, Uttarakhand, India

Ravi Teja Kothuru
University of San Diego, Applied Artificial Intelligence, San Diego, CA, United States

Deepika Koundal
University of Eastern Finland, A.I. Virtanen Institute for Molecular Sciences, Kuopio, Finland; UPES, Dehradun, Uttarakhand, India

Gurpreet Kour
Department of Electronics and Communication Engineering, Model Institute of Engineering and Technology, Jammu, Jammu and Kashmir, India

Ashwani Kumar
School of Computer Science Engineering and Technology, Bennett University, Greater Noida, Uttar Pradesh, India

Manish Kumar
School of Computer Science & Engineering, IILM University, Gurugram, Haryana, India

Sanjeev Kumar
AI Cluster, School of Computer Science, UPES, Dehradun, Uttarakhand, India

Vijay Kumar
Dr. B.R. Ambedkar National Institute of Technology, Department of Information Technology, Jalandhar, Punjab, India

Priyanka Kumari
CGC University, Department of Electronics and Communication Engineering, School of Engineering and Technology (SET), Mohali, Punjab, India

L. Leo Prasanth
Vel Tech Rangarajan Dr. Sagunthala R&D Institute of Science and Technology, Department of Computer Science and Engineering, Chennai, Tamil Nadu, India

Edson Bernardo da Conceicao Mahangue
Department of Computer Science & Engineering (CSE), Sharda School of Computing Science & Engineering (SSCSE), Sharda University, Greater Noida, Uttar Pradesh, India

Arun Malik
Department of Computer Science and Engineering, Lovely Professional University, Phagwara, Punjab, India; School of Computer Science and Engineering, Lovely Professional University, Phagwara, Punjab, India

Rachel Nallathamby
Vel Tech Rangarajan Dr. Sagunthala R&D Institute of Science and Technology, Department of Computer Science and Engineering, Chennai, Tamil Nadu, India

I. Nandhini
Vel Tech Rangarajan Dr. Sagunthala R&D Institute of Science and Technology, Department of Computer Science and Engineering, Chennai, Tamil Nadu, India

Peace Ayo Olaoluwa
University of Ilorin, Department of Physiology, Ilorin, Nigeria

Raghavendra S. Patil
KLE Institute of Technology, Hubli, Karnataka, India

Godwyn Paulson Pitchaimani
Escuela Superior de Ingeniería y Arquitectura (ESIA), Unidad Ticomán, Instituto Politécnico Nacional (IPN), Gustavo A. Madero, Mexico

Laxmi Pujari
KLE Institute of Technology, Hubli, Karnataka, India

Asha Rani
Lovely Professional University, Phagwara, Punjab, India

Ritesh Rastogi
Department of Information Technology, Noida Institute of Engineering & Technology, Greater Noida, Uttar Pradesh, India

Preeti Saini
Manav Rachna International Institute of Research and Studies, School of Allied Health Sciences, Department of Physiotherapy, Faridabad, Haryana, India

Priyanka Sethi
Manav Rachna International Institute of Research and Studies, School of Allied Health Sciences, Department of Physiotherapy, Faridabad, Haryana, India; Sarvodaya Hospital, Department of Physiotherapy, Faridabad, Haryana, India

Ajay Sharma
School of Computer Science and Engineering, Lovely Professional University, Phagwara, Punjab, India; Department of Computer Science and Engineering, Lovely Professional University, Phagwara, Punjab, India

Avinash Kumar Sharma
Department of Computer Science & Engineering (CSE), Sharda School of Computing Science & Engineering (SSCSE), Sharda University, Greater Noida, Uttar Pradesh, India

Shamneesh Sharma
School of Computer Science and Engineering, Geeta University, Panipat, Haryana, India

Ajmer Singh
Lovely Professional University, Phagwara, Punjab, India

R. Sivaramakrisknan
Vel Tech Rangarajan Dr. Sagunthala R&D Institute of Science and Technology, Department of Computer Science and Engineering, Chennai, Tamil Nadu, India

Amit Kumar Tyagi
National Forensic Sciences University, School of Law Forensic Justice Policy Studies, Gandhinagar, Gujarat, India

PART I

Foundations

CHAPTER 1

Introduction to neuroscience imaging modalities: fMRI, EEG, MEG, and their challenges

Laxmi Pujari and Raghavendra S. Patil
KLE Institute of Technology, Hubli, Karnataka, India

1. Introduction

The human brain, comprising approximately 86 billion neurons interconnected by an estimated 100 trillion synapses, represents the most complex biological structure in the known universe. Understanding how this vast network generates perception, cognition, emotion, and behavior—and how its function is disrupted in neurological and psychiatric disease—constitutes one of the defining scientific challenges of the 21st century. Progress toward this understanding has accelerated dramatically over the past 4 decades, driven in large part by the development and refinement of non-invasive neuroimaging technologies that allow researchers and clinicians to observe brain activity in living participants without surgical intervention.

Among the array of neuroimaging tools available today, three modalities have achieved the greatest clinical and research adoption: functional Magnetic Resonance Imaging (fMRI), Electroencephalography (EEG), and Magnetoencephalography (MEG). Each is grounded in a distinct biophysical principle. fMRI capitalizes on the hemodynamic response to neural activity, exploiting the differential magnetic properties of oxygenated and deoxygenated hemoglobin to map cerebral blood flow changes that serve as proxies for underlying neural firing. EEG records the summed postsynaptic electrical potentials of large populations of cortical neurons using scalp electrodes, capturing brain electrical dynamics with millisecond precision. MEG detects the magnetic fields generated by intracellular ionic currents that accompany neural firing, offering a complementary electrophysiological window on cortical activity with somewhat better source localization than EEG.

The clinical relevance of these modalities spans a broad spectrum of conditions. fMRI is routinely used in pre-surgical planning to map eloquent cortex, in psychiatric research to characterize network-level dysfunction in depression and schizophrenia, and in cognitive neuroscience to delineate the neural substrates of language, memory, and executive function. EEG remains the gold standard for epilepsy diagnosis and monitoring, polysomnography, and neonatal neurological assessment, and is the dominant

Generative Artificial Intelligence for Neuroimaging
ISBN 978-0-443-45010-5
https://doi.org/10.1016/B978-0-443-45010-5.00008-3

technology in brain–computer interface research. MEG is employed in specialized tertiary centers for pre-surgical epilepsy evaluation and functional mapping, particularly where EEG localization is insufficient.

Despite decades of development and widespread clinical adoption, each modality continues to face significant technical limitations. These limitations have motivated two complementary research strategies: engineering of next-generation hardware platforms that push the boundaries of resolution and portability, and development of multimodal acquisition and analysis frameworks that combine the complementary strengths of multiple modalities within a single study. The rapid maturation of machine learning—and particularly deep learning—as analytical tools has further accelerated both strategies by enabling extraction of clinically meaningful patterns from high-dimensional neuroimaging datasets.

This chapter is organized into 10 principal sections covering the historical and scientific background of the field, the methodology of this review, six extended clinical case studies, a comparative technical analysis, a results and discussion synthesis, a survey of advances from 2015 to 2025 with all supporting figures, a catalog of major open neuroimaging datasets, prospective future directions, and a comprehensive conclusion.

2. Background and literature review

Deciphering the mechanisms of the human brain has historically represented one of the greatest intellectual challenges in science and medicine. Prior to the advent of modern neuroimaging, understanding of brain function relied primarily on three categories of evidence: behavioral observations of individuals with localized brain lesions, invasive electrophysiological recordings in animal models, and post-mortem neuropathological analysis. Each approach offered valuable but partial insights, constrained by the impossibility of directly observing dynamic neural activity in an intact, functioning human brain.

The earliest widely adopted neuroimaging technique was Electroencephalography, pioneered by the German psychiatrist Hans Berger, who in 1929 published the first recordings of human brain electrical activity from the scalp (Berger, 1929). Berger identified the characteristic alpha and beta rhythms that remain clinically significant today. EEG spread rapidly through European neurology and psychiatry, establishing itself as the primary tool for diagnosing epilepsy and studying sleep disorders. Its millisecond temporal sensitivity to neural electrical dynamics gave it an enduring advantage for studying rapid cognitive processes such as sensory perception, attention, and language comprehension.

Magnetic Resonance Imaging, introduced in clinical practice during the 1980s, initially provided only structural information about brain anatomy. The pivotal extension to functional imaging was achieved in the early 1990s through the discovery of the Blood Oxygen Level Dependent (BOLD) contrast mechanism by Ogawa and colleagues (Ogawa et al., 1990). Their demonstration that deoxyhaemoglobin behaves as an endogenous contrast agent—detectable by MRI because of its paramagnetic properties—opened the possibility of mapping brain activation non-invasively and with high spatial resolution. fMRI grew with extraordinary speed throughout the 1990s and 2000s, generating thousands of studies illuminating the neural correlates of virtually every cognitive and affective domain.

Magnetoencephalography emerged as a research modality during the 1970s following the development of superconducting quantum interference devices (SQUIDs), which demonstrated that the extremely weak magnetic fields generated by human brain activity—on the order of femtotesla—could be detected by cryogenic sensors (Hari & Salmelin, 2012). MEG measures primarily tangentially oriented cortical sources, which generate magnetic fields that penetrate the skull and scalp with minimal distortion, unlike the electrical fields recorded by EEG. This characteristic confers a source localization advantage that has made MEG especially valuable in the presurgical evaluation of drug-resistant epilepsy and in functional mapping of language and sensorimotor areas (Baillet, 2017).

A dominant theme in the contemporary neuroimaging literature is the recognition that no single modality provides a complete picture of brain function, and that strategic integration of multiple techniques yields insights inaccessible to any one technology alone. Huster and colleagues (Huster et al., 2012) provided a comprehensive methodological review of simultaneous EEG-fMRI, demonstrating that the millisecond temporal resolution of EEG and the millimeter spatial resolution of fMRI are genuinely complementary. Michel and Murray (2012) similarly argued for the broader adoption of high-density EEG as a functional brain imaging tool, proposing that, when combined with appropriate source localization methods, EEG can provide spatiotemporal brain maps that complement fMRI findings.

The past decade has witnessed a further acceleration of progress driven by three intersecting developments: advances in hardware, standardization of data formats and processing pipelines, and the application of machine learning to neuroimaging data. The Brain Imaging Data Structure (BIDS) standard, introduced by Gorgolewski and colleagues (Gorgolewski et al., 2016), established a community-wide framework for organizing, documenting, and sharing neuroimaging datasets, substantially improving reproducibility. The development of optically pumped magnetometer-based MEG systems by Boto and colleagues (Boto et al., 2018) demonstrated that wearable, room-temperature MEG is feasible, with sensitivity comparable to SQUID-based systems, opening transformative possibilities for mobile brain imaging during natural movement.

3. Methodology

This chapter employs a systematic narrative review methodology grounded in critical synthesis of the peer-reviewed neuroimaging literature. The analytical framework examines fMRI, EEG, and MEG through four consistent dimensions: biophysical operating principles; data acquisition hardware and signal processing pipelines; established and emerging clinical applications; and documented technical limitations and proposed mitigations. This structure enables rigorous cross-modality comparison while preserving the technical specificity necessary for a graduate-level reference.

3.1 Literature search strategy

Source material was retrieved systematically from PubMed, IEEE Xplore, ScienceDirect, and Google Scholar using Boolean combinations of modality names (functional MRI, EEG, MEG, OPM-MEG) with methodological and clinical topic terms (BOLD signal, source localization, artifact correction, epilepsy, pre-surgical mapping, brain-computer interface, deep learning, transformer, BIDS, fMRIPrep). The search was restricted to publications from 1990 to 2025, with priority given to systematic reviews, meta-analyses, landmark empirical studies, and methodological papers published in high-impact journals. Authoritative textbooks and published book chapters were consulted for foundational technical material (Niedermeyer & da Silva, 2005).

3.2 Inclusion and exclusion criteria

Studies were included if they addressed technical specifications, physiological mechanisms, data quality, or clinical outcomes of fMRI, EEG, or MEG; were published in English or French in peer-reviewed venues; and presented original data, systematic review findings, or authoritative technical guidance. Studies were excluded if they were conference abstracts without associated peer-reviewed publication, if they reported only non-human animal data without direct translational relevance, or if they could not be verified as methodologically sound.

3.3 Comparative analysis framework

Comparative analysis was structured around five standardized performance dimensions: spatial resolution, temporal resolution, acquisition cost, system portability, and demonstrated clinical utility. These dimensions reflect the practical decision criteria used by clinicians and researchers when selecting a neuroimaging modality for a given application. A structured comparison table was developed to enable direct visual inspection of these metrics across the three modalities, supplemented by extended prose discussion of nuances not capturable in tabular format.

3.4 Multimodal integration review

The review extended beyond individual modality assessment to examine multimodal imaging frameworks that combine two or more technologies within a single study. Published evidence on concurrent EEG-fMRI and MEG-fMRI acquisition was evaluated to characterize the practical and analytical challenges of merging complementary data streams. The survey of recent advances specifically examined the period 2015–2025 to capture the most clinically and scientifically consequential developments, including ultra-high-field MRI, OPM-MEG, BIDS standardization, fMRIPrep, ICA-AROMA denoising, and transformer-based neural decoding.

4. Case studies and clinical applications

The practical value of neuroimaging technologies is most effectively illustrated through documented clinical cases in which fMRI, EEG, and MEG have been applied to resolve diagnostic questions, guide therapeutic interventions, or monitor treatment response. The following six extended case studies demonstrate the complementary clinical roles of the three modalities and the incremental diagnostic value achievable through multimodal integration.

4.1 Case study 1: Pre-surgical mapping in drug-resistant temporal lobe epilepsy

A 17-year-old patient was referred to a tertiary epilepsy surgery program following failure of four antiepileptic drug regimens. Video-EEG telemetry recorded over 72 h identified interictal epileptiform discharges and ictal onset predominantly at left anterior temporal electrodes. MEG was performed using a 306-channel SQUID-based whole-head system. Dipole source modeling identified a cluster of equivalent current dipoles localized to the left anterior mesial temporal cortex with three-dimensional spatial precision superior to that achievable from scalp EEG alone. The MEG cluster was co-registered with the patient's structural MRI to generate a magnetic source imaging (MSI) map, confirming that the irritative zone was anterior and medial to Wernicke's area. Task-based fMRI using a validated language paradigm identified left-lateralized activation in Broca's and Wernicke's areas, delineating the functional boundary between the proposed resection margin and critical language cortex. On the basis of integrated EEG, MEG, and fMRI data, a stereotactic anterior temporal lobectomy was performed sparing the posterior language area. The patient achieved Engel Class I seizure freedom at 24-month follow-up with no postoperative language deficit.

4.2 Case study 2: Early detection of Alzheimer's disease using resting-state fMRI

A 65-year-old female academic with a first-degree family history of Alzheimer's disease presented with subjective cognitive decline over 18 months. Neuropsychological testing revealed performance 1.5 standard deviations below age-adjusted norms on delayed verbal recall. Structural MRI showed no significant hippocampal atrophy. Resting-state fMRI was performed to evaluate functional connectivity within the Default Mode Network (DMN). Independent component analysis identified reduced functional connectivity between the posterior cingulate cortex and precuneus—a pattern documented as an early and sensitive biomarker of Alzheimer's-related neurodegeneration even before structural atrophy. Following amyloid PET positivity, the patient was diagnosed with prodromal Alzheimer's disease and enrolled in a multimodal prevention program. Follow-up resting-state fMRI at 12 months demonstrated stabilization of DMN connectivity metrics, and neuropsychological testing showed no further decline.

4.3 Case study 3: EEG-guided neurorehabilitation following mild traumatic brain injury

A 29-year-old professional footballer sustained a mild traumatic brain injury (mTBI) during competitive play. Structural CT and MRI performed within 24 h were unremarkable. The player reported persistent post-concussive symptoms including concentration difficulty, emotional lability, and sleep disturbance. Quantitative EEG performed at 4 weeks post-injury identified significant increases in frontal theta band power and reductions in alpha power relative to age-matched normative data—a pattern associated with post-concussive cognitive dysfunction. Event-related potential analysis revealed prolonged P300 latency during an auditory oddball paradigm, indicating slowed information processing speed. These findings guided a structured 8-week cognitive rehabilitation program. Serial qEEG recordings at 2-week intervals documented progressive normalization of the theta: alpha ratio and P300 latency, paralleling subjective symptom resolution. Medical clearance for graduated return to training was provided at 10 weeks, supported by objective electrophysiological evidence of neural recovery.

4.4 Case study 4: Multimodal EEG–fMRI in treatment-resistant depression

A 34-year-old male with 7-year treatment-resistant major depressive disorder underwent simultaneous EEG-fMRI during an emotional face-processing paradigm. fMRI analysis revealed marked hypoactivation of the left dorsolateral prefrontal cortex (DLPFC) and hyperactivation of the amygdala to negative facial expressions, consistent with impaired top-down prefrontal modulation of limbic affective responses.

Concurrent EEG identified frontal alpha asymmetry—greater right frontal alpha power relative to left—a validated neurophysiological marker of depressive disorders. EEG-informed fMRI analysis revealed inverse coupling between frontal alpha power and DLPFC activity. On the basis of this multimodal characterization, 20 sessions of high-frequency repetitive transcranial magnetic stimulation (rTMS) targeting the left DLPFC were administered. Post-treatment fMRI demonstrated normalized DLPFC activation and attenuated amygdala reactivity, accompanied by significant improvement in standardized depression rating scores.

4.5 Case study 5: MEG and fMRI language mapping for low-grade glioma resection

A 45-year-old right-handed male was referred for neurosurgical evaluation following MRI detection of a T2-hyperintense, non-enhancing lesion of 3.2 cm diameter in the left posterior superior temporal gyrus, adjacent to Wernicke's language area. Biopsy confirmed a WHO Grade 2 diffuse astrocytoma. Task-based fMRI using three validated language paradigms (silent word generation, auditory sentence comprehension, object naming) identified robust left-lateralized activation in Wernicke's area, Broca's area, and the supplementary motor area. MEG using beamformer source analysis distinguished metabolically active language-functional cortex within and adjacent to the tumor from electrically silent tumor core. The integrated fMRI/MEG functional map was used as a neuro navigation overlay during intraoperative cortical stimulation mapping, enabling gross total resection with clear margins from critical language areas. Postoperative assessment confirmed intact speech and language function.

4.6 Case study 6: fMRI monitoring of neural plasticity following ischemic stroke

A 58-year-old right-handed male suffered an acute ischemic stroke affecting the left middle cerebral artery territory, producing a left posterior parietal and superior temporal infarct with moderate Broca-type aphasia and right hemispatial neglect. Serial task-based fMRI was performed at 1 month, 3 months, and 6 months post-stroke during language paradigms. At 1 month, fMRI demonstrated near-absent activation in the infarcted left posterior temporal cortex with emerging compensatory right-hemisphere engagement. Following a targeted melodic intonation therapy program specifically designed to recruit right hemisphere language areas, fMRI at 3 months showed significantly increased activation in right posterior temporal and inferior frontal regions. At 6 months, right hemisphere language network activation was robust, correlating with the patient's substantially recovered verbal communication. This longitudinal case demonstrates fMRI's unique capacity to visualize neural reorganization after brain injury and to guide personalized rehabilitation strategies.

4.7 Technical overview of neuroimaging modalities

4.7.1 Functional Magnetic Resonance Imaging (fMRI)

4.7.1.1 Biophysical principles and BOLD signal generation

fMRI relies on the neurovascular coupling mechanism; whereby local neural activity elicits a rapid and regionally specific increase in cerebral blood flow (CBF) that exceeds the concurrent increase in cerebral metabolic rate of oxygen consumption ($CMRO_2$). This hemodynamic overshoot transiently increases the local ratio of oxyhemoglobin to deoxyhaemoglobin. Since deoxyhaemoglobin is paramagnetic and oxyhemoglobin is diamagnetic, this shift alters the local magnetic field homogeneity in a manner detectable by gradient-echo MRI sequences sensitive to T2 contrast. The resulting BOLD signal constitutes an indirect, hemodynamically smoothed readout of underlying neural activity, with a characteristic temporal profile—the hemodynamic response function (HRF)—those peaks approximately 5–6 s after stimulus onset and returns to baseline over 15–20 s.

Standard task-based fMRI employs block or event-related designs in which participants perform cognitive or perceptual tasks while undergoing MRI scanning. Statistical parametric mapping software packages model the expected BOLD response as a convolution of the HRF with the task design matrix, estimating activation maps using general linear models and performing group-level random effects inference. Resting-state fMRI, acquired during passive wakefulness without a task, characterizes spontaneous low-frequency BOLD fluctuations (<0.1 Hz) reflecting intrinsic functional connectivity within brain networks including the Default Mode, Salience, and Executive Control networks.

4.7.1.2 Advantages and limitations of fMRI

Advantages: High spatial resolution (1–3 mm); whole-brain coverage; non-invasive and safe for repeated use; effective in functional localization for pre-surgical planning and research; resting-state variants require no participant task performance.

Limitations: Low temporal resolution (seconds) due to hemodynamic response lag; measures indirect neural activity proxy; sensitive to motion artifacts and B0 field inhomogeneity; requires expensive scanner infrastructure and specialist expertise; contraindicated with metallic implants.

4.7.2 Electroencephalography (EEG)

4.7.2.1 Biophysical principles and signal origin

EEG measures scalp potential differences arising from synchronous postsynaptic currents in populations of cortical pyramidal neurons oriented perpendicular to the cortical surface. The recorded signal reflects primarily the summation of excitatory and inhibitory postsynaptic potentials across thousands to millions of neurons rather than action potentials. Because the skull and scalp act as volume conductors with spatially distributed and non-uniform conductivity, the spatial resolution of EEG is limited by signal smearing, typically to 1–3 cm at best with high-density arrays and advanced source localization algorithms.

Standard clinical EEG uses the international 10–20 electrode placement system, providing 19–21 electrodes arranged according to proportional skull measurements from anatomical landmarks. Research-grade high-density EEG employs 64–256 electrodes enabling improved spatial sampling and source analysis. EEG amplifiers record potential differences with bandpass filtering from 0.1 Hz to 70–100 Hz and sampling rates of 500–2000 Hz for clinical use.

Advantages: Excellent temporal resolution (milliseconds); portable and relatively inexpensive; widely used in clinical diagnosis; suitable for ambulatory and long-term monitoring; no contraindications.

Limitations: Poor spatial resolution due to volume conduction; difficult to localize deep brain sources; susceptible to electrical noise and movement artifacts; reference electrode choice affects signal interpretation.

4.7.3 Magnetoencephalography (MEG)

4.7.3.1 Biophysical principles and SQUID technology

MEG detects the extracranial magnetic fields generated by intracellular ionic currents—primarily the same postsynaptic currents that produce the EEG signal—but with a critical geometric selectivity: MEG is most sensitive to tangentially oriented cortical sources (located in sulci), while radially oriented sources generate closed magnetic field loops that partially cancel at the sensor level. The magnetic fields produced by brain activity are on the order of 10–100 fT, requiring cryogenically cooled superconducting quantum interference devices (SQUIDs), typically operated at liquid helium temperature (4.2 K), to achieve the necessary measurement sensitivity. Whole-head MEG systems typically contain 102–306 sensor channels arranged in gradiometer configurations.

Advantages: High temporal resolution comparable to EEG; better source localization than EEG for tangential cortical sources; non-invasive and operationally silent; sensitive to cortical oscillatory dynamics.

Limitations: Extremely high cost (£2–4 M); requires magnetically shielded room; sensitive to magnetic interference; less effective for deep subcortical sources; limited clinical availability.

5. Comparative analysis of imaging modalities

A structured comparison of fMRI, EEG, and MEG across key performance dimensions is presented in Table 1.1 and illustrated in Fig. 1.1. No single modality achieves simultaneously high spatial and temporal resolution; rather, the three technologies occupy

Table 1.1 Structured comparison of fMRI, EEG, and MEG across nine key performance and clinical dimensions.

Feature	fMRI	EEG	MEG	Clinical significance
Spatial resolution	High: 1–3 mm	Low: ~1–3 cm	Moderate–High: ~5–10 mm	fMRI best for anatomical localization
Temporal resolution	Low: 1–6 s (HRF)	Excellent: 1–2 ms	Excellent: 1–2 ms	EEG/MEG critical for dynamic tracking
Signal origin	Hemodynamic BOLD	Electrical potentials	Magnetic fields	Different physiological bases
Acquisition cost	High: £1–3 M	Low: £5–50 K	Very high: £2–4 M	MEG most resource-intensive modality
Portability	Fixed (MRI suite)	Highly portable	Fixed (shielded room)	EEG only bedside/ambulatory option
Depth sensitivity	Excellent (all regions)	Reduced (deep sources)	Limited (deep sources)	fMRI best for subcortical imaging
Primary clinical role	Pre-surgical mapping, connectivity	Epilepsy, sleep, BCI	Pre-surgical epilepsy, language	Complementary clinical niches
Noise sensitivity	Motion, B0 field inhomogeneity	Electrical artifacts, muscle	Magnetic interference	Each requires specific mitigation strategy
Regulatory status	CE/FDA cleared (clinical)	CE/FDA cleared (clinical)	CE/FDA cleared (clinical/ research)	All approved for appropriate clinical use

complementary positions in the resolution space that motivate multimodal integration strategies.

Fig. 1.1 offers the highest temporal resolution at limited spatial precision; fMRI (lower right) provides superior spatial detail with constrained temporal bandwidth; MEG occupies an intermediate position combining millisecond temporal fidelity with improved source localization. Multimodal combinations exploit these complementary profiles to achieve both high spatial and temporal resolution in a single study.

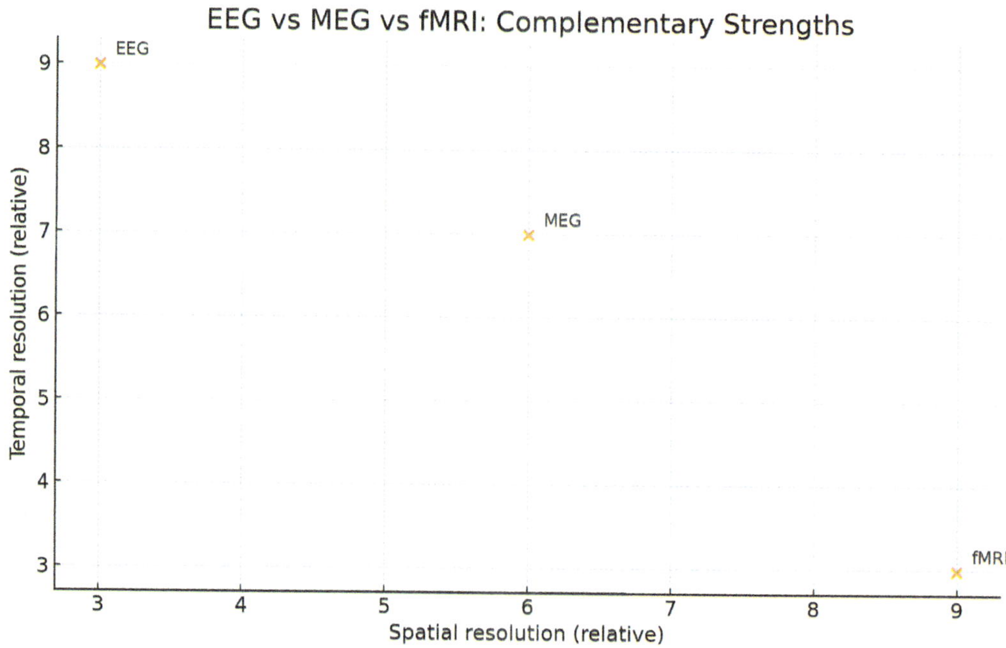

Figure 1.1 Complementary spatial and temporal resolution profiles of EEG, MEG, and fMRI.

6. Results and discussion

Systematic comparative analysis of fMRI, EEG, and MEG revealed a consistent and clinically meaningful pattern of complementary strengths and trade-offs across all performance dimensions, corroborating the prevailing view in the neuroimaging literature that these technologies are best regarded as a suite of tools to be strategically combined rather than competing alternatives.

fMRI demonstrated unequivocal superiority in spatial resolution among the three modalities, with modern 3T systems achieving isotropic voxel sizes of 2 mm routinely, and ultra-high-field 7T systems enabling sub-millimeter resolution that resolves cortical layers and columns. This spatial precision renders fMRI irreplaceable for applications where anatomical localization is paramount, including pre-surgical mapping of eloquent motor, sensory, and language cortex; characterization of spatially specific resting-state network disruptions in neurodegenerative disease; and detection of focal hemodynamic biomarkers of tumor infiltration or post-stroke reorganization.

EEG and MEG, in contrast, both operate at millisecond temporal scales that directly reflect the temporal structure of neural electrical activity without the intervening hemodynamic filter. This temporal fidelity is clinically critical in epileptology, where the morphology, propagation, and temporal sequence of epileptiform discharges carry diagnostic information that hemodynamic surrogates cannot capture, and in cognitive neuroscience, where event-related potentials and oscillatory dynamics encode information about the timing and sequence of neural computations.

The critical distinction between EEG and MEG in clinical and research practice is spatial resolution. EEG source localization is substantially hampered by volume conduction through the skull and scalp, typically achieving localization accuracy of 1–3 cm even with high-density arrays and sophisticated inverse methods. MEG, because the skull is magnetically transparent, achieves localization accuracy of 5–10 mm for superficial cortical sources, making it the preferred electrophysiological modality when spatial precision is required alongside temporal fidelity.

Multimodal integration emerged as the most powerful investigative strategy across all case studies reviewed. Simultaneous EEG-fMRI enables study designs in which EEG-identified neural events are used as parametric regressors in fMRI analysis, unlocking analyses that neither modality permits independently. The integration of machine learning with multimodal neuroimaging data has opened new frontiers in automated diagnosis, with convolutional neural networks applied to fMRI connectivity matrices demonstrating diagnostic accuracy for Alzheimer's disease, schizophrenia, and autism spectrum disorder approaching that of expert clinicians. Transformer-based architectures applied to raw EEG time series have achieved state-of-the-art performance on seizure detection, sleep staging, and BCI control tasks (Barachant et al., 2012; Debener et al., 2012; Haufe et al., 2014; Kostas et al., 2021; Lawhern et al., 2018; Li et al., 2023).

In applied clinical contexts, EEG retained its dominant role in epilepsy monitoring and polysomnography, driven by low cost, widespread availability, and portability. fMRI found primary application in pre-surgical functional mapping and psychiatric research protocols, while MEG was reserved for specialized indications requiring precise localization of epileptic foci or eloquent cortex. These findings reinforce the principle that modality selection should be deliberately matched to the diagnostic question, patient context, and available resources.

7. Advances in neuroimaging (2015–2025)

The decade from 2015 to 2025 witnessed an extraordinary acceleration of progress across all three neuroimaging modalities, driven by parallel advances in hardware engineering, computational methods, data standardization, and open science infrastructure.

7.1 Standards, pipelines, and open data

The Brain Imaging Data Structure (BIDS) standard, introduced in 2016 (Gorgolewski et al., 2016), established a community-wide, file-system-level convention for organizing neuroimaging data that enables automated validation, interoperability across processing pipelines, and archival sharing following FAIR principles (Fig. 1.2). Extensions of BIDS

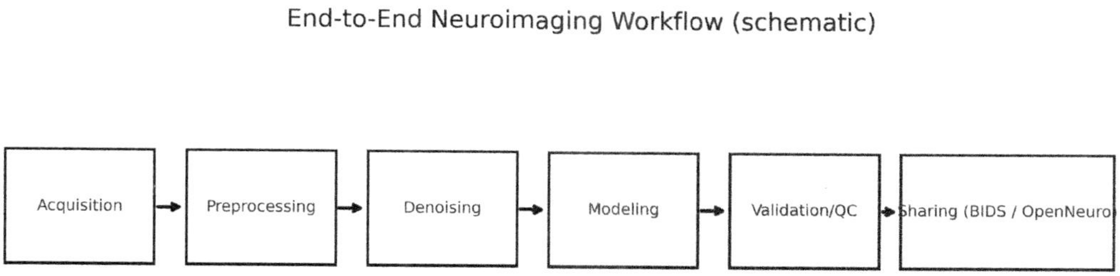

Figure 1.2 End-to-end neuroimaging workflow from data acquisition to FAIR-compliant open sharing.

to EEG (EEG-BIDS (Pernet et al., 2019)) and MEG (MEG-BIDS (Niso et al., 2018)) subsequently unified the data organization standard across all three major modalities. The fMRIPrep preprocessing pipeline (Esteban et al., 2019) consolidated best-practice processing steps into a modular, containerized workflow generating comprehensive quality reports, dramatically reducing between-site analytical variability. The OpenNeuro platform (Markiewicz et al., 2021) has grown to host thousands of openly accessible datasets, serving as both a data sharing repository and a benchmark platform for analytical methods development.

7.2 Wearable OPM-MEG: A paradigm shift in MEG

The development of practical optically pumped magnetometers (OPMs) operating at room temperature has precipitated a paradigm shift in MEG technology. Unlike SQUID sensors requiring immersion in liquid helium at 4.2 K and housing within a rigid cryogenic dewar separated from the scalp by 2–3 cm of insulating vacuum, OPM sensors exploit the quantum properties of alkali metal vapor to achieve femtotesla magnetic sensitivity at room temperature. This enables the sensors to be mounted directly on the scalp in individually conformable headgear (Fig. 1.3), reducing source-to-sensor distance from approximately 3 cm in conventional SQUID-MEG to effectively 0 mm, substantially increasing signal amplitude and spatial resolution. Boto and colleagues' landmark 2018 demonstration (Boto et al., 2018) of wearable OPM-MEG capable of recording high-quality neural signals during walking and ball-catching provided proof-of-concept that MEG could be performed outside the stationary paradigm that had constrained the modality since its inception.

7.3 Artifact mitigation in concurrent EEG-fMRI

The five major artifact categories (Fig. 1.4) are: (1) Gradient artifacts caused by rapidly changing MRI magnetic fields inducing voltages in EEG leads; (2) Ballistocardiogram (BCG) artifacts from cardiac-synchronized electrode displacement; (3) Cable loop induction from EEG leads forming conductive loops in the MRI bore; (4) Physiological

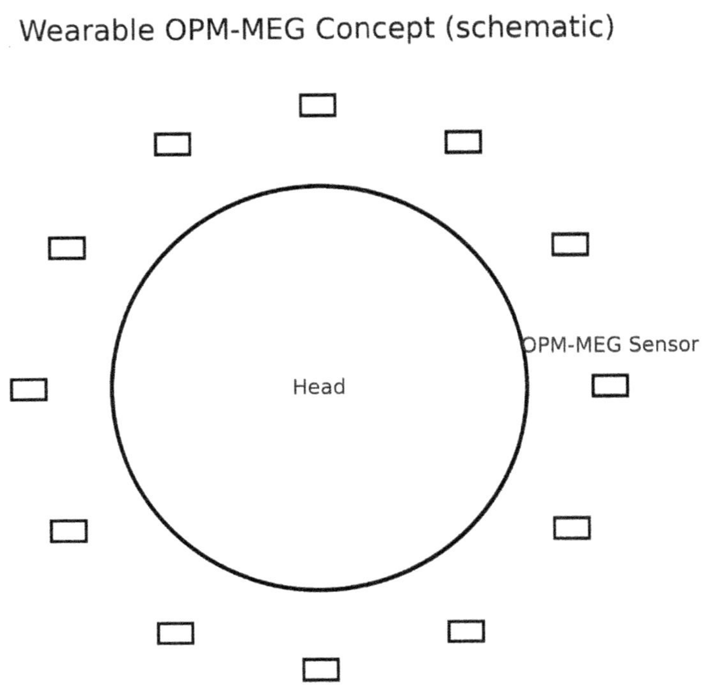

Figure 1.3 Schematic of a wearable Optically Pumped Magnetometer-MEG (OPM-MEG) system.

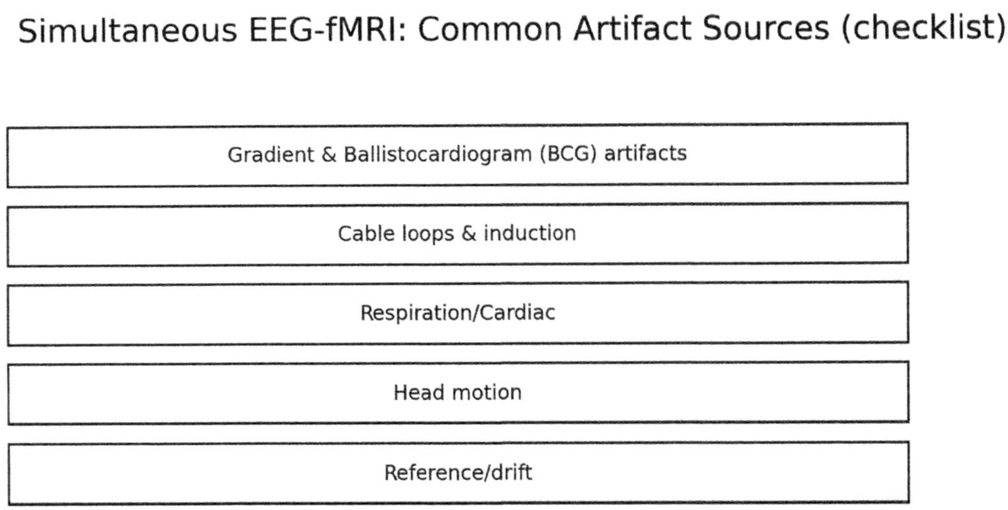

Figure 1.4 Checklist of principal artifact sources encountered during simultaneous EEG-fMRI acquisitions.

noise from respiratory and cardiac cycles; (5) Reference electrode drift from impedance fluctuations or heating. Each requires specific correction strategies including average artifact subtraction, adaptive filtering, and physiological signal regression.

Motion and physiological artifacts remain the central technical challenge in neuroimaging data quality. ICA-AROMA (Pruim et al., 2015) introduced a principled, classifier-assisted approach to identifying and removing motion-related independent components from fMRI data. Multi-echo fMRI enables principled separation of genuine neural BOLD signal from physiological noise using TE-dependence models. For simultaneous EEG-fMRI, the gradient artifact—caused by rapidly switching MRI field gradients inducing voltages in EEG electrode leads orders of magnitude larger than the neural EEG signal—and the ballistocardiogram (BCG) artifact require specialized correction. Average artifact subtraction combined with adaptive filtering has become standard for gradient artifact correction, while BCG correction typically employs independent component analysis (Bullock et al., 2021).

7.4 EEG electrode standardization: The 10–20 system

The 10–20 electrode placement system, first standardized by Jasper in 1958 and formally adopted by the International Federation of Clinical Neurophysiology, provides the universal spatial framework within which EEG recordings are performed and compared globally (Fig. 1.5). The system derives its name from the principle that electrode positions are defined by measurements of 10% and 20% of the total skull distance between paired anatomical landmarks. This proportional scheme ensures that electrode positions scale appropriately with head size, maintaining consistent anatomical coverage across individuals from infants to adults. Extended high-density montages based on the 10–10 and 10–5 variants provide 64–256 electrode positions, enabling the spatial sampling density required for reliable EEG source analysis, beamforming, and brain-computer interface applications.

7.5 Machine learning and AI-driven neuroimaging

The application of machine learning to neuroimaging has grown from isolated proof-of-concept demonstrations to a central methodological stream producing clinically actionable tools. For fMRI, graph convolutional networks and transformer architectures applied to functional connectivity matrices have demonstrated diagnostic performance for Alzheimer's disease, schizophrenia, and autism spectrum disorder approaching expert clinical classification in independent validation cohorts. For EEG, architectures specifically adapted to the temporal and spectral structure of EEG signals—including EEGNet (Lawhern et al., 2018), ShallowConvNet, and transformer-based models—have achieved reliable automated performance on seizure detection, sleep staging, and BCI control tasks (Barachant et al., 2012; Debener et al., 2012; Haufe et al., 2014; Kostas

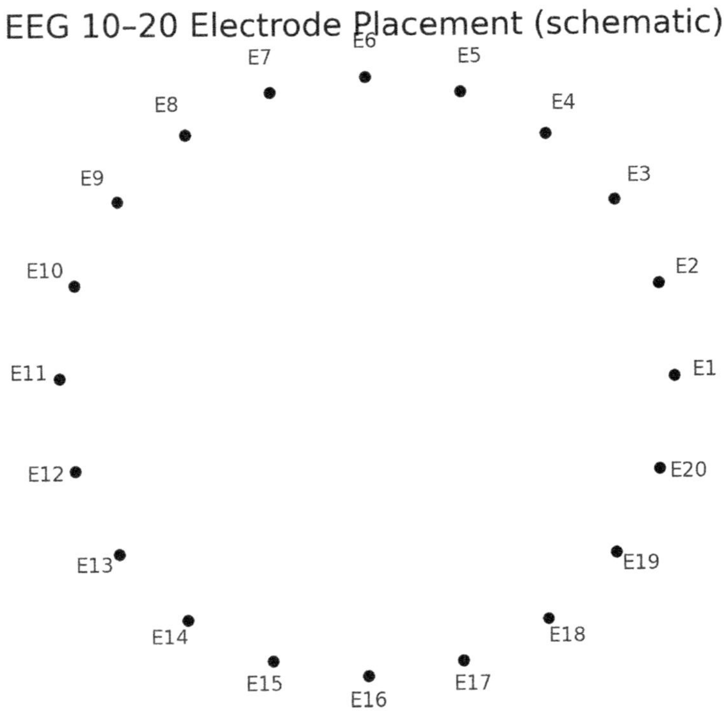

Figure 1.5 The international 10–20 EEG electrode placement system.

et al., 2021; Lawhern et al., 2018; Li et al., 2023; Roy et al., 2019; Schirrmeister et al., 2017). Transformer-based models leveraging self-attention mechanisms have achieved state-of-the-art results on multiple EEG benchmark tasks and are increasingly applied to cross-subject and cross-session transfer learning, a critical requirement for clinical deployment.

8. Clinical neuroimaging datasets (2015–2025)

The establishment of large-scale, openly accessible neuroimaging datasets has been transformative for the field, enabling pooled analyses with statistical power unavailable to individual research groups, facilitating benchmarking of analytical methods, and accelerating the development of machine learning models trained on clinically representative data. Table 1.2 catalogs the most significant datasets relevant to fMRI, EEG, and MEG research.

Fig. 1.6 explains about the data acquisition from patients and research participants via initiatives such as ADNI, HCP, and UK Biobank; (2) Preprocessing—BIDS-compliant pipeline-based cleaning, normalization, and artifact rejection; (3) AI/ML Analysis—convolutional neural networks for spatial imaging data and recurrent networks or

Table 1.2 Major publicly available clinical neuroimaging datasets (2015–2025) relevant to fMRI, EEG, and MEG research.

Dataset	Modalities	N (subjects)	Clinical focus	Access	Key reference
ADNI (Alzheimer's disease neuroimaging initiative)	MRI, fMRI, PET, CSF	>1500	AD progression, MCI conversion	Public (LONI portal)	Petersen et al. (2010)
HCP (human connectome project)	MRI, dMRI, fMRI, MEG, EEG	1200+	Healthy connectomics	Public (ConnectomeDB)	Van Essen et al. (2013)
UK biobank imaging	MRI, fMRI, dMRI	100,000+	Population health, dementia, stroke	Restricted (application)	Miller et al. (2016)
OpenNeuro/EEG-BIDS	EEG, fMRI, MEG	Variable (500+ datasets)	Epilepsy, sleep, psychiatric	Open (CC0 license)	Markiewicz et al. (2021)
OPM-MEG epilepsy pilot	MEG (OPM sensors)	<100	Epilepsy pre-surgical mapping	Limited research access	Boto et al. (2018)
ABIDE (autism brain imaging)	fMRI, sMRI	1112	Autism spectrum disorder	Public (NITRC)	Di Martino et al. (2014)
ADHD-200	fMRI, sMRI	973	ADHD diagnosis	Public (NITRC)	The ADHD-200 Consortium (2012)
TUH EEG corpus	EEG (clinical)	16,000+	Epilepsy, seizure detection	Public (Temple Univ.)	Obeid and Picone (2016)

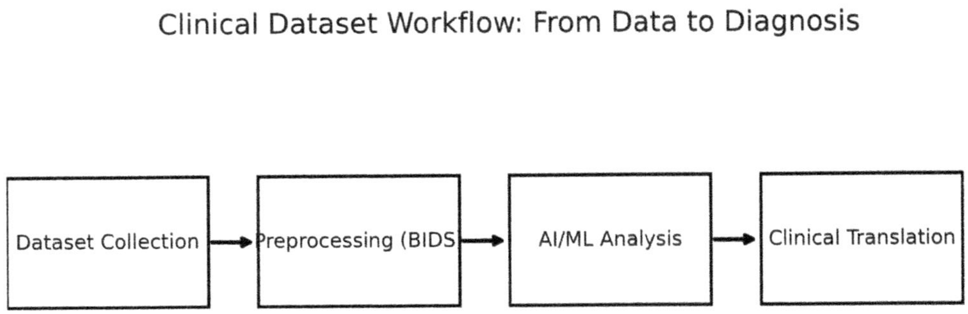

Figure 1.6 Four-stage clinical neuroimaging dataset workflow, from dataset collection through to clinical translation.

transformers for time-series EEG, enabling disease detection, seizure prediction, and biomarker discovery; (4) Clinical Translation—conversion of data-driven insights into diagnostic tools, surgical planning resources, and continuous monitoring systems.

The Alzheimer's Disease Neuroimaging Initiative (ADNI) has been particularly impactful as a longitudinal biomarker study providing multimodal data from participants spanning normal cognition, mild cognitive impairment, and Alzheimer's disease. ADNI data have enabled development and validation of imaging biomarkers for disease staging and progression prediction, training of machine learning diagnostic models, and characterization of the temporal sequence of biomarker changes in the AD continuum. The Human Connectome Project established the definitive normative reference for structural and functional brain connectivity in healthy young adults, employing ultra-high-resolution multiband fMRI at 3 and 7T alongside diffusion MRI tractography and resting-state MEG and EEG, producing multi-modal parcellation atlases of the human cortex that serve as references for clinical network analysis.

The UK Biobank imaging sub-study, with its planned 100,000-participant MRI and fMRI dataset linked to comprehensive health, genetic, and lifestyle records, is creating unprecedented statistical power to identify imaging biomarkers of brain aging, dementia risk, stroke, and mental health conditions at the population level. The OpenNeuro platform (Markiewicz et al., 2021), tightly integrated with BIDS, hosts thousands of openly accessible datasets encompassing fMRI, EEG, MEG, and PET modalities, serving as both a data sharing repository and a benchmark platform for analytical methods development.

9. Future directions

The next decade of neuroimaging research is likely to be defined by the convergence of four transformative trends: continued hardware innovation extending resolution and portability; the maturation of AI-driven analysis from research tools to validated clinical decision support systems; the scaling of multimodal acquisition and data integration frameworks; and the development of robust ethical and governance infrastructure commensurate with the unprecedented sensitivity of detailed brain data.

9.1 Next-generation hardware platforms

Ultra-high-field MRI beyond 7T—including emerging 14T systems under development—promises sub-millimeter fMRI resolution capable of resolving individual cortical columns and the laminar organization of cortical microcircuits. For MEG, the consolidation of OPM sensor technology into reliable, commercially viable wearable systems will extend magnetoencephalography from specialized research centers to general neurology and psychiatry departments. The development of triaxial OPM arrays with on-scalp active magnetic field control will further improve sensitivity and enable recordings in unshielded environments, dramatically expanding clinical accessibility. For EEG, advances in dry electrode technology and miniaturized wireless amplifiers are producing systems capable of high-density recording without conductive gel, suitable for long-term ambulatory monitoring and home-based neurofeedback.

9.2 Artificial intelligence and precision neuroimaging

The integration of deep learning with large-scale neuroimaging datasets is creating the foundations for precision neuroimaging: the ability to generate individualized predictions of disease trajectory, treatment response, and cognitive reserve from a single or longitudinal neuroimaging session. Foundation models for neuroscience—large-scale transformer architectures pre-trained on aggregated EEG and fMRI datasets—are beginning to demonstrate cross-task and cross-subject generalizability that earlier task-specific models could not achieve. Federated learning frameworks enabling training across distributed clinical sites without centralizing sensitive patient data will accelerate the development of clinically deployable AI neuroimaging tools while addressing data governance and privacy concerns.

9.3 Multimodal integration and personalized medicine

Simultaneous acquisition of multiple neuroimaging modalities—EEG-fMRI, MEG-fMRI, and potentially trimodal systems—will become increasingly feasible as hardware and analytical tools mature. The development of principled statistical frameworks for fusing heterogeneous data streams with different spatial and temporal properties, noise characteristics, and units of measurement is an active and important methodological frontier. In clinical practice, the realization of personalized neuroimaging—in which each patient's unique functional network organization and neurophysiological biomarker profile guides individualized diagnostic and therapeutic decisions—represents the ultimate translational aspiration of 3 decades of neuroimaging research.

9.4 Ethics, data governance, and equitable access

As neuroimaging data become increasingly detailed, linked to genetic and phenotypic information, and interpretable by AI systems capable of inferring sensitive personal attributes, the ethical dimensions of brain data collection, storage, and use require urgent attention. The principles of informed consent, data minimization, purpose limitation, and the right to erasure enshrined in frameworks such as the GDPR must be implemented in neuroimaging research practice in technically robust and genuinely participant-centered ways. Questions of equity—whether the benefits of advanced neuroimaging technologies are accessible across socioeconomic, geographic, and demographic divides—will determine whether the transformative potential of these technologies is realized for all people or only for those with access to the world's best-resourced medical centers.

10. Conclusion

The chapter has provided a comprehensive and critically integrated examination of the three principal non-invasive neuroimaging modalities in contemporary neuroscience and clinical medicine: functional Magnetic Resonance Imaging (fMRI), Electroencephalography (EEG), and Magnetoencephalography (MEG). Each technology embodies a distinct biophysical approach to the common challenge of observing neural activity in the living human brain, and each has achieved a unique and irreplaceable position in the clinical and scientific landscape through decades of technical refinement and application.

fMRI, by exploiting the neurovascular coupling mechanism and the BOLD contrast of deoxyhaemoglobin, provides spatial resolution unmatched among non-invasive neuroimaging modalities. This enables precise anatomical mapping of brain activation during cognitive, perceptual, and motor tasks and characterization of large-scale resting-state functional networks disrupted in neurological and psychiatric disease. Its temporal resolution, constrained by the hemodynamic response function to seconds rather than milliseconds, is a fundamental limitation that motivates combination with electrophysiological modalities. Despite this limitation, fMRI has become the dominant research tool in cognitive neuroscience and an increasingly central element of clinical neurological practice, particularly in pre-surgical functional mapping and the emerging field of neuroimaging biomarkers for precision medicine.

EEG, with its direct and millisecond-resolution readout of cortical electrical dynamics, remains indispensable in epilepsy diagnosis and monitoring, sleep medicine, neonatal neurology, and brain-computer interface research. Its portability and relatively low cost make it the only practical option for bedside, ambulatory, and home-based brain monitoring. Advances in high-density electrode arrays, dry electrode technology, and deep learning-based signal analysis are continuously expanding the diagnostic and research utility of EEG, while simultaneous EEG-fMRI acquisition frameworks leverage the complementary temporal and spatial strengths of the two modalities in increasingly routine fashion.

MEG occupies a specialized but high-value niche in the neuroimaging ecosystem, offering electrophysiological recordings with temporal resolution equivalent to EEG but substantially improved source localization for tangential cortical sources attributable to the magnetic transparency of the skull. Its primary clinical applications in pre-surgical mapping of eloquent cortex and epileptic foci are well-validated. The advent of wearable OPM-MEG systems marks a potentially transformative development that could extend MEG's clinical reach far beyond the small number of specialized centers that currently operate conventional cryogenic systems.

The overarching message of this review is that fMRI, EEG, and MEG are not competing technologies but complementary instruments that, when strategically combined, provide a spatiotemporal window on brain function of unparalleled comprehensiveness. The rapid maturation of open science infrastructure—BIDS standards, reproducible preprocessing pipelines, open-access datasets, federated learning frameworks—is creating the conditions for multimodal neuroimaging to transition from a research tool to a precision clinical instrument capable of providing individualized, mechanistically informed guidance for diagnosis, prognosis, and treatment selection across the full spectrum of neurological and psychiatric conditions.

Future progress will require sustained investment not only in hardware and algorithms but in the ethical, governance, and equity frameworks necessary to ensure that the benefits of advanced neuroimaging are equitably distributed and that the unprecedented sensitivity of detailed brain data is protected with commensurate rigor. The decade ahead offers extraordinary scientific opportunities; realizing them in ways that genuinely serve patients and society will require the neuroimaging community to engage these broader responsibilities with the same creativity and rigor it brings to technical innovation.

References

Baillet, S. (2017). Magnetoencephalography for brain electrophysiology and imaging. *Nature Neuroscience, 20*(3), 327–339. https://doi.org/10.1038/nn.4504

Barachant, A., Bonnet, S., Congedo, M., et al. (2012). Multiclass brain-computer interface classification by Riemannian geometry. *IEEE Transactions on Biomedical Engineering, 59*(4), 920–928. https://doi.org/10.1109/TBME.2011.2172210

Berger, H. (1929). Über das Elektrenkephalogramm des Menschen. *Archiv für Psychiatrie und Nervenkrankheiten, 87*, 527–570. https://doi.org/10.1007/BF01797193

Boto, E., Holmes, N., Leggett, J., et al. (2018). Moving magnetoencephalography towards real-world applications with a wearable system. *Nature, 555*, 657–661. https://doi.org/10.1038/nature26147

Bullock, M., Jackson, J., Kreuzer, M., et al. (2021). Artifact reduction in simultaneous EEG-fMRI: A systematic review. *Frontiers in Neurology, 12*, 622719. https://doi.org/10.3389/fneur.2021.622719
Debener, S., Minow, F., Emkes, R., et al. (2012). How about taking a low-cost, small, and wireless EEG for a walk? *Psychophysiology, 49*(11), 1617–1621. https://doi.org/10.1111/j.1469-8986.2012.01471.x
Di Martino, A., Yan, C.-G., Li, Q., et al. (2014). The autism brain imaging data exchange: Towards a large-scale evaluation of the intrinsic brain architecture in autism. *Molecular Psychiatry, 19*(6), 659–667. https://doi.org/10.1038/mp.2013.78
Esteban, O., Markiewicz, C., Blair, R. W., et al. (2019). fMRIPrep: A robust preprocessing pipeline for functional MRI. *Nature Methods, 16*, 111–116. https://doi.org/10.1038/s41592-018-0235-4
Gorgolewski, K. J., Auer, T., Calhoun, V. D., et al. (2016). The Brain Imaging Data Structure (BIDS): A standard for organizing and describing outputs of neuroimaging experiments. *Scientific Data, 3*, 160044. https://doi.org/10.1038/sdata.2016.44
Hari, R., & Salmelin, R. (2012). Magnetoencephalography: From SQUIDs to neuroscience. *NeuroImage, 61*(2), 386–396. https://doi.org/10.1016/j.neuroimage.2011.11.074
Haufe, S., Meinecke, F., Görgen, K., et al. (2014). On the interpretation of weight vectors of linear models in multivariate neuroimaging. *NeuroImage, 87*, 96–110. https://doi.org/10.1016/j.neuroimage.2013.10.067
Huster, R. J., Debener, S., Eichele, T., & Herrmann, C. S. (2012). Methods for simultaneous EEG-fMRI: An updated review. *Frontiers in Human Neuroscience, 6*, 106. https://doi.org/10.3389/fnhum.2012.00106
Kostas, D., Aroca-Ouellette, S., Rudzicz, F., et al. (2021). BENDR: Using transformers and contrastive self-supervised learning to learn from massive amounts of EEG data. *Frontiers in Human Neuroscience, 15*. https://doi.org/10.3389/fnhum.2021.653659
Lawhern, V. J., Solon, A. J., Waytowich, N. R., et al. (2018). EEGNet: A compact convolutional neural network for EEG-based brain-computer interfaces. *Journal of Neural Engineering, 15*(5), 056013. https://doi.org/10.1088/1741-2552/aace8c
Li, X., Cao, D., Wei, L., et al. (2023). Deep learning in EEG: Advance of the last ten-year critical period. *IEEE Transactions on Cognitive and Developmental Systems, 15*(2), 410–432. https://doi.org/10.1109/TCDS.2022.3183479
Markiewicz, C. J., Gorgolewski, K. J., Feingold, F., et al. (2021). The OpenNeuro resource for sharing of neuroscience data. *eLife, 10*, e71774. https://doi.org/10.7554/eLife.71774
Michel, C. M., & Murray, M. M. (2012). Towards the utilization of EEG as a brain imaging tool. *NeuroImage, 61*(2), 371–385. https://doi.org/10.1016/j.neuroimage.2011.12.039
Miller, K. L., Alfaro-Almagro, F., Bangerter, N. K., et al. (2016). Multimodal population brain imaging in the UK Biobank prospective epidemiological study. *Nature Neuroscience, 19*(11), 1523–1536. https://doi.org/10.1038/nn.4393
Niedermeyer, E., & da Silva, F. L. (2005). *Electroencephalography: Basic principles, clinical applications, and related fields* (5th ed.). Lippincott Williams & Wilkins.
Niso, G., Tadel, F., Bock, E., et al. (2018). MEG-BIDS: The brain imaging data structure extended to magnetoencephalography. *Scientific Data, 5*, 180110. https://doi.org/10.1038/sdata.2018.110
Obeid, I., & Picone, J. (2016). The temple university hospital EEG data corpus. *Frontiers in Neuroscience, 10*, 196. https://doi.org/10.3389/fnins.2016.00196
Ogawa, S., Lee, T. M., Kay, A. R., & Tank, D. W. (1990). Brain magnetic resonance imaging with contrast dependent on blood oxygenation. *Proceedings of the National Academy of Sciences, 87*(24), 9868–9872. https://doi.org/10.1073/pnas.87.24.9868
Pernet, C. R., Appelhoff, S., Gorgolewski, K. J., et al. (2019). EEG-BIDS, an extension to the brain imaging data structure for electroencephalography. *Scientific Data, 6*, 103. https://doi.org/10.1038/s41597-019-0104-8
Petersen, R. C., Aisen, P. S., Beckett, L. A., et al. (2010). Alzheimer's Disease Neuroimaging Initiative (ADNI): Clinical characterization. *Neurology, 74*(3), 201–209. https://doi.org/10.1212/WNL.0b013e3181cb3e25

Pruim, R. H. R., Mennes, M., van Rooij, D., Llera, A., Buitelaar, J. K., & Beckmann, C. F. (2015). ICA-AROMA: A robust ICA-based strategy for removing motion artifacts from fMRI data. *NeuroImage, 112*, 267–277. https://doi.org/10.1016/j.neuroimage.2015.02.064

Roy, Y., Banville, H., Albuquerque, I., et al. (2019). Deep learning-based electroencephalography analysis: A systematic review. *Journal of Neural Engineering, 16*(5), 051001. https://doi.org/10.1088/1741-2552/ab260c

Schirrmeister, R. T., Springenberg, J. T., Fiederer, L. D. J., et al. (2017). Deep learning with convolutional neural networks for EEG decoding and visualization. *Human Brain Mapping, 38*(11), 5391–5420. https://doi.org/10.1002/hbm.23730

The ADHD-200 Consortium. (2012). The ADHD-200 Consortium: A model to advance the translational potential of neuroimaging in clinical neuroscience. *Frontiers in Systems Neuroscience, 6*, 62. https://doi.org/10.3389/fnsys.2012.00062

Van Essen, D. C., Smith, S. M., Barch, D. M., et al. (2013). The WU-Minn Human Connectome Project: An overview. *NeuroImage, 80*, 62–79. https://doi.org/10.1016/j.neuroimage.2013.05.041

Further reading

Brickwedde, M., Keilholz, S. K., Iivanainen, J., et al. (2024). Applications of OPM-MEG for translational neuroscience: A perspective. *Translational Psychiatry, 14*(1), 1–12. Article No. 341. https://doi.org/10.1038/s41398-024-03047-y

Bzdok, D., Altman, N., & Krzywinski, M. (2018). Statistics versus machine learning. *Nature Methods, 15*, 233–234. https://doi.org/10.1038/nmeth.4642

Caetano, G., Mayeli, P., & Hampson, M. (2023). NeuXus: Real-time artifact reduction in simultaneous EEG-fMRI. *NeuroImage, 280*, Article 120353. https://doi.org/10.1016/j.neuroimage.2023.120068

Chen, J. E., & Glover, G. H. (2015). BOLD fractional contribution to resting-state functional connectivity above 0.1 Hz. *NeuroImage, 107*, 207–218. https://doi.org/10.1016/j.neuroimage.2014.12.012

Marek, S., Tervo-Clemmens, B., Calabro, F. J., et al. (2022). Reproducible brain-wide association studies require thousands of individuals. *Nature, 603*, 654–660. https://doi.org/10.1038/s41586-022-04492-9

Noble, S., Scheinost, D., Shen, X., et al. (2019). A decade of test-retest reliability of functional connectivity: A systematic review and meta-analysis. *NeuroImage, 203*, 116157. https://doi.org/10.1016/j.neuroimage.2019.116157

Rier, L., Michelmann, S., Ritz, H., et al. (2023). Test-retest reliability of the human connectome: An OPM-MEG study. *Imaging Neuroscience, 1*. https://doi.org/10.1162/imag_a_00020

CHAPTER 2

A primer on generative AI: Key concepts and architectures

Edson Bernardo da Conceicao Mahangue[1], Avinash Kumar Sharma[1], Ashwani Kumar[2] and Amit Kumar Tyagi[3]
[1]Department of Computer Science & Engineering (CSE), Sharda School of Computing Science & Engineering (SSCSE), Sharda University, Greater Noida, Uttar Pradesh, India; [2]School of Computer Science Engineering and Technology, Bennett University, Greater Noida, Uttar Pradesh, India; [3]National Forensic Sciences University, School of Law Forensic Justice Policy Studies, Gandhinagar, Gujarat, India

1. Introduction

Artificial intelligence (AI) has revolutionized modern technology, by the improvement of communication and information sharing through automation, real-time data processing, and intelligent systems, which reduces the costs and increases efficiency.

Through the combination of intelligent algorithms and data-driven automation, AI has improved the communication and information-sharing mechanisms across various domains making them more accessible and less costly. This improvement is primarily accomplished through the automation of routine tasks, real-time data processing abilities, and the employment of adaptive intelligent systems. The implementation of these systems based in AI not only modernizes operations and reduces operational costs but also improves overall efficiency and accuracy, thus allowing organizations and individuals to achieve more in less time.

Among the subfield of AI, deep learning (DL) is known as one of the most prominent and powerful subset, it also has gathered substantial attention due to its supreme ability to process vast amounts of unstructured data. Deep learning, which is inspired by the structure and function of the human brain, employs artificial neural networks (ANNs) with multiple layers to perform complex tasks such as image recognition, speech processing, and decision-making. The most impactful application of deep learning is found in the domain of generative AI, which has been defined by its proficiency to produce novel content, including text, images, audio, and even video. Different from traditional AI models that are primarily focused on classification or prediction, these models learn patterns in data to create new, original outputs that mimic human creativity.

GANs, which are subfield of generative AI, operate through a dual-model framework consisting of a generator and a discriminator, where the generator creates data and the discriminator evaluates its authenticity. This antagonistic process leads to the generation of gradually realistic outputs over time. On the other hand, transformer models, which are the other hand of generative AI, are the base of systems such as

Generative Artificial Intelligence for Neuroimaging
ISBN 978-0-443-45010-5
https://doi.org/10.1016/B978-0-443-45010-5.00006-X

OpenAI's GPT series or Google's BERT, outshine in processing sequential data, making them remarkably effective in natural language understanding and generation. These models have allowed significant progress in fields such as natural language processing (NLP), where they are used for tasks including language translation, summarization, and chatbot development.

The influence of generative AI can be seen across a wide spectrum of industries. In healthcare, it is used to generate synthetic medical data, project drug molecules, and even assist in diagnostic procedures. In entertainment, where AI-driven tools helps to produce realistic visual effects, representative video game environments, and dynamic content personalization. Within software engineering, these models are used to transform code generation, software test, and documentation. The use of applications combined with these models not only improves productivity but also unlocks new potentials for innovation and personalization.

Given the extensive capabilities and suggestions of generative AI, become essential to address fundamental concerns such as explainability and interpretability of these models. Understanding how and why generative models make decisions or produce specific outputs is important for building trust, safeguarding ethical use, and maintaining transparency, particularly in high-stakes environments. This chapter delivers a comprehensive overview of the foundational concepts, key architectures, and practical considerations in the field of generative AI, with a particular focus on improving its explainability and interpretability to ensure responsible and effective deployment in real-world applications.

2. Generative AI

Generative AI has been described as the subset of AI systems, which is designed to create new content by learning patterns from existing data (Caselli, 2025).

It can be also defined as the computational methods that can create an original, meaningful content such as text, images, or audio in basis of patterns learned from training data collected where it is inserted (Feuerriegel & Janiesch, 2024). There rapid adoption can be seen in tools such as DALL·E 2, GPT-4, and Copilot, which they are transforming how people work and communicate.

Different from traditional AI models, which are typically concentrated on narrow tasks such as classification, regression, or prediction, with the aim of analyzing input data and returning a specific output or decision. Generative AI purpose has been defined as to mimic the innovative aspects of human intelligence.

It ensures so by building data that is not simply a replication of the training input but is instead a novel fusion, which maintains the characteristics and consistency of human-generated content.

These systems are capable of separately producing a wide range of content types, including natural language text, realistic images, synthetic audio, animations, and even functional code.

For example, in the realm of natural language handling, generative models can compose coherent articles, generate dialog for virtual assistants, or translate text across languages with human-like fluency. In visual domains, they can generate detailed creations, photorealistic images, or design prototypes from textual descriptions.

In music and sound engineering, they are used to compose melodies or restructure voices. To completely understand generative AI, it is important to begin with the foundational principles of AI.

2.1 Related terms and key concepts

A comprehensive understanding of generative AI involves a previous and clear grasp of the introductory concepts and related terms that support its development.

2.2 Artificial intelligence

An extensive field of computer science worried with construction machines that can perform tasks that typically require human intelligence. These include learning, reasoning, problem-solving, perception, and language understanding.

2.3 Machine learning

A subset of AI where machines increase their performance on a task over time by learning from data, rather than being explicitly programmed.

2.3.1 Deep learning

A specialized form of ML uses artificial neural networks (ANN) with multiple layers to automatically learn representations of data. It is the primary technique behind most generative AI systems. The interaction among these main concepts is illustrated in Fig. 2.1.

Generative AI can thus be understood as a subset of AI, grounded in deep learning, which uses models trained on large datasets to create new content by applying transformation-based techniques (Zhang et al., 2025).

This capability is enhanced by the implementation of sophisticated deep learning architectures, which learn from immense datasets and replicate the patterns and structures originate within them. By employing unique mechanisms to generate data, these models distinguish among them, and they can be classified by their mechanisms and applications as listed in the following:

- **Generative adversarial networks (GANs)** consist of two opposing neural networks a generator and a discriminator, which work together to produce extremely realistic artificial data, such as images and videos.

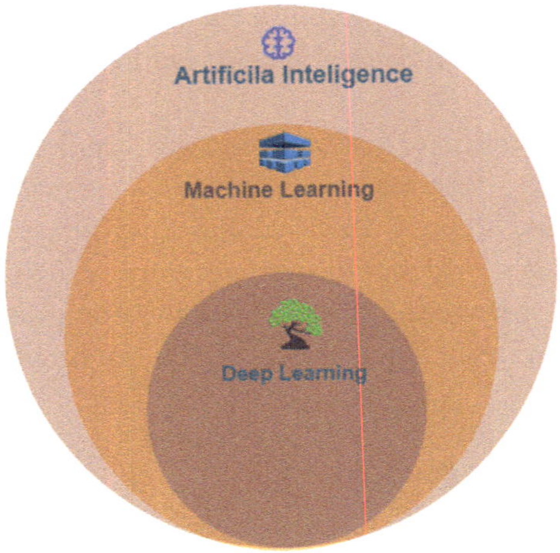

Figure 2.1 AI versus ML versus deep learning.

- **Variational autoencoders (VAEs)** use probabilistic encoding and decoding to generate new data instances that preserve the arithmetical properties of the original dataset, often applied in image generation and anomaly detection.
- **Autoregressive models** create content one step at a time, with each output conditioned on previous elements, making them especially effective in natural language processing tasks.
- **Diffusion models** start with random noise and iteratively refine it to produce coherent images, excelling in high-resolution image synthesis.
- **Transformers**, originally designed for sequence-to-sequence tasks, have become initial in modern generative AI due to their scalability and capacity to handle large datasets across various modalities.

Composed, these models represent the core technological pillars driving generative AI's rapid development.

3. Generative adversarial networks

GANs have been described as a type of deep learning models, which their models are trained based in a confrontational process. They are based in a specific architecture, which defines how it interprets inputs and produces outputs, what makes it a key factor in its overall performance and its suitability for particular tasks (Bandi et al., 2023).

It is a model that is made up of two important components: the generator and the discriminator. These two interact in a competitive process, where the generator tries to produce realistic data while the discriminator evaluates its authenticity. With many experiences, this rivalry helps the generator improve its ability to create increasingly realistic outputs (Pahuja & Singh, 2025). The all process is summarized into these two main neural network (Kim et al., 2020). The aim of these elements is to produce the most realistic images, videos, or audios based on given data. The role of each is described as follows.

3.1 Generator

The primary role is to generate credible data from fake and noise data. This type of neural network is responsible to make difficult to differentiate the real data distribution from a fake one (Tahmid et al., 2020).

3.2 Discriminator

The discriminator has the role of discriminating whether the input data is real data or a fake data, which has been produced by the generator. The discriminator is trained to classify well, while the generator is trained to betray the discriminator. The working principle of these two components is described in Fig. 2.2.

The working principle of GAN's models can also be explained through mathematical expression as shown in the following equation. The equation shows the value function for GANs, where the generator G tries to minimize the function while the discriminator D tries to maximize it. In this equation, the variable G generates fake data G(z) to trick D, and this finally tries to distinguish between real data x and fake data (Goodfellow et al., 2020).

$$\underset{G}{\text{Min}}\ \underset{D}{\max} V(D, G) = E_{\{x \sim p_{\text{data}(x)}\}}[\log D(x)] + E_{\{z \sim p_{z(z)}\}}[\log(1 - D(G(z)))] \quad (2.1)$$

where:

- $D(x)$: The discriminator's estimate of the probability that real data instance x is real.
- $G(z)$: The generator's output when given random input z (noise).
- $D(G(z))$: The discriminator's estimate of the probability that a fake instance $G(z)$ is real.
- $pdata(x)$: The distribution of the real data.
- $pz(z)$: The distribution of the generator's input noise (usually uniform or Gaussian).

3.3 Key application of generative AI

Generative AI has originated diverse applications across multiple domains, meaningfully improving capabilities and automating complex tasks. It is also used in the text domain, by controlling tools such as chatbots, automated content writing systems, text summarization, and multilingual translation; these models offer improved communication and productivity.

The integration of generative AI in education has outperformed both learning approaches and instructional methods. Since this technology focuses on creating new

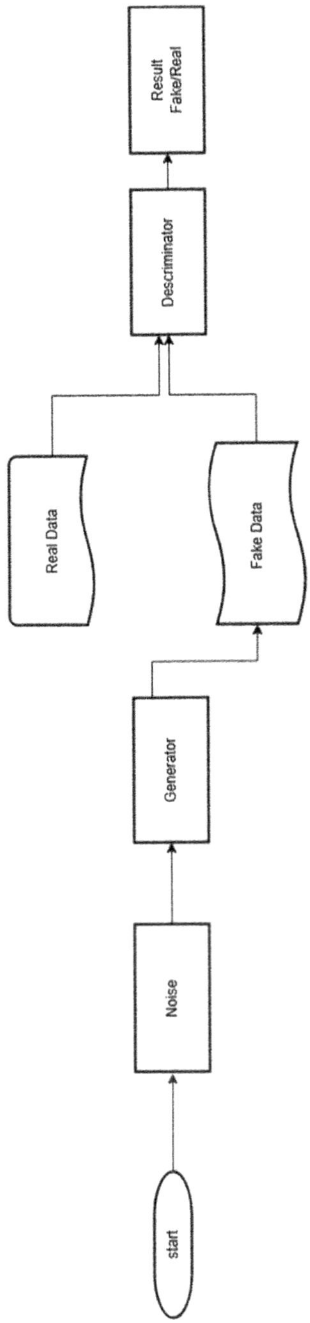

Figure 2.2 GAN working principle.

content by reworking and adapting existing material, it helps the students to learn new concepts in an immersive, interactive, and engaging manner (Labib et al., 2025).

On other hand in the image domain, generative AI is used to create AI-based artwork, style transfer, realistic face generation, and advanced image editing, enabling creativity and visual design innovation.

The application of these models in the field of audio can be seen in music composition, voice synthesis, and the creation of sound effects, converting the way audio content has been produced. Likewise, in video, generative AI enables the creation of deepfakes, automated animations, and realistic virtual avatars, impacting both entertainment and digital media production.

In healthcare, it is used for innovations such as the generation of drug molecules and predictions of protein folding, accelerating research and personalized medicine.

Within education, generative systems are used to improve intelligent teaching platforms, the mechanized quiz creation, and custom lesson plans generation, improving personalized learning experiences.

In software development, these models simplify code generation, autocompletion, and bug detection, thereby refining the software quality and development speed. The gaming industry influences generative AI for practical content generation and dynamic character or dialog creation, offering more immersive and receptive gaming experiences.

Lastly, in finance, it helps produce synthetic data for fraud detection and simulate market behaviors, supportive risk analysis, and secure financial operations. The summarization of these key points can be found in Table 2.1.

Table 2.1 has been presented a selection of recent studies that apply GANs across various domains, emphasizing their specific objectives and the primary gaps identified in each work. These examples underline the diverse applications of GANs from drug discovery to deepfake detection while also highlighting the persistent challenges researchers face.

As summary can be said that GANs represent a powerful class of deep learning models structured around a competitive training process between two neural networks: the generator and the discriminator. The training dynamic between these two networks is antagonistic: the generator continuously improves its outputs to fool the discriminator, while the discriminator becomes more accurate at detecting fakes.

When they are compared to generative models, it can easily be understood that a generative model is a statistical approach that aims to estimate the underlying probability distribution from observed data samples, while in the context of generative AI, they allow these models to produce new, randomly generated examples that follow the same learned distribution (Ronge et al., 2025).

This conflict results in both networks progressively improving their capabilities. The mathematical basis of this interaction is represented in the value function of GANs,

Table 2.1 Key application of generative AI.

Domain	Applications
Text	Chatbots, content writing, summarization, translation
Images	AI art, style transfer, face generation, image editing
Audio	Music arrangement, voice synthesis, sound effects
Video	Deepfake generation, animation, virtual avatars
Healthcare	Drug molecule generation, protein folding
Education	Intelligent tutoring systems, automatic quiz, and lesson plan generation
Software Dev	Code generation, code completion, bug detection
Gaming	Procedural content generation, character/dialog generation
Finance	Synthetic data for fraud detection, market simulations

where the generator pursues to minimize the loss, and the discriminator aims to maximize it.

Through this min-max game, GANs learn to produce highly realistic synthetic data. Fig. 2.2 illustrates this working principle, and Eq. (2.1) reinforces the fundamental optimization objective, highlighting the core probabilistic interactions between real data distributions and generated content.

The practical consequences of GANs and wider generative AI technologies are extensive, covering multiple sectors and transforming how content is shaped and tasks are automated. As described in detail in Table 2.1, generative AI has been widely applied in the text domain, the visual arts and design sector, in audio and video, healthcare, educational software, gaming fields to improve their objectives. Additionally, Table 2.2 provides a comparative overview of recent GAN-based

Table 2.2 GAN models comparison.

Author	Model	Objective	Gap analysis
Caselli (2025)	GAN	Drug molecule generation, medicine application	High data requirements for training
Zhang et al. (2025)	GAN	Tasks automation in a hotel	Risk of misuse for deepfakes, misinformation, etc.
Kim et al. (2020)	GAN	N/A	High resource consumption during training
Goodfellow et al. (2020)	GAN	Descriptive	Lack of reliable, universal metrics
Deng et al. (2022)	GAN	Deepfake detection	Lack of reliable, universal metrics
Trabelsi et al. (2022)	GAN	Deepfake detection	Poor performance when evaluated in the unseen dataset

research studies, extracting various purposes such as drug design, automation, and deepfake detection. Despite their aptitude, these models still face notable challenges, including high computational demands, data requirements, and the absence of universal performance metrics.

4. Variational autoencoders

VAEs have been known by the capability of generating data from learned probability distribution and maintaining the output identical to the training set. They are widely considered as a variation of the original autoencoder, with an extra function of being capable to generate from a latent vector of a specific data (Mandava, 2024; Mansouri & Lachiri, 2020). VAEs compress data into a latent representation and then decode it to produce new samples. Different traditional autoencoders increase a probabilistic layer, which allows further variability and originality in output. This can be represented mathematically

$$p\theta(x, z) = p\theta(x \mid z) \cdot p\theta(z) \tag{2.2}$$

where

- $p\theta(x,z)$ signifies the joint probability of both the observed variable x and the latent variable z.
- $p\theta(x|z)$ represents the conditional probability of x given z. It acts as a decoder network.
- $p\theta(z)$ represents the prior distribution of the latent variable z.

In general, $p\theta(x|z)$, $p\theta(z)$, and θ represent correspondingly the generative model, the preceding distribution, and the model parameters. The mathematical expression is possible to identify that these models are trained to minimize the probability by the application of the following equation:

$$p\theta(x) = \int p\theta(z) \cdot p\theta(x \mid z)\, dz \tag{2.3}$$

These elements represent the difference between the VAEs and the traditional autoencoders, being the first more tending to data generation than autoencoder as shown in Fig. 2.3.

4.1 Key application of VAE

VAEs are a class of generative models commonly recognized for their ability to learn efficient latent representations of data. This feature allows VAEs to be effectively applied across various domains, each advancing from the model's capacity to encode and reconstruct data with high fidelity.

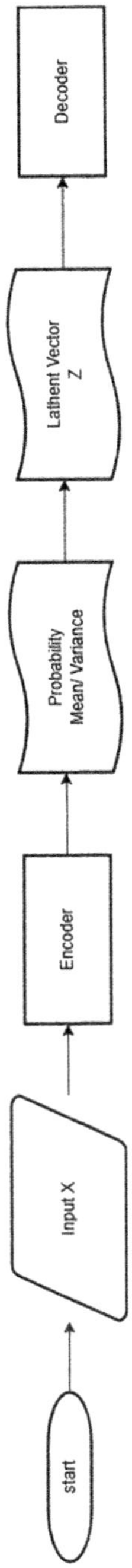

Figure 2.3 VAE architecture.

In computer vision, the use of these models is seen on image generation, where they learn latent features from training data and use this compressed representation to produce new, realistic images. This is predominantly useful in tasks such as synthetic data generation, image improvement, and creative visual design (Ramalakshmi & Asha, 2024).

For anomaly detection in fields such as cybersecurity and industrial IoT, VAEs are trained on normal data behavior to identify deviations. By rebuilding data from its hidden representation, the model can detect anomalies such as network intrusions or system faults when the restoration error is significantly higher than expected.

In multimedia processing, these models support data compression by encoding complex data into compact latent vectors, which can then be decoded to reconstruct the original input with minimal data loss. This has been described as the key point beneficial for reducing storage and transmission requirements in images, videos, and audio files (Asperti et al., 2021).

In the domain of bioinformatics and healthcare, VAEs have been applied in drug discovery to generate novel molecular structures with chosen properties. These models can also be used to discover the latent space of chemical compounds, suggest new molecules that may not exist in current databases but meet specific biological or chemical criteria.

VAEs also play a key role in speech production, where they are used to produce natural-sounding speech from latent audio features. This use contributes to the development of voice assistants, text-to-speech systems, and personalized speech generation tools.

Lastly, in natural language processing (NLP), VAEs have been employed in text generation tasks such as rephrasing, summarization, and translation. By capturing the fundamental structure of language in latent form, these models can generate fluent and contextually appropriate text outputs, enhancing automation and human–computer interaction (Tables 2.3, 2.4, 2.5 and 2.6).

4.2 Autoregressive models

Autoregressive (AR) models have been described as foundational class of generative models, which build sequences by generating one data point at a time, where each new element is conditioned on the previous ones. This sequential nature makes them especially suited for tasks involving structured data such as text, audio, or time series, where the order of information is crucial.

In essence, an autoregressive model guesses the joint probability of a sequence $x=(x1, x2, \ldots, xT)$ $x = (x1, x2, \ldots, xT)$ $x = (x1, x2, \ldots, xT)$ by factorizing it into a product of conditional probabilities. This application allows the model to capture dependencies across a sequence, with each output depending on the context built from the previously generated outputs.

Table 2.3 Key application of VAE.

Application and domain	Description
Image generation—*Computer vision*	Producing new images from learned latent images.
Anomaly detection—*Cybersecurity/industrial IoT*	Detecting deviances from normal patterns, useful in intrusion detection and fault diagnosis.
Data compression—*Multimedia processing*	Encoding data into compact latent vectors and reconstructing it with minimal loss.
Drug discovery—*Bioinformatics/healthcare*	Producing novel molecular structures with preferred chemical properties.
Speech synthesis—*Audio processing/NLP*	Producing natural-sounding language from latent variables.
Text generation—*NLP*	Generating expressive text sequences, particularly for tasks such as rephrasing or summarization.

Table 2.4 Key application and domain of autoregressive models.

Application and domain	Description
Language modeling—*Natural language processing*	Prediction of the next word/token in a sequence (e.g., GPT models).
Text generation—*Creative writing/chatbots*	Generation of coherent and human-like text based on prompts.
Speech synthesis—*Audio AI/assistive tech*	Production of realistic speech audio from text (e.g., WaveNet).
Image generation—*Computer vision*	Pixel-by-pixel image synthesis (e.g., PixelRNN, PixelCNN).
Time series forecasting—*Finance/weather/IoT*	Prediction of future values based on past temporal data (e.g., ARIMA, RNNs).
Music generation—*Creative AI/entertainment*	Composition of music notes or audio tracks using past sequences.
Stock price prediction—*Finance/market analysis*	Modulation and forecasting asset prices based on historical data.
Sensor data modeling—*Industrial IoT/robotics*	Prediction of sensor readings for monitoring or control applications.

Table 2.5 Key application of diffusion models.

Application and domain	Description
Image generation—*Computer vision/creative AI*	Creation of high-quality, realistic images from noise (e.g., DALL·E 2, stable diffusion).
Text-to-image synthesis—*Multimodal AI*	Generation of images from natural language descriptions (prompts).
Video generation—*Media and entertainment*	Creation of consistent, high-fidelity video frames from sequential noise inputs.
Audio synthesis—*Speech processing/music creation*	Generation of speech and music with natural sound quality (e.g., DiffWave, AudioLM).
3D object generation—*Computer graphics/robotics*	Production of realistic 3D shapes or scenes from diffusion-based processes.
Molecular design—*Drug discovery/chemistry*	Crafty molecules with specific properties by denoising latent chemical space.
Super-resolution—*Image processing*	Improvement of low-resolution images using learned diffusion-based refinement.
Image inpainting—*Computer vision/restoration*	Rich in missing or corrupted parts of images with coherent content.

Table 2.6 Key application of transformers.

Application and domain	Description
Machine translation—*Natural language processing*	Translation of text between languages using attention-based sequence models (e.g., Google translate).
Text summarization—*NLP/media analysis*	Generation of concise summaries from long texts or documents.
Text classification—*NLP/sentiment analysis*	Categorization of text into predefined labels (e.g., spam detection, sentiment).
Named entity recognition (NER)—*Information extraction*	Identification and classification of proper nouns (people, places, organizations).
Question answering—*Conversational AI/search engines*	Production of precise answers from unstructured text sources.
Code generation—*Software development*	Generation of code snippets or completing programs (e.g., GitHub Copilot).
Protein folding—*Bioinformatics/drug discovery*	Prediction of 3D structure of proteins from amino acid sequences (e.g., AlphaFold).
Vision transformers (ViT)—*Computer vision*	Image classification and segmentation using attention mechanisms.

Famous examples: GPT series, BERT, T5, DALL·E.

The training progression involves maximizing the probability of the observed sequence by learning these conditional probabilities.

A noticeable example of autoregressive modeling is the generative pretrained transformer (GPT) family. GPT models utilize a transformer decoder architecture, which incorporates self-attention mechanisms to learn circumstantial representations from large-scale textual data. During generation, the model produces text one token at a time, with each token conditioned on all previous tokens in the sequence. This permits for comprehensible and contextually rich text generation, making GPT and similar models extremely effective in tasks such as dialog generation, translation, summarization, and code synthesis.

Autoregressive models are not limited to the use in NLP. They are also applied in areas such as speech synthesis, where each audio frame is produced based on the previously generated ones, and image generation, where pixels or image patches are predicted sequentially. Despite their effectiveness, autoregressive models have been known for being computationally intensive during inference due to their sequential nature, making generation slower compared to parallel models such as VAEs or GANs. However, their ability to capture long-range habits with high fidelity makes them indispensable in modern generative modeling.

These models generate one data point at a time, with each output depending on previously generated outputs. Examples include GPT (generative pretrained transformer).

4.2.1 Key application of autoregressive

Autoregressive models are extensively used in NLP, particularly in language modeling, where their primary task is to predict the next word or token in a sentence based on previously seen words. This essential principle underlies the performance of models, which have been demonstrated noteworthy success in understanding and generating human-like text.

Building on this capability, text generation solicitations such as creative writing tools, conversational agents, and chatbots. The use autoregressive models help to generate comprehensible and contextually appropriate text responses from prompts, thereby improving human–computer interaction.

In the domain of audio AI and assistive technologies, autoregressive models are employed for speech synthesis.

Autoregressive models also have been applied in time series forecasting, where they are used to predict future values based on historical trends. This is particularly valuable in domains such as finance, weather prediction, and IoT-based monitoring, where models such as ARIMA and RNNs can forecast stock prices, weather conditions, or scheme performance.

In music generation, autoregressive models can learn from present compositions to generate new musical sequences or full audio tracks, making them useful in entertainment, video game design, and automated composition tools. Lastly, in sensor data modeling, especially in industrial IoT and robotics, these models can be used to predict future sensor outputs to aid in system diagnostics, predictive maintenance, and real-time control systems.

Together, these applications highlight the flexibility of autoregressive models across both structured and unstructured data domains. Their strength lies in their ability to learn sequential dependencies, making them essential tools in tasks where the order of information is critical to meaning and accuracy.

4.3 Diffusion models

They have been known as the type of generative AI, which transforms noise into a representative sample data through the employment of generative probabilistic models (Sallami & Ben Fredj, 2024). These models generate data by continuously filtering random noise into comprehensible output, directed by a learned reverse diffusion process (Zhang et al., n.d.).

In a simple manner, a diffusion model involves two stages: a forward process and a reverse process. Where during the forward stage, a clean data sample is gradually corrupted by adding Gaussian noise step by step until, over time, it becomes nearly indistinguishable from pure noise. On the other hand, in the reverse stage, a neural network is trained to progressively remove this noise, reconstructing a clean sample and effectively learning to generate new data from the noise (Chen et al., 2024).

Diffusion models, when compared to VAE, they have a different working principle, acting on bases of high-dimensional latent variable and predefined methodology for learning process.

These models work based on two fundamental mathematical frameworks, which use the same method of adding and removing noise to generate new samples. The processes are

- the forward diffusion process
- the backward denoising process

Diffusion models are widely known as the most prominent text-to-image generation, attracting interest due to their potential to provide unlimited and distinct images (Pattankudi et al., 2024). The sample production of a diffusion model can be expressed mathematically as follows:

$$q(xt \mid xt - 1) = N\left(xt; \sqrt{\alpha t} \cdot xt - 1, (1 - \alpha t) \cdot I\right) \tag{2.4}$$

where:

- xt represents the noise of data t.
- $xt - 1$ represents the previous stage of t.
- αt is a scaling factor that measure the noise level.

- *I* it represents the matrix.
- $N(\mu,\Sigma)$ signifies a Gaussian distribution with mean μ and covariance Σ.

The two mathematical frameworks can be expressed as follows:

For the forward process from $x0$ *to* xT mathematically is expressed by:

$$q(xT \mid x0) = \prod(t = 1 \text{ to } T)\ q(xt \mid xt - 1) \tag{2.5}$$

The same process can be expressed graphically as shown in Fig. 2.4.

On the other hand, the backward denoising process is represented through:

$$p\theta(xt\ - 1 \mid xt) = N(xt - 1;\ \mu\theta(xt, t), \Sigma\theta(xt, t)) \tag{2.6}$$

where

- $\mu\theta(xt,t)$ represents the mean predicted by the model.
- $\Sigma\theta(xt,t)$ represents the predicted covariance.

With this, the complete equation for reverse process can be denoted by

$$p\theta(x0 \mid xT) = \prod(t = 1 \text{ to } T)\ p\theta(xt - 1 \mid xt) \tag{2.7}$$

Diffusion models have appeared as powerful generative models in AI; they have been known for their capability to generate highly realistic data through a process of gradually denoising random noise. One of their most noticeable applications is in image generation within computer vision and creative AI. Models such as DALL·E 2 and stable diffusion demonstrate how noise can be iteratively refined to produce visually rich and photorealistic images.

In a narrowly related domain, text-to-image synthesis combines multimodal AI capabilities, using diffusion models to translate textual prompts into comprehensible and detailed visual representations, efficiently connecting the gap between language and vision (Yadav, 2024). These advances are widely used in art generation, digital content creation, and creative industries.

In the field of video generation, diffusion models are applied to synthesize high-fidelity video content by consistently denoising frame sequences, ensuring temporal coherence and realistic motion—a valuable tool in media production and virtual simulations. Similarly, in audio synthesis, models such as DiffWave and AudioLM use diffusion processes to generate lifelike speech and music, enhancing quality and expressiveness in sound design and speech applications. Diffusion models are also making strides in 3D object generation, particularly in robotics and computer graphics, by generating structured three-dimensional shapes and environments with spatial realism from noisy latent variables.

Another important application lies in molecular design for drug discovery and chemistry, where these models have been used to generate molecular structures with targeted properties by exploring and refining latent chemical space.

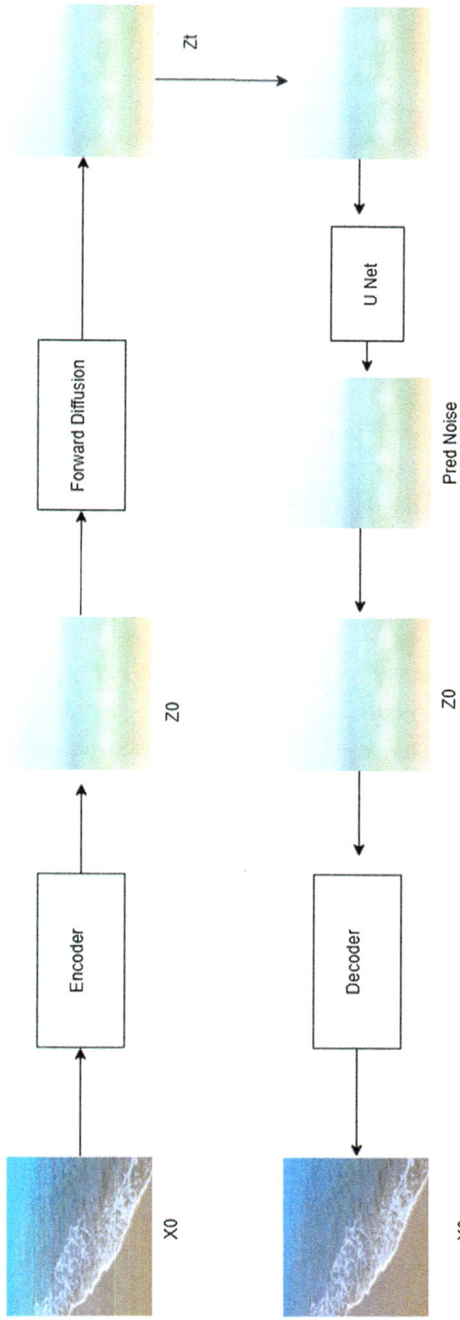

Figure 2.4 Diffusion model that uses the forward process to add the noise to image.

In image processing, diffusion models have been used in superresolution, where low-quality images are improved to higher resolutions through a learned improvement process, recovering fine details lost in the original. Mutually, these applications demonstrate the flexibility and transformative potential of diffusion models crosswise a wide range of scientific, creative, and industrial domains.

4.4 Transformers

First introduced in the paper "Attention is All You Need" (2017), transformers revolutionized sequence modeling by using self-attention mechanisms, making them highly effective for language and image generation tasks.

Transformers have modernized various domains of AI by introducing powerful attention-based mechanisms that outrival at understanding and generating structured data (Cho et al., 2017). These models have been used in NPL through machine translation, where models such as Google Translate utilize transformer architectures to convert text between languages with remarkable fluency and accuracy (Firdous & Masood, 2026). Beside translation, text summarization has advanced expressively; transformers can summarize lengthy documents into comprehensible summaries, which is especially useful in media, legal, and academic fields.

Transformers have also been described by playing a crucial role in text classification, where they are used to allocate predefined labels to textual data. Applications comprise sentiment analysis, spam detection, and topic categorization, and essential tasks in content equability and customer feedback analysis (Liu & Li, 2026; Bellini & Bignami, n.d.).

In question answering models such as BERT and GPT, the use of transformers facilitates machines to interpret questions and generate precise answers from vast text corpora, improving search engines and conversational AI.

Away from text, transformers are also applied in code generation, where models such as GitHub Copilot influence learned programming patterns to assist developers by generating code snippets, completing functions, or identifying bugs.

In bioinformatics, transformers such as AlphaFold have been used to predict the 3D structure of proteins from their amino acid sequences, which represents an innovation in drug discovery and molecular biology.

Additionally, in computer vision, vision transformers (ViT) apply the attention mechanism to image data, enabling accurate classification and segmentation tasks without depending on traditional convolutional networks. These applications validate the broad adaptability of transformer models across linguistic, biological, and visual domains, making them a cornerstone of modern AI development.

4.5 Advantages of generative AI

Generative AI is renovating a wide range of industries through their ability to systematize, personalize, simulate, and democratize content and process creation. Through the analyses of the core areas where generative AI has demonstrated significant impact, the advantages can be seen.

4.5.1 Content automation

Generative AI modernizes and accelerates content production, principally in sectors such as media, marketing, and education. By the automation of tasks, which previously required significant manual effort such as article writing, graphic design, and instructional material development, the use of generative AI can now be performed in seconds using these models. For example:

- In media, such tools generate reports, headlines, and summaries.
- In marketing, AI assistances in the creation of tailored advertising copy, product descriptions, and campaign visuals.
- In education, it supports the formulation of quizzes, personalized study materials, and automated grading tools.

With this incorporation, it achieved the faster production cycles, consistent quality, and scalable operations.

4.5.2 Cost reduction

The automation of repetitive and time-consuming tasks, generative models decrease the need for human labor, resulting in direct cost savings:

- **Creative industries** in this field save time and money by automating variations of designs, content drafts, or promotional materials.
- **Customer service** in this, the use of chatbots which are trained with predefined answers minimizes the human intervention needs, improving operational efficiency.
- **Gaming and media** companies can use AI to procedurally generate visual and audio assets instead of creating each manually.

4.5.3 Personalization

Generative AI surpasses at creating tailored content and experiences, allowing for high levels of personalization across various sectors:

- Educational platforms.
- E-commerce systems.
- Healthcare and wellness apps.

4.5.4 Scientific discovery

Generative models allow innovations in scientific research and engineering by

- simulating experiments which are too costly, time-consuming, or dangerous to perform physically;
- generating synthetic datasets in fields such chemistry, biology, and physics for model training and hypothesis testing; and
- scheming novel molecular complexes or materials using GANs and VAEs, aiding in drug discovery and materials science.

4.5.5 Accessibility

The use of generative AI enables users without specialized skills to involve in activities such as

- designing graphics, using tools like DALL·E or Canva AI;
- composing music, with platforms like Amper Music or Soundraw; and
- coding applications, via AI pair-programming tools like GitHub Copilot or text-to-code generators.

4.6 Societal impact of generative AI

4.6.1 Positive impacts

- Democratization of creativity: With the use of Generative AI, it is possible to see individuals without technical or artistic proficiency which through the use of it can generate professional-grade content.
- Enhanced accessibility: By enabling access of generative AI, which can produce text-to-speech and image captioning tools, the improvement is seen in the information access by people with disabilities.
- Faster innovation: Generative models benefit scientists in simulating scenarios, generate hypotheses, and explore unknowns giving space for faster innovation process.

4.6.2 Risks and challenges

- Generative AI can be used to misinformation and deepfakes, by the production of realistic fake content, which can deceive viewers and corrode trust in digital media.
- Production of fake news, identity fraud, and political manipulation.

5. Conclusion

The progression of generative AI has marked a paradigm modification in how technology is applied across industries, from creative automation to high-impact scientific exploration. Through the use of models such GANs, VAEs, diffusion models, and transformers, machines are now skillful of producing original and contextually relevant content, which would require time in human creativity. Their extensive adoption of this models has resulted in faster workflows, personalized experiences, reduced barriers for nonexperts, and accelerated innovation in fields such as education, media, healthcare, and engineering. However, with this extensive use of AI, challenges remain comprising ethical concerns, resource consumption, data dependency, and the lack of universal assessment metrics. Finally, generative AI holds the potential not only to enhance human productivity but also to redefine creativity, discovery, and access in the digital age.

References

Asperti, A., Evangelista, D., & Loli, E. (2021). A survey on variational autoencoders from a green AI perspective. *SN Computer Science, 2*(4), 1–23. https://doi.org/10.1007/s42979-021-00702-9

Bandi, A., Venkata, P., Ramesh, S., Eswar, Y., Pratap, V., & Kuchi, K. (2023). The power of generative AI : A review of requirements, models, input—Output formats. evaluation metrics, and challenges. *Future Internet, 15*(8), 260. https://doi.org/10.3390/fi15080260

Bellini, V., & Bignami, E. G. Comment generative pre-trained transformer 4 (GPT-4) in clinical settings. The Lancet Digital Health, 7(1), e6–e7. https://doi.org/10.1016/j.landig.2024.12.002.

Caselli, T. (2025). Understanding generative AI to harness its potentials and minimize risks: A perspective. *European Journal of Radiology, 184*, Article 111951. https://doi.org/10.1016/j.ejrad.2025.111951

Chen, M., Mei, S., Fan, J., & Wang, M. (2024). Opportunities and challenges of diffusion models for generative AI. *National Science Review, 11*. https://doi.org/10.1093/nsr/nwae348

Cho, A., et al. (2017). *Transformer explainer: Interactive learning of text-generative models* (pp. 29625–29627).

Deng, Z., Zhang, B., He, S., & Wang, Y. (2022). Deepfake detection method based on face edge bands. In *2022 9th international conference on digital home (ICDH)* (pp. 251–256).

Feuerriegel, S., & Janiesch, C. (2024). Generative AI. 66(1), 111–126. https://doi.org/10.1007/s12599-023-00834-7.

Firdous, A., & Masood, W. (2026). *Generative AI-based image captioning system using vision transformers and large language models* (pp. 67–73). https://doi.org/10.59256/ijire.20260702008

Goodfellow, I., et al. (2020). Generative adversarial networks. *Communications of the ACM, 63*(11), 139–144.

Kim, J., Tae, D., & Seok, J. (2020). A survey of missing data imputation using generative adversarial networks. In *2020 international conference on artificial intelligence in information and communication (ICAIIC)* (pp. 454–456).

Labib, W., Abdelsattar, A., Abowardah, E., & Abdelalim, M. (2025). The role of generative AI in architecture education from students' perspectives—A cross-sectional descriptive and correlational study. 1–37.

Liu, X., & Li, Z. (2026). *Optimization and application of generative AI algorithm based on transformer architecture in adaptive learning* (pp. 1–22).

Mandava, M. (2024). Leveraging variational autoencoders to identify genes involved with cancer metastasis. In *2024 IEEE MIT undergraduate research technology conference (URTC)* (pp. 1–5). https://doi.org/10.1109/URTC65039.2024.10937517

Mansouri, N., & Lachiri, Z. (2020). Laughter synthesis: A comparison between variational autoencoder and autoencoder. In *2020 5th international conference on advanced technologies for signal and image processing (ATSIP)* (pp. 1–6). https://doi.org/10.1109/ATSIP49331.2020.9231607

Pahuja, S., & Singh, A. (2025). ScienceDirect comprehensive review of generative artificial intelligence: Comprehensive review of generative artificial intelligence: Mechanisms, models, and applications mechanisms, a models, and b applications. *Procedia Computer Science, 258*, 3731–3740. https://doi.org/10.1016/j.procs.2025.04.628

Pattankudi, M. S. Z., Attar, A. R., Jewargi, K., Uppin, S., Khanapure, A., & Kulkarni, U. (2024). AI driven LLM integrated diffusion models for image enhancement- A survey. In *2024 IEEE conference on engineering informatics (ICEI)* (pp. 1–8). https://doi.org/10.1109/ICEI64305.2024.10912400

Ramalakshmi, S., & Asha, G. (2024). Exploring generative AI: Models, applications, and challenges in data synthesis. *Asian Journal of Research in Computer Science, 17*(12), 123–136.

Ronge, R., Maier, M., & Rathgeber, B. (2025). Towards a definition of generative artificial intelligence. *Philosophy & Technology*, 1–26. https://doi.org/10.1007/s13347-025-00863-y

Sallami, M. N., & Ben Fredj, O. (2024). Survey on diffusion models. In *2024 international symposium of systems, advanced technologies and knowledge (ISSATK)* (pp. 1–7). https://doi.org/10.1109/ISSATK62463.2024.10808957

Tahmid, M., Alam, S., & kalim Akram, M. (2020). Comparative analysis of generative adversarial networks and their variants. In *2020 23rd international conference on computer and information technology (ICCIT)* (pp. 1–6).

Trabelsi, A., Pic, M. M., & Dugelay, J.-L. (2022). Improving deepfake detection by mixing top solutions of the DFDC. In *2022 30th European signal processing conference (EUSIPCO)* (pp. 643–647).

Yadav, A. B. (2024). Generative AI in the era of transformers: Revolutionizing natural language processing with LLMs. *Journal of Image Processing and Intelligent Remote Sensing*, (2), 54–61.

Zhang, H., Xiang, Z., & Zach, F. J. (2025). Generative AI vs. humans in online hotel review management: A task-technology fit perspective. *Tourism Management, 110*, Article 105187. https://doi.org/10.1016/j.tourman.2025.105187

Zhang, C., Zhang, M., Kweon, I.S., & Kim, J. Text-to-image diffusion models in generative AI: A survey, pp. 1–40.

CHAPTER 3

Data handling and preprocessing in neuroimaging for generative AI. Addressing noise, artifacts, and variability

Priyanka Kumari[1], Akshay Juneja[2], Vijay Kumar[3] and Prateek Bhadauria[4]
[1]CGC University, Department of Electronics and Communication Engineering, School of Engineering and Technology (SET), Mohali, Punjab, India; [2]COER University, Computer Science and Engineering, College of Smart Computing, Roorkee, Uttarakhand, India; [3]Dr. B.R. Ambedkar National Institute of Technology, Department of Information Technology, Jalandhar, Punjab, India; [4]MITS-DU Gwalior, Department of Electronics Engineering, Gwalior, Madhya Pradesh, India

1. Introduction

Neuroimaging technologies are used by doctors and radiologists to study the brain structure and function. Each technology represents unique data characteristics that are used along with artificial intelligence (AI) in different applications. The datasets of these modalities are trained and validated using deep learning models for neuroscience research and clinical applications (Mårtensson et al., 2020).

Structural magnetic resonance imaging (sMRI) presents high-resolution three-dimensional brain anatomy images, such that their sizes vary from 0.5 to 1 mm^3. sMRI data is processed using convolutional neural networks (CNNs), such as U-Net architectures (Gautam & Singh, 2025). The automated segmentation of different parts such as the hippocampus or ventricles is performed. Generative adversarial networks (GANs) are used by researchers to generate sMRI images to address data scarcity issues in rare disease research. However, there are several challenges such as intensity inhomogeneity across scans and motion artifacts (Islam et al., 2024).

Functional MRI (fMRI) depicts brain activity via blood oxygenation level-dependent (BOLD) signals. It produces time-series data that reflect neural activity. fMRI has high spatial and low temporal resolution. Various 3D CNNs and recurrent neural networks are designed for different fMRI data applications to predict clinical outcomes. Variational autoencoders (VAEs) are used to generate synthetic outcomes for limited datasets using data augmentation (Al-zahrani et al., 2025).

Diffusion MRI (dMRI) is detected using white matter architecture as it measures the directional diffusion of water molecules. It enables reconstruction of fiber path using tractography, which produces complex 3D representations of structural connectivity. Recently, the disease classification is performed by researchers using graph neural

Generative Artificial Intelligence for Neuroimaging
ISBN 978-0-443-45010-5
https://doi.org/10.1016/B978-0-443-45010-5.00014-9

networks (GNNs) that consist of structural connections. However, dMRI data is preprocessed to reduce eddy current, which reduces the efficiency of the architecture (Zhang et al., 2025b).

Electroencephalography (EEG) and magnetoencephalography (MEG) have unparalleled temporal resolution, which measures the neural activity in milliseconds (ms). These techniques utilize electric or magnetic fields generated by neurons (Chaudhary, 2025). The data augmentation is performed on EEG data using GANs to increase the size of dataset, while transformer architectures detect pathological patterns such as epileptic spikes (Naghipour et al., 2026).

This section presents the various data handling and preprocessing techniques for different medical data modalities.

1.1 Data handling

The generative adversarial networks (GANs) are preprocessed to avoid training instability. GANs are trained on neuroimaging modalities database such that the data is normalized to prevent the generator from producing unrealistic intensity distributions (Saad et al., 2024). Motion correction generates high-resolution structural images, as minor artifacts in the training dataset lead to image blurring. In recent years, spatial normalization and skull stripping of input images generated using GAN architectures have improved the anatomical plausibility of generated brains. The data handling of temporal correlations in fMRI time series aids GANs to produce realistic BOLD signal dynamics instead of noise patterns (Qiang et al., 2023).

VAEs preserves variance and reduces technical noise. The disentangled latent spaces extracted from VAEs depend on distributed input data. The harmonization across scanners and protocols is essential for training. For instance, synthetic MRI images are generated using VAEs. It is followed by the bias correction of the training data that helps the model to focus on anatomical features instead of scanner-specific intensity variations. Similarly, physiological noise is filtered to ensure that VAE has captured the neural-relevant signal components rather than the artifacts to generate functional data (Nowroozilarki et al., 2024).

Diffusion models are sensitive to input data quality due to their iterative denoising process. The diffused MRI data including eddy current correction and tensor estimation improved the quality of synthetic fiber tracts generated by diffusion models. The data handling of temporal autocorrelations in fMRI inputs aided diffusion models to produce realistic time series instead of independent noise patterns for functional data generation (Vetter et al., 2024).

The preprocessing pipeline considers modality-specific requirements. sMRI needs improved artifact removal algorithm, while functional data generation needs proper data handling of temporal correlations. The emerging techniques such as test-time adaptation are automated adjusting preprocessing based on model performance, which reduces the need for manual pipeline optimization (Liang et al., 2025).

1.2 Preprocessing techniques

The high-quality neuroimaging data is essential to achieve high accuracy. It relies on clean, structured, and standardized input to perform pattern recognition with higher accuracy. Preprocessing is a step-by-step procedure to enhance the quality of neuroimaging data. The distortions and noise are reduced along with alignment of scans across the subjects. The different preprocessing techniques frequently used in the neuroimaging pipeline are depicted in Fig. 3.1.

1.2.1 Advanced motion and distortion correction

Head motion and scanner-induced distortions are the significant sources of noise in neuroimaging data. FMRIPrep and SPM12 are some preprocessing tools that used a combination of transformations, field map correction, and susceptibility distortion correction to ensure high accuracy of spatial alignment. Furthermore, ML-based artifact detection is adopted to automatically classify and remove motion-related components in fMRI datasets (Raval et al., 2022).

1.2.2 Spatial normalization and anatomical alignment

The spatial normalization is essential to facilitate cross-subject comparisons and data integration. Further, recent techniques employ nonlinear registration methods that allow for high-precision anatomical alignment that are even used in neurodegenerative disease cohorts (Mårtensson et al., 2020).

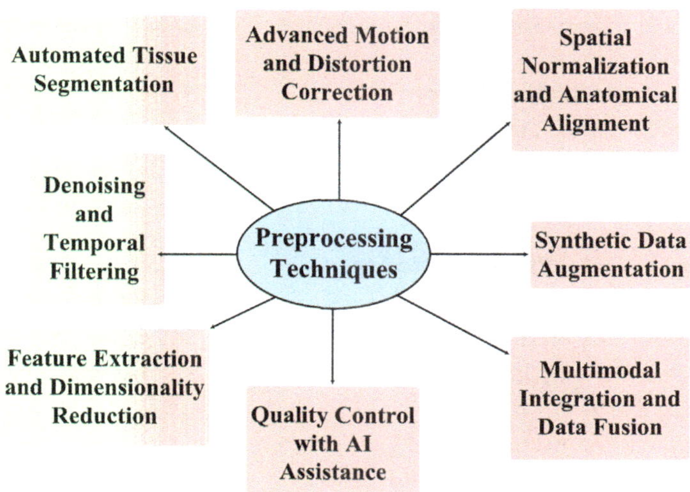

Figure 3.1 ***Preprocessing techniques for neuroimaging data.*** Preprocessing techniques.

1.2.3 Automated tissue segmentation

The precise segmentation of gray matter, white matter, cerebrospinal fluid (CSF), and other structures is a prerequisite for region-based analysis and localized generation tasks. The DL-based segmentation tools, such as FastSurfer, SynthSeg, and U-Net variants, outperforms traditional algorithms in terms of speed and accuracy. Moreover, brain parcellation using atlases (e.g., AAL, Desikan-Killiany, or Schaefer) enables region-specific modeling in generative applications such as lesion synthesis or connectivity mapping (Zhang et al., 2024).

1.2.4 Denoising and temporal filtering

The neuroimaging data is affected by physiological noise, scanner drift, and background fluctuations. The advanced temporal filtering, such as bandpass filtering and sliding-window denoising, ensures that important signal patterns are retained for generative models as they are sensitive to temporal dynamics (Zhang et al., 2025a).

1.2.5 Feature extraction and dimensionality reduction

The high-dimensional data is compressed into lower-dimensional before feeding neuroimaging data into generative models. These techniques are independent component analysis (ICA), principal component analysis (PCA), graph-based representations, and ROI-based averaging. These methods reduce computational time and enhance interpretability of the learned features (Paul et al., 2024).

1.2.6 Multimodal integration and data fusion

Generative AI models rely on multiple data types such as combining structural MRI with fMRI and DTI. Advanced preprocessing frameworks enhance multimodal registration, feature-level fusion, and shared latent space representation (Duan et al., 2024).

1.2.7 Synthetic data augmentation

The challenges related to data scarcity and class imbalance are reduced using GAN-based data augmentation. The pipeline simulates realistic variations and maintains anatomical plausibility (Polavarapu et al., 2026).

1.2.8 Quality control with AI assistance

The preprocessing pipelines include AI-driven quality control (QC) to correct suboptimal scans. MRIQC, Qoala-T, and DeepQC provide automated assessments of scan quality-based feature extraction. It reduces human workload and ensures reliability (Klapwijk et al., 2019).

In this chapter, the preprocessing techniques are validated on preexisting publicly available brain tumor MRI dataset: https://www.kaggle.com/datasets/masoudnickparvar/brain-tumor-mri-dataset.

The key contributions of this chapter are as follows:

- The various sources of noise in the medical data along with their mitigation techniques are presented.
- The artifact types and their mitigation techniques are discussed.
- The different sources introducing variability in the data along with their minimization techniques are presented.

The different types of denoising techniques are discussed in Section 3.2, followed by artifact removal techniques in Section 3.3. Further, Section 3.4 presents several variability analysis techniques in neuroimaging data. Section 3.5 presents the results and discussion. Finally, concluding remarks are presented in Section 3.6.

2. Denoising in neuroimaging data

Effective denoising is essential to enhance the signal-to-noise ratio (SNR) and ensure reliability. This section presents the various sources of noise and different denoising techniques (Herthum & Hetzer, 2024).

2.1 Sources of noise

The various noise sources are categorized as thermal noise, physiological noise, and system-related artifacts.

Thermal noise occurs due to the random movement of electrons in the imaging hardware. It is important in high-field MRI systems and low-SNR modalities such as diffusion MRI. It depicts a Gaussian distribution and affects all frequency ranges (Ayde et al., 2024).

Physiological noise occurs due to biological processes such as cardiac and brain pulsations, and respiratory movements due to blood flow. For instance, heartbeat and breathing introduce periodic artifacts in fMRI that can produce distortion in neural activities. Similarly, EEGs are distorted electromyograms and electro-oculograms (EOG) (López et al., 2023).

There are few miscellaneous noise sources such as motion artifacts (i.e., sudden head movements by patients during scans) and system-related noises (i.e., gradient coil vibrations and magnetic field inhomogeneities). Therefore, it is difficult to acquire clean, high-fidelity medical data.

2.2 Denoising techniques

The several denoising techniques for noise mitigation as follows:

2.2.1 Conventional techniques

Gaussian filtering performs convolution of the image with a Gaussian kernel, which smoothen the data by calculating the average of neighboring pixel intensities. Although it is found effective to reduce high-frequency thermal noise, it also generated blurred structural details. Thus, it is not found suitable for high spatial resolution applications (Vasilache et al., 2009).

$$F(X,\Omega) = \frac{1}{(2\pi\sigma^2)} e^{-\frac{X^2+\Omega^2}{2\sigma^2}} \tag{3.1}$$

where F represents the Gaussian function value at (X,Ω) pixel position. X and Ω represent horizontal and vertical spatial coordinates. σ represents the controlling function that decides the region to be preprocessed. Fig. 3.2 represents medical image denoising using Gaussian filtering.

Wavelet filtering decomposes the image into different frequency components using high and low pass filters, and performs noise removal from specific bands. It is found more effective than Gaussian filtering to preserve edges. Thus, it is more suitable to process MRI and CT data. However, wavelet denoising needs appropriate selection of threshold values and wavelet type, depending upon the dataset used (Noorbasha et al., 2026). Fig. 3.3 depicts wavelet filtering on medical data.

Figure 3.2 ***Gaussian denoising on medical image.*** Gaussian denoising filter is used.

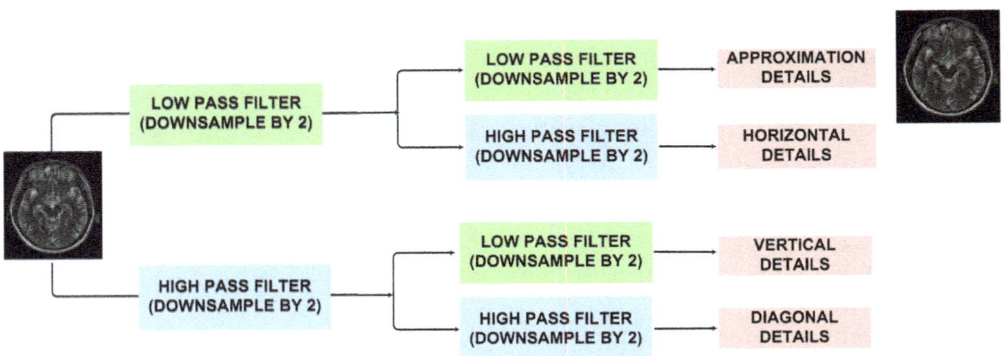

Figure 3.3 ***Wavelet denoising on medical image.*** Only approximate feature obtained using wavelet denoising is considered.

2.2.2 AI-based techniques

In the recent years, deep learning has revolutionized neuroimaging denoising for large datasets. For instance, autoencoders are trained to reconstruct clear images from noisy inputs. It is achieved by compressing data into a latent space, followed by its decoding. Denoising autoencoders (DAEs) are designed to remove noise from fMRI and diffusion MRI.

CNNs have been frequently used in denoising tasks over the years. The deep learning models such as DnCNN (denoising CNN) and UNet perform feature extraction from noisy images and generate denoised outputs. They outperformed traditional methods by capturing complex noise distributions and process anatomical details. Further, diffusion models have also been proven effective for denoising. They iteratively refine noisy images to perform reverse diffusion process, to compute high-quality reconstructions. Although they are computationally intensive, they minimize the noise and artifacts, and found useful in low-SNR modalities and low-field MRI. Fig. 3.4 depicts deep learning–based image denoising. Also, Table 3.1 presents various denoising techniques for neuroimaging data.

3. Artifact removal in neuroimaging data

The artifacts are the patterns that are integral part of acquisition processes or subject physiology. This section discusses the different artifacts in neuroimaging, followed by their correction techniques.

3.1 Types of artifacts

The artifacts occur due to hardware limitations, patients' movement, and tissue properties. Motion artifacts are caused due to involuntary head movements of patients during scanning. In MRI and fMRI, even minor movements (such as 1–2 mm) introduce blurring, ghosting, or misalignment between slices (Tripathi et al., 2024). Fig. 3.5 depicts the motion artifacts obtained in MRI data. Scanner drift refers to the signal instability over a period of time, due to rise in temperature of hardware, or magnetic field fluctuations. In long fMRI sessions, drift performs similar to low-frequency neural activity (Richer et al., 2024).

Susceptibility artifacts are present near tissue–air interfaces (such as sinuses and ear canals) where magnetic field inhomogeneities distort spatial encoding. They majorly occur in echo-planar imaging for fMRI and diffusion MRI.

Ghosting artifact is a shifted duplicate of the original modality along the phase-encoding direction. It originates from periodic signal variations that occur due to pulsatile blood flow or inconsistent gradient performance (Tripathi et al., 2024). Fig. 3.6 presents the ghosting artifacts.

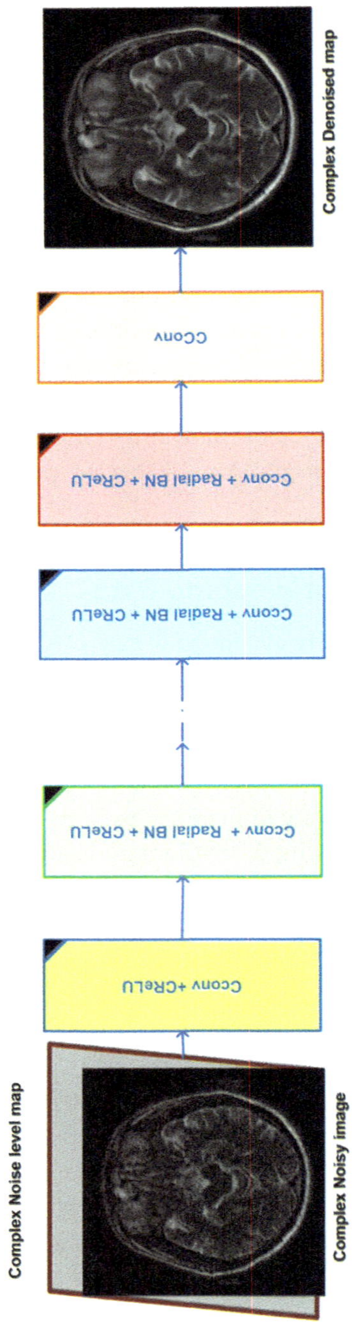

Figure 3.4 ***MRI image denoising using deep learning.*** MRI image denoising.

Table 3.1 Denoising techniques for neuroimaging data.

Algorithm	Advantages	Disadvantages	Limitations	Applications
Conventional techniques				
Gaussian filter	• Simple • High computation speed	• Oversmoothing • Loss of fine details • Fixed kernel size	• Not adaptive to varying noise levels • Poor edge preservation	• Initial preprocessing • Low-resolution scans
Wavelet filter	• Multiresolution analysis • Edge preservation • Minimizes non-Gaussian noise	• Complexity in parameter tuning • Artifacts from thresholding	• Wavelet basis selection • Sensitive to data • Not efficient for 3D dataset	• High-resolution MRI • fMRI time series
Generative AI-based techniques				
Autoencoders	• Learns data-specific features • Handles complex noise patterns • Can be unsupervised	• Requires representative training data • Black-box nature	• Oversmoothness in unique features • Memory intensive	• Multimodal datasets • Large-scale studies
CNN	• Preserves anatomical details • End-to-end training	• Needs paired clean/noisy data • Large parameter count	• GPU-dependent • Hallucination	• Clinical quality MRI • Diffusion imaging
Diffusion models	• Efficient for large noise • Progressive refinement	• Extremely resource-intensive • Slow inference	• Requires careful scheduling	• Ultralow SNR data • PET reconstruction

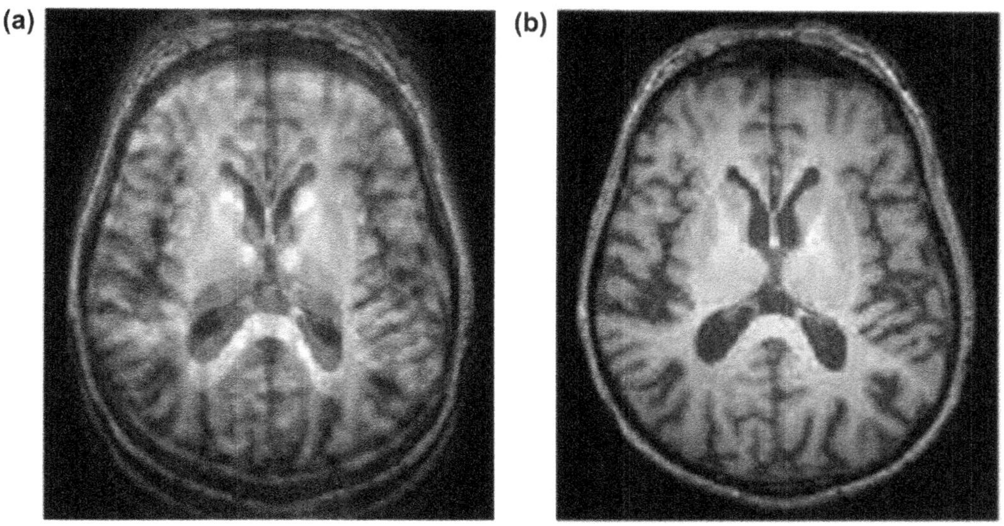

Figure 3.5 ***Motion artifacts in MRI image (A) image with motion artifacts (B) corrected image.*** Motion artifacts in MRI image.

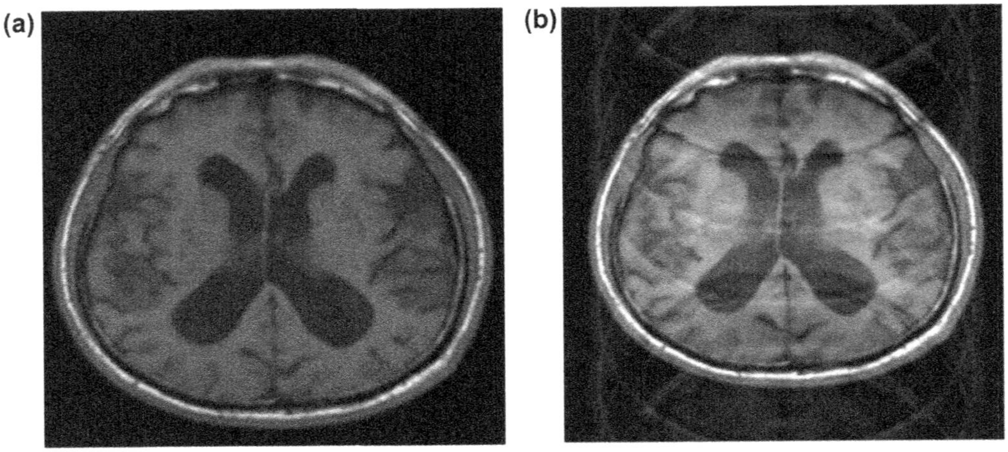

Figure 3.6 ***Ghosting artifacts in MRI image (A) original image (B) image with ghosting artifact.*** Ghosting artifacts in MRI image.

3.2 Artifacts correction techniques

This section discusses various conventional and AI-based artifacts correction techniques.

3.2.1 Conventional techniques

Gaussian filtering is also used to mitigate motion artifacts by smoothing localized distortions. However, the spatial resolution is reduced. For instance, in sMRI, a small kernel ($\sigma = 1$–2 mm) suppresses high-frequency artifacts without affecting the fine details (Lorenz et al., 2024).

Wavelet filtering is more efficient than Gaussian filtering to minimize ghosting artifacts. The images are divided into different frequency bands using combinations of low and high pass filters to preserve edges.

3.2.2 AI-based techniques

CycleGAN transforms MRI images with artifact distortions into clear images. It performs feature extraction of unpaired mappings between low- and high-quality datasets.

Patch-based GANs are used to mitigate localized artifacts (i.e., motion streaks). It only refines regions of interest (ROI) (Rana et al., 2024; Tanim et al., 2024).

Transformers with attention mechanisms are found effective in designing long-range dependencies. These models correct spatially extended artifacts. For instance, Swin transformers are used to acquire both local and global features (i.e., local artifact patterns and global context) and process the images. Vision transformer-based denoisers utilize self-attention modules to differentiate between artifacts and anatomical features.

Table 3.2 Artifact removal techniques for neuroimaging data.

Technique	Advantages	Disadvantages	Applications	Implementation considerations
		Conventional techniques		
Gaussian filtering	• Reduces high-frequency artifacts	• Blurs fine details • Fixed kernel limitations	• Mild motion artifacts • sMRI preprocessing	• Kernel size of 1–2 mm found optimal for neuroimaging
Wavelet denoising	• Multiscale artifact removal • Better edge preservation	• Complex parameter tuning • Wavelet basis selection sensitive	• Ghosting artifacts • fMRI time series	• Soft thresholding
AI-based techniques				
GAN	• Unpaired training • Mitigate large artifacts	• Probability of hallucinations • Unstable training	• Cross-scanner harmonization • Severe motion artifacts	• Needs representative artifact examples
Swin transformers	• Captures long-range dependencies • Hierarchical processing	• High memory requirements • Needs large datasets	• Ghosting correction • Whole-brain artifact patterns	• Pretraining on natural images helps
UNet++ with Attention	• Combines local/global features	• Complex architecture • Difficult to interpret	• Diffusion MRI artifacts • Multicontrast correction	• Attention gates improve lesion preservation
Federated learning models	• Multisite training • Preserves data privacy	• Communication overhead • Heterogenous data challenges	• Large-scale studies • Clinical collaborations	• Needs standardized preprocessing

Some researchers proposed hybrid approaches that combine the advantages of CNNs and transformers. For instance, UNet++ is combined with attention gates to suppress motion artifacts while preserving the edges (Jasmin et al., 2024).

Table 3.2 presents various artifact removal techniques for neuroimaging data.

4. Variability in neuroimaging data

This section discusses the variability sources and data harmonization techniques (Pinto et al., 2020).

4.1 Sources of variability

The variability in neuroimaging data originates due to different subjects and scanning equipment, and inconsistencies in acquisition protocols.

There is natural biological variability among different individuals based on brain size, cortical folding patterns, and the white matter tracts distribution. It is observed that different individuals of similar age and geographical location have up to 15%–20% variation in cortical thickness. In fMRI, the variability increases due to cognitive strategies, neural efficiency, and baseline metabolic activity.

Scanner variations refer to technical variability that reflects in biological signals. The different MRI manufacturers (such as Siemens, GE, and Philips) or different articles from the same manufacturer have different image contrasts due to variations in pulse sequences, gradient performance, and radiofrequency coils. The machines with higher field strengths generate high signal-to-noise ratio but also introduce more artifacts.

Protocol inconsistencies are present in and across imaging sites. The performance metrics such as voxel size, repetition time, echo time, flip angle, and bandwidth affects efficiency of the machine. The performance of multisite machines leads to increase in variance and reduction in statistical power.

4.2 Harmonization techniques

Harmonization techniques reduce technical variability across neuroimaging datasets to preserve biological differences (Pinto et al., 2020).

4.2.1 Conventional techniques

Combating batch effects (ComBat) is proven to be an effective technique for multisite harmonization. It is designed to process genomic data. It uses empirical Bayes estimation to mitigate site-specific biases without affecting any biological variability. The site effects act as additive and multiplicative to find mean differences and variance differences, respectively.

It aligns the distributions across scanners irrespective of subject matching. The performance of sMRI is calculated using evaluation metrics such as cortical thickness and volumetric measures, such that 30%–50% of cross-site variance is achieved (Jasmin et al., 2024).

Z-score normalization transforms image intensities to standard values obtained from a reference population. It minimizes the gross differences between contrast and brightness. Another normalization technique designed specific to MRI such as white stripe normalization is also utilized as it uses robust intensity references from white matter tissue.

4.2.2 *AI-based techniques*

The recent machine learning algorithms are found effective to solve nonlinear relationships and higher-dimensional data. The domain adaptation methodologies compute scanner-invariant features. The deep learning models such as domain-adversarial neural networks (DANNs) uses a dual optimization strategy, that is, a generator and a discriminator. The generator produces similar features across various scanners, while a discriminator is used to find the scanner source. The network neglects the scanner-specific information during adversarial training, while retaining the biological features. It is commonly applied to fMRI data and has achieved 15%–20% higher accuracy in cross-scanner classification as compared with unharmonized data.

Another technique used to perform the data harmonization is style transfer using VAEs. The anatomical information is separated from the scanner-specific appearance in their latent representations. The features of one scan are combined with the style of another. The synthetic images are generated using different scanners and preserve the underlying biology. The researchers have designed 3D VAEs to normalize T1-weighted MRI across vendors. The generated synthetic images have achieved more than 90% similarity with real scans.

Federated learning has achieved traction for multisite models in which data sharing is restricted. The models are trained locally at each site and periodically aggregated instead of data centralization. Thus, harmonization is achieved without any direct access to the raw data. It is observed that the federated versions of ComBat have achieved 80%–90% of the performance of centralized harmonization without affecting data confidentiality.

Table 3.3 presents various variability minimization techniques for neuroimaging data.

5. Conclusion and future scope

The conclusion and future scope are presented in this section.

5.1 Conclusion

The data handling and preprocessing are performed on neuroimaging data obtained from different modalities. This chapter presented various algorithms for mitigation of noise, artifacts, and variability, which enhances the performance of GANs, VAEs, and diffusion models. Although conventional techniques are relevant for preprocessing, the AI-driven methods are found superior in handling complex and nonlinear distortions as they preserve biological signals. The preprocessing pipelines are integrated with generative architectures to improve synthetic data quality and model stability. The Gaussian and wavelet filters are used to mitigate noise and artifacts from the images of different modalities. Further, CNNs, autoencoders, and wavelet CNNs are also used

Table 3.3 Variability minimization techniques for neuroimaging data.

Technique	Advantages	Disadvantages	Applications	Implementation considerations
		Sources of variability		
Biological (intersubject differences)	• Depicts biological diversity • Essential for personalized medicine	• Inaccurate group analyses • Requires large sample sizes	• Genetic studies • Precision neuroimaging	• Individual differences are preserved during normalization
Technical (scanner variations)	• Reflects technological progress • Different clinical strengths	• Introduces nonbiological variance • Vendor-specific artifacts	• Multicenter trials • Longitudinal studies	• Phantom scans recommended for QC
Protocol (inconsistencies)	• Allows protocol optimization • Site-specific adaptations	• Affects quantitative comparisons • Impacts reproducibility	• Retrospective studies • Data pooling	• Standard protocols are preferred
Conventional harmonization				
ComBat	• Handles additive/multiplicative effects • Preserves biological variance	• Assumes parametric distributions • Needs site metadata	• Structural MRI metrics • Multisite studies	• Availability of combat harmonization package
Z-score normalization	• Simple implementation • No training required	• Sensitive to outliers • Linear only	• Single-site intensity standardization	• White matter improves robustness
AI-based harmonization				
Domain-adversarial networks	• Learns nonlinear mappings • Scanner-invariant features	• Needs representative data • Complex training	• fMRI connectivity studies • Disease classification	• Gradient reversal layer critical
VAE style transfer	• Generates synthetic images • Separates content/style	• Memory intensive	• Cross-vendor MRI harmonization	• Latent space regularization needed
Federated learning	• Preserves data privacy • Multi-site collaboration	• Communication overhead • Heterogenous data	• Clinical data partnerships • Global studies	• Differential privacy may be required

to mitigate noises from modalities. Moreover, CycleGAN, patch-based GANs, swin transformers, combination of UNet and attention, and federated learning models are used to mitigate artifacts. Also, ComBat, Z-score normalization, and VAEs are used to minimize variability in neuroimaging data.

5.2 Future scope

The proposed work can be extended by future researchers as follows: Real-time adaptive processing systems can be designed using lightweight neural networks to enable instant artifact correction while scanning. Also, multimodal harmonization techniques can be improved to integrate structural, functional, and electrophysiological data streams. The federated learning frameworks can be combined with built-in quality control mechanisms to overcome ongoing limitations in multicenter data sharing. Moreover, standard evaluation metrics must be designed for processed outputs. In future, professionals of different domains such as AI researchers, imaging physicists, and practicing neurologists can work together to design new methodologies and provide temporal and spatial advantages over conventional approaches.

Conflict of interests

Authors have no conflict of interests.

AI tool

No AI tool has been used to prepare this manuscript.

References

Al-zahrani, M. S., Alsaade, F. W., & Alsaadi, F. E. (2025). Variational autoencoder-guided data augmentation for imbalanced medical diagnostics Author links open overlay panel. *Results in Engineering, 27*. https://doi.org/10.1016/j.rineng.2025.107011

Ayde, R., Vornehm, M., Zhao, Y., Wu, E. X., & Sarracanie, M. (2024). MRI at low field: A review of software solutions for improving SNR. *NMR in Biomedicine, 38*(1). https://doi.org/10.1002/nbm.5268

Chaudhary, U. (2025). Foundation of brain-computer interface technology non-invasive brain signal acquisition techniques: Exploring EEG, EOG, fNIRS, fMRI, MEG, and fUS. In *Expanding senses using neurotechnology. Vol. 1*. Cham: Springer. https://doi.org/10.1007/978-3-031-76081-5_2

Jasmin, N. H., Rizhan Wan Idris, W. M., Hilmi, N. A. M., Zulkifli, M. I., Roslan, N. A., & Mohd Azhar, S. N. A. (2024). Interactive 3D models and animations for cardiac imaging education. *Journal of Medical Imaging and Radiation Sciences, 55*(3), Article 101547. https://doi.org/10.1016/j.jmir.2024.101547

Duan, J., Xiong, J., Li, Y., & Ding, W. (2024). Deep learning based multimodal biomedical data fusion: An overview and comparative review. *Information Fusion, 112*, Article 102536. https://doi.org/10.1016/j.inffus.2024.102536

Gautam, P., & Singh, M. (2025). A multidimensional U-net architecture for Alzheimer's disease classification using T1 weighted brain structural MRI. *Computers and Electrical Engineering, 128*(Part A). https://doi.org/10.1016/j.compeleceng.2025.110687

Herthum, H., & Hetzer, S. (2024). Tensor denoising of quantitative multi-parameter mapping. *Magnetic Resonance in Medicine, 92*(1), 145–157. https://doi.org/10.1002/mrm.30050

Islam, S., Aziz, Md. T., Nabil, H. R., Rahman Jim, J., Mridha, M. F., Mohsin Kabir, Md, Asai, N., & Shin, J. (2024). Generative Adversarial Networks (GANs) in medical imaging: Advancements, applications, and challenges. *IEEE Access, 12*, 35728–35753. https://doi.org/10.1109/ACCESS.2024.3370848

Noorbasha, S. K., Arunkumar Gudivadab, A., Raveendra, P., Murali Krishna, Y., & Aravindraj, E. (2026). Skewness based discreate wavelet transform and swarm decomposition for the automatic elimination of ocular artifacts from EEG signals. *Computer Methods in Biomechanics and Biomedical Engineering*, 1–14. https://doi.org/10.1080/10255842.2026.2636254

Klapwijk, E. T., van de Kamp, F., van der Meulen, M., Peters, S., & Wierenga, L. M. (2019). Qoala-T: A supervised-learning tool for quality control of FreeSurfer segmented MRI data. *NeuroImage, 189*, 116–129. https://doi.org/10.1016/j.neuroimage.2019.01.014

Polavarapu, B. L., Dinesh Reddy, V., Kumar Morampudi, M., Hussain, Md M., & Abdul, A. (2026). A feature fusion framework for improved autism spectrum disorder prediction using sMRI and phenotype information. *Journal of Sensor and Actuator Networks, 15*, 1–21. https://doi.org/10.3390/jsan15010021

Liang, J., He, R., & Tan, T. (2025). A comprehensive survey on test-time adaptation under distribution shifts. *International Journal of Computer Vision, 133*, 31–64. https://doi.org/10.1007/s11263-024-02181-w

López, A., Villar, J. R., Fernández, M., & Ferrero, F. J. (2023). Comparison of classification techniques for the control of EOG-based HCIs. *Biomedical Signal Processing and Control, 80*, Article 104263. https://doi.org/10.1016/j.bspc.2022.104263

Lorenz, E. A., Su, X., & Skjæret-Maroni, N. (2024). A review of combined functional neuroimaging and motion capture for motor rehabilitation. *Journal of NeuroEngineering and Rehabilitation, 21*(1). https://doi.org/10.1186/s12984-023-01294-6

Mårtensson, G., Ferreira, D., Granberg, T., Cavallin, L., Oppedal, K., Padovani, A., Rektorova, I., Bonanni, L., Pardini, M., Kramberger, M. G., Taylor, J.-P., Hort, J., Snædal, J., Kulisevsky, J., Blanc, F., Antonini, A., Mecocci, P., Vellas, B., Tsolaki, M., … Westman, E. (2020). The reliability of a deep learning model in clinical out-of-distribution MRI data: A multicohort study. *Medical Image Analysis, 66*. https://doi.org/10.1016/j.media.2020.101714

Naghipour, J., Ghazizadeh, R., & Hadi, M. (2026). GAN-based deep learning strategy for advanced EEG focal and non-focal classification in epilepsy. *Signal Image and Video Processing, 20*. https://doi.org/10.1007/s11760-026-05155-2

Nowroozilarki, Z., Mortazavi, B. J., & Jafari, R. (2024). Variational autoencoders for biomedical signal morphology clustering and noise detection. *IEEE Journal of Biomedical and Health Informatics, 28*(1), 169–180. https://doi.org/10.1109/JBHI.2023.3320585

Paul, S. G., Saha, A., Hasan, Md. Z., Noori, S. R. H., & Ahmed, M. (2024). A systematic review of graph neural network in healthcare-based applications: Recent advances, trends, and future directions. *IEEE Access, 12*, 15145–15170. https://doi.org/10.1109/ACCESS.2024.3354809

Pinto, M. S., Paolella, R., Billiet, T., Van Dyck, P., Guns, P. J., Jeurissen, B., Ribbens, A., den Dekker, A. J., & Sijbers, J. (2020). Harmonization of brain diffusion MRI: Concepts and methods. *Frontiers in Neuroscience, 14*. https://doi.org/10.3389/fnins.2020.00396

Qiang, N., Gao, J., Dong, Q., Yue, H., Liang, H., Liu, L., Yu, J., Hu, J., Zhang, S., Bao, G., Sun, Y., Liu, Z., Liu, T., Jin, L., Song, H., & Zhao, S. (2023). Functional brain network identification and fMRI augmentation using a VAE-GAN framework. *Computers in Biology and Medicine, 165*. https://doi.org/10.1016/j.compbiomed.2023.107395

Rana, S. U., Alam, M. N., Akter, S., & Likhon, M. N. (2024). KidneyMultiNet: A web-based automatic system for kidney disease detection using hybrid machine learning model from CT scan images. *Biomedical Materials & Devices*, 1–12. https://doi.org/10.1007/s44174-024-00231-y

Raval, V., Nguyen, K. P., Pinho, M., Dewey, R. B., Jr., Trivedi, M., & Montillo, A. A. (2022). Pitfalls and recommended strategies and metrics for suppressing motion artifacts in functional MRI. *Neuroinformatics, 20*. https://doi.org/10.1007/s12021-022-09565-8

Richer, N., Cortney Bradford, J., & Ferris, D. P. (2024). Mobile neuroimaging: What we have learned about the neural control of human walking, with an emphasis on EEG-based research. *Neuroscience & Biobehavioral Reviews, 162*, Article 105718. https://doi.org/10.1016/j.neubiorev.2024.105718

Saad, M. M., O'Reilly, R., & Husain Rehmani, M. (2024). A survey on training challenges in generative adversarial networks for biomedical image analysis. *Artificial Intelligence Review, 57*. https://doi.org/10.1007/s10462-023-10624-y

Tanim, S. A., Alam, G. M. I., Shrestha, T. E., Islam, M., Jahan, F., & Nur, K. (2024). Breast cancer diagnosis with XAI-integrated deep learning approach. In *International conference on innovation and intelligence for informatics, computing, and technologies (3ICT)* (pp. 659–665). Bangladesh: Institute of Electrical and Electronics Engineers Inc.. https://doi.org/10.1109/3ict64318.2024.10824574

Tripathi, V. R., Tibdewal, M. N., & Mishra, R. (2024). A survey on motion artifact correction in magnetic resonance imaging for improved diagnostics. *SN Computer Science, 5*(3). https://doi.org/10.1007/s42979-023-02596-1

Vasilache, S., Ward, K., Cockrell, C., Ha, J., & Najarian, K. (2009). Unified wavelet and Gaussian filtering for segmentation of CT images; application in segmentation of bone in pelvic CT images. *BMC Medical Informatics and Decision Making, 9*. https://doi.org/10.1186/1472-6947-9-S1-S8

Vetter, J., Macke, J. H., & Gao, R. (2024). Generating realistic neurophysiological time series with denoising diffusion probabilistic models. *Patterns, 5*(9). https://doi.org/10.1016/j.patter.2024.101047

Zhang, X., Chen, M., Li, Y., Gao, J., Sun, Y., Liu, F., & Wei, S. (2025a). A dual-path interactive denoising autoencoder for removing mixed noise in multi-lead electrocardiogram signals. *Computers and Electrical Engineering, 123*(Part B). https://doi.org/10.1016/j.compeleceng.2025.110177

Zhang, F., Cho, K. I. K., Seitz-Holland, J., Ning, L., Haitz Legarreta, J., & Rathi, Y. (2024). DDParcel: Deep learning anatomical brain parcellation from diffusion MRI. *IEEE Transactions on Medical Imaging, 43*(3), 1191–1202. https://doi.org/10.1109/TMI.2023.3331691

Zhang, Y., Kong, X., He, W., & Xu, Z. (2025b). Eddy current characterization using efficient finite element analysis for Z-gradient coils in a low-field MRI scanner. *IEEE Transactions on Instrumentation and Measurement, 74*, 1–9. http://10.1109/TIM.2025.3544743.

PART II

Methodological advancements

CHAPTER 4

Generative adversarial networks (GANs) for neuroimaging networks

Ankur Chaudhary[1], **Manish Kumar**[2] **and Ritesh Rastogi**[1]
[1]Department of Information Technology, Noida Institute of Engineering & Technology, Greater Noida, Uttar Pradesh, India; [2]School of Computer Science & Engineering, IILM University, Gurugram, Haryana, India

1. Introduction

In the last 10 years, there have been incredible changes in the field of neuroimaging due to the development and improvement of various imaging technologies such as MRI, positron emission tomography (PET), CT, and functional magnetic resonance imaging (fMRI). With the increased spatial and temporal resolution provided by these methodologies, it has become possible to see what is going on in the living human brain at an ever deeper level; this also includes its structural and functional organization. As a result, many steps forward have been taken in comprehending diverse neurological and psychiatric problems such as Alzheimer's disease, brain tumors, epilepsy, multiple sclerosis, etc. But on the other hand, the evolution of these imaging modalities led to an exponential growth of data volume, dimensions, and complexity. Currently, there are massive neuroimaging databases provided nationally or internationally, offered by some consortia such as ADNI and Human Connectome Project (HCP). These datasets comprise multimodal information taken over multiple time points with additional information about patients' age and sex, and often include demographic, clinical, cognitive, and genomic data. It remains difficult for researchers to extract relevant patterns and clinically effective information from such complex high-dimensional data, which is also very different in nature.

Despite their continued relevance, customary statistical methods and conventional machine learning algorithms often prove inadequate at handling the nonlinear and intricate features inherent in brain scans. This gap is increasingly being filled by incorporating artificial intelligence (AI) into medical imaging workflows. Within AI research, the field known as deep learning has gained much attention lately for its great potential in solving a variety of problems without needing manually designed solutions called features or descriptors that would otherwise be used if one employed traditional image processing techniques, for example, histogram equalization followed by edge detection and filter bank analysis, and convolutional neural networks (CNNs). A good example of this is the CNNs, which are already showing exceptional results across a wide range of neuroimaging applications, including image segmentation, anomaly detection, disease

Generative Artificial Intelligence for Neuroimaging
ISBN 978-0-443-45010-5
https://doi.org/10.1016/B978-0-443-45010-5.01004-2

classification, and prognosis estimation, among others. Many other architectures exist today within the deep learning realm, but there are some, such as generative adversarial networks (GANs), whose unmatched capability lies in producing authentic digital pictures (Goodfellow et al., 2014).

In their 2014 publication, Goodfellow et al. described a GAN as two neural networks, one playing the role of a generator and the other a discriminator, which are both trained by means of competition with each other, more particularly in an adversarial fashion. The work is meant to see how close the generated data can get to looking like real samples while the discriminator tries hard to tell if what it sees is not real data but only some fake data. As time goes by, this contradictory process yields very good outputs that capture the underlying data structure without any explicit modeling assumptions. Deep learning is seen in a different light with the emergence of GANs for neuroimaging. In fact, they were applied successfully for the expansion of training datasets, solving issues related to small sizes and unequal class distribution through generations of realistic brain MRIs or CT scans known as augmented artificial intelligence (AAI). It has become possible to generate PET images from MRI data using a modality translation approach based on CycleGANs and conditional GANs (cGANs) such that even if certain scans are unavailable, one could still obtain some information on multiple cross-cutting modalities. Denoising, superresolution, and artifact correction, are examples of its applications aimed at improving image quality even when obtained under unfavorable conditions in patients with neurological disorders. On top of that, GANs are now employed in creating virtual patients that simulate personalized disease progression; for example, they can generate an older version of one's brain scan & predict how lesions will change over time.

Various recent forms of GANs have tackled numerous technical challenges encountered within specific contexts of application, too. An example being Wasserstein GANs (WGANs), which enhance training stability through the adoption of a less sensitive objective function, while another case relates to StyleGANs that support controllable and interpretable synthesis of brain images. CycleGANs belong to a class of their own because they can perform image-to-image translations without needing matching examples: this technology finds special application areas within the healthcare sector due to the unavailability of a typically paired dataset. Therefore, these improvements rendered GANs more adaptable and less vulnerable, particularly in the neuroimaging field. Nonetheless, much work remains to be done in this promising area. The anatomical validity of these artificial pictures remains uncertain and must therefore undergo very careful confirmation before being applied clinically either for prognosis or diagnosis. Mode collapse could also occur, whereby generators just learn a few patterns of images, hence decreasing diversity in synthetic datasets. Besides, ethics-related issues concerning synthetic data usage, its possible abuse, patient anonymization, as well as explainable AI in the decision support systems context should not be underestimated prior to clinical deployment of these models (Seeliger et al., 2017).

In this chapter, we will look at how GANs have changed with time and are now used in neuroimaging just like other advanced technologies. To achieve this objective, the chapter discusses the technical basics, practical use, and possible applications of GANs for neuroimaging. By going through some of the fundamental as well as most recent concepts, we hope to show that these models can help us see and understand data in a new way, particularly data emerging from neuroimaging, which may be very important both within traditional healthcare practices as well as research on such issues. With advancements witnessed in the field, it is anticipated that GAN-centered approaches will gain importance in customized healthcare, accurate diagnostics, and computational neuroscience, among other emerging fields.

In Fig. 4.1, we can see the important roles of GANs in neuroimaging. It is evident in the figure that these networks can be used in generating fake brain pictures to boost other tasks, such as conversion across different modalities, for instance, from MRI scans into PET images. To be specific, GANs increase the level at which a diagnosis is correct or identifies some facts, help in recognizing when there is something wrong, imitate the way diseases evolve and fill up any blank spots on an image. Such capabilities facilitate intensive as well as individualized examinations under limited clinical data.

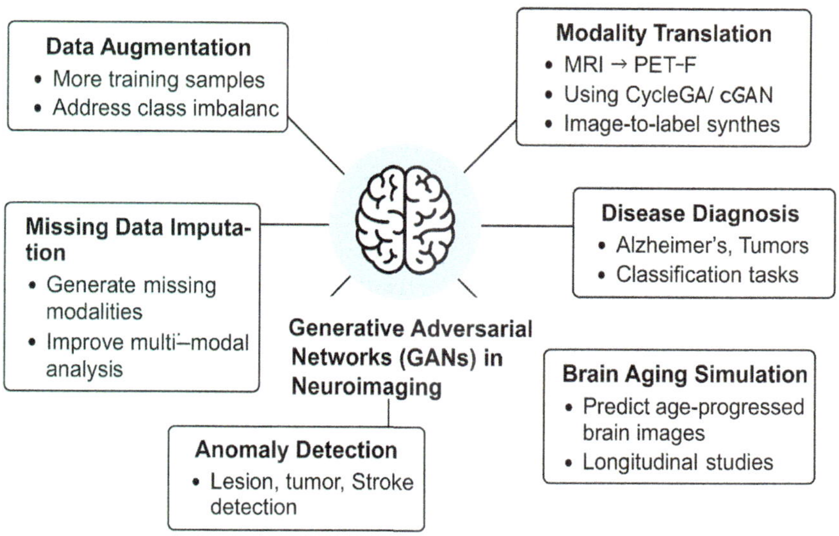

Figure 4.1 Overview of GAN applications in neuroimaging.

2. Fundamentals of generative adversarial networks

2.1 Architecture of GANs: Generator versus discriminator

Goodfellow et al. in 2014 introduced an innovative adversarial training approach for data generation that is centered on the interaction between two neural networks: the Generator (G) and the Discriminator (D). The idea is based on GAN architecture (Goodfellow et al., 2014).

2.1.1 The generator (G)

The generator is an upsampling deep neural network that takes a random input from a prior distribution (commonly Gaussian or uniform) and transforms it into a fake data sample, which resembles the real data and can be represented as $G(z)$. It works on tricking the discriminator that the produced sample is authentic with respect to its objectives. Generators in terms of architecture usually employ some form of up-sampling strategy; examples include transposed convolutions or nearest neighbor interpolation, particularly when dealing with tasks such as generating brain scans or reconstructing PET images. In the neuroimaging context, it is possible for the generator to consider certain attributes (e.g., age, disease, imaging type) to yield very realistic contextually appropriate outputs, for example, synthesizing MRIs for Alzheimer patients or creating MRI-based PET scan images.

2.1.2 The discriminator (D)

The discriminator works by classifying as either genuine or fake. It is supplied with real data $x \sim p\text{data}(x)$ extracted from the training set and with false data $G(z)$ produced by a generator, then it gives a prediction whether it is real or fake. The objective of the discriminator is to increase classification accuracy, that is, to correctly determine whether the input provided to it was taken from the real data samples or generated by G. For this purpose, discriminators are usually designed using CNNs, which can capture both local and global structures in inputs such as textures or anatomy present in medical images. This discriminator can generate images that appear real as well as biologically plausible because, within its many applications, it learns the fine details of anatomy.

2.2 Training methodology: Adversarial process

The training of GANs follows an adversarial framework where there are two neural networks: one is the generator, while the other is the discriminator. Training involves both models at the same time with contradicting goals, creating a two-player game theory. Specifically, the generator is trained to produce synthetic samples that approximate the true data distribution, while the discriminator is trained to tell fake and real samples apart. The discriminator provides a probability score for whether a particular input is real or fake. Thus learning the structure of the target data distribution. This adversarial interaction is defined by a mathematical objective, typically in the form of:

$$L = \log(D(x)) + \log(1 - D(G(z))) \tag{2.1}$$

In this expression:

- x is a real data sample
- z is a noise vector sampled from a prior distribution (e.g., Gaussian)
- $D(x)$ is the discriminator's output when fed a real sample
- $D(G(z))$ is its output when fed a fake sample generated by G.

Here, training involves giving the discriminator feedback to enhance its capacity for telling what is real from what is not, then doing the same for the generator, but in this case, on improving its fake samples. With this kind of training, it is possible for the GAN to get better and better at a certain point where it cannot discriminate between real data and fake data anymore.

This kind of learning is essential in creating models that can work well with data, which may be very complex and high-dimensional, too. It is for this reason that GANs are now widely used in tasks such as creating new images or improving the existing ones. Nevertheless, training GANs has also remained difficult and unreliable. The first one is known as mode collapse, whereby the generator starts producing very few examples because it ignores most of the real data. Another one is vanishing gradients, a common challenge experienced when the discriminator provides weak feedback because of being too strong itself.

Several improved GAN frameworks have been proposed to overcome these issues. An example includes the Wasserstein GAN (WGAN), which differs from the initially employed Jensen–Shannon divergence loss function by using Wasserstein distance as a metric of dissimilarity. This change results in smoother gradient flow with improved convergence properties during training; hence, WGANs are more resilient & easier to train in practice.

2.2.1 Adversarial training in neuroimaging

Adversarial training in neuroimaging makes it possible to create high-quality brain pictures that are close to what they should look like anatomically, even when there is not much data. Therefore GANs are very appropriate in adding data into research with few patients, as well as aiding in multiple issues, such as cross-disease progression evolution analysis. To be specific, Xia et al. (2021) employed a Transformer-GAN architecture to learn the dynamics of structural changes from longitudinal MRI data, proving that generative models could simulate age-related atrophy profiles (Seeliger et al., 2017). The use of such technologies has extended into medical practice, whereby GANs are used for tasks such as denoising, superresolution, and cross-modality imaging, among others, because of their ability to improve the images through numerous iterations, which involve both generation and discrimination processes.

In addition to this, GANs' capacity for producing realistic nonidentifiable synthesized neuroimages promotes confidentiality-preserved information exchange and AI model creation under data-limited situations. In conclusion, adversarial training stands out as a valuable introduction that addresses longstanding issues experienced within neuroimaging, such as little data for analysis, and how well findings can apply to different groups.

2.3 Evaluation metrics for GAN performance

Assessing how good GANs are in neuroimaging can be quite difficult because they are supposed to appear real and, at the same time, be biologically valid. While classification models use predetermined metrics such as accuracy or area under the curve (AUC) for evaluation, this is not applicable for GANs; rather than comparing at a pixel level alone, with some other issues such as the structure of images or even clinical data, it requires evaluation of the closeness of generated data to real data. The text provided below outlines some common evaluation approaches employed in research (Islam & Zhang, 2020):

- **Inception Score (IS):** The Inception score (IS) is a widely used metric for assessing the quality and diversity of synthetic images. While this measure is quite reliable within the field of computer vision, its applicability for neuroimaging is questionable, given that natural image databases are not entirely suitable for medical imaging purposes.
- **Fréchet Inception Distance (FID):** In the comparison, the mean and covariance are used to compare the characteristic distribution of real and generated data. The closer these two datasets are, the lower their FIDs. Nevertheless, in neuroimaging, specific medical encoders must be employed rather than general Inception networks for the significance of this measure.
- **Structural Similarity Index (SSIM):** SSIM is commonly applied in medical imaging for assessing perceptual similarity as well as structural changes, which may include comparison of intensity levels, contrast, and texture details. Thus, making it very important for assessing synthetic brain images.
- **Peak Signal-to-Noise Ratio (PSNR):** The peak signal-to-noise ratio (PSNR) is one of the fundamental metrics for image fidelity, and it is used for measuring pixel-wise errors between the original image and reconstructed image. This metric is simple to calculate; however, it does not consider the human eye characteristics and, as such, it might wrongly portray anatomical malformations.
- **Visual Turing Test/Expert Evaluation:** Numerous neuroimaging research involves specialists who determine if computer-generated pictures can be told from

the real ones. This human-in-the-loop approach continues to be vitally important for confirming clinical realism and anatomical fidelity in various imaging studies.

- **Downstream Task Performance:** The effectiveness of a GAN can be best tested in neuroimaging by applying its results to some downstream jobs, such as disease classification and segmentation. This will help in understanding whether the GAN has added any value to the model used. If the accuracy improves on a dataset that has been enhanced with GAN samples, then it means that these samples are of high quality and useful as far as the training process is concerned.

GANs' evaluation in neuroimaging goes beyond the use of conventional image quality metrics. It is very difficult for the generator networks to produce clinical neuroimages with relevant data that could be easily masked out. As such, it is recommended for researchers to employ a combined measure of evaluation that comprises quantitative image similarity metrics, expert-based qualitative reviews, and performance metrics based on applications. By doing this, we make sure that the artificial data appears real on the one hand and is biologically and clinically informative on the other.

2.4 Popular GAN variants in neuroimaging

The first GAN presented a revolutionary approach to data generation, which could imitate even highly intricate real-world patterns. Nonetheless, the original formulation encounters various difficulties, such as unstable training, lack of control over outputs, and vulnerability to data distribution. Several GAN variations have been developed to overcome these challenges and make them applicable in specialized areas. Specifically for neuroimaging, these different forms are very important for creating images that can be controlled, of high quality, and do not change much even when used in different clinical scenarios. This write-up examines some of the common GANs in relation to neuroimaging (Wang et al., 2023).

2.4.1 Conditional GAN (cGAN)

The original GAN architecture can be improved using conditional GANs, which make use of additional information y (such as labels, age, diagnosis, modality) as input for both the generator and discriminator. With this extra conditioning, the model can now produce data that follows the semantics of the input data. In neuroimaging, cGANs find application in:

- Generation of synthetic MRIs or PET scans intended for specific diseases or demographic groups.
- Formulating disease evolution models, such as the transformation from young to elderly brain structures in the neuroimaging of certain patients.
- Training on one type of labeled data and translating it into other modalities (for example, "label-to-image" problem.

For instance, researchers applied cGANs to generate MRI-derived PET images in Alzheimer's studies, providing a low-cost PET substitute (Goodfellow et al., 2014).

2.4.2 CycleGAN

CycleGANs were created specifically for performing image-to-image translations without the need for paired input data. This property is very important in medical imaging since it is usually impossible to find paired data from the same patient obtained on various imaging modalities. It consists of two generators and two discriminators with an additional cycle-consistency loss that makes sure that after translating from domain A to B and back to A, we obtain an image close to x again.

In neuroimaging, CycleGANs find applications in tasks like:

- Conversion between T1-weighted and T2-weighted MRIs.
- Generating missing modalities (e.g., FLAIR or diffusion-weighted imaging).
- Creating artificial cross-sectional data changes related to age or illness in disparate groups (Seeliger et al., 2017).

2.4.3 Wasserstein GAN (WGAN)

The WGAN is different from the original GAN because it uses the Wasserstein distance (Earth-Mover's Distance) instead of the Jensen–Shannon divergence. Because of this fact, training stability is enhanced, and gradients are smoother. As a result, these types of GANs can be very useful for dealing with volumetric brain data.

In neuroimaging applications, WGANs are highly valuable since:

- The generator has to keep in mind very small details, such as cortical thickness or where exactly one part ends, and another begins, when it creates anatomical pathology pictures.
- A situation may arise when modes fall down, which will have a significant impact on clinical applicability.
- On top of that, scientists want to obtain reliable results even with incomplete information (small/imbalanced samples) (Nguyen et al., 2018).

2.4.4 Least squares GAN (LSGAN)

Least squares generative adversarial networks (LSGANs) are an adaptation of the traditional GAN loss function. The adaptation involves replacing binary cross-entropy with a less square loss. By doing this, there is a decreased probability for the gradients to vanish, and therefore the generator is pushed better toward producing samples that are close to the true data distribution. Even though they are not as widely used as WGANs or CycleGANs, LSGANs provide some key advantages and are recommended for training when one intends:

- Removal of noise and artifacts from medical images.

- Tasks that require high pixel similarity, such as fMRI signal denoising or upscaling low-quality pictures.

2.4.5 StyleGAN

The introduction of Style-based generative network (StyleGAN) has greatly revolutionized image synthesis with its ability to disentangle high-level attributes from low-level ones through adaptive instance normalization (AdaIN). By doing this, it becomes possible to manipulate every detail of an image, such as its shape and texture, among others. Although initially designed for the generation of human faces, researchers are now looking at how they can apply StyleGAN to produce:

- High-quality anatomical MRIs that look real and come in many different forms or types.
- Conditional generation of brain morphology based on certain clinical profiles (age, gender, and disease).
- Exploring controllable latent spaces to analyze between-group differences in brain structures (Islam & Zhang, 2020).

2.4.6 StyleGAN

Pix2Pix is a type of GAN used for image-to-image translations, which are supervised and need paired data for training. The generator in Pix2Pix learns the mapping from input to output images by combining the L1 loss function with adversarial loss. In neuroimaging, Pix2Pix would work well for tasks that involve paired images, such as:

- Conversion of segmentation to anatomical brain images.
- Changing CT scans into MRIs or vice versa with appropriate pairs taken.
- Localization of brain tumors and mask creation used in scheduling surgery plans.

Even though it is rarely applied with clinical data due to the need for pairs, Pix2Pix can still be considered an important research tool that is employed within various deep learning frameworks for synthesizing artificial information.

All these versions have their special strengths that are suitable for different types of images. The choice of the most suitable GAN architecture may depend on the size of the dataset as well as the availability of paired or unpaired data, what kind of output control is needed & where this is applied. Such options gain increased importance within neuroimaging since their results are commonly applied within clinical diagnostics.

3. Applications for GANs in neuroimaging

GANs are very flexible, and this has made them useful in almost every part or sector of neuroimaging. It is now possible for GANs to enhance the quality of brain scans, simulate diseases, and carry out cross-modality synthesis among many other applications, which have been found both within research and clinical practice. This is because they can produce fake pictures that are just like real ones, containing details at their appropriate places while also being very clear or sharp, thus overcoming several conventional challenges in neuroimaging, for example, low data volume, nonavailability of some modalities, and confidentiality, among others. These techniques not only improve current imaging systems, but they also create opportunities for early detection, customized treatment plans, and monitoring of changes in the human brain over time (Yan et al., 2022) (Figs. 4.2, 4.3, and 4.4).

3.1 Data augmentation in limited and imbalanced datasets

The issue of having datasets that are imbalanced and limited is very common in neuroimaging. This problem can be solved using GANs, which have been found to generate artificial samples. Through the generation of fake brain images with closely distributed class properties to those of the authentic data, GANs aid in resolving class imbalance while increasing available training samples. As a result, it becomes more immune and applicable even to less common cases such as initial-stage Alzheimer's or tumors (Ali et al., 2022).

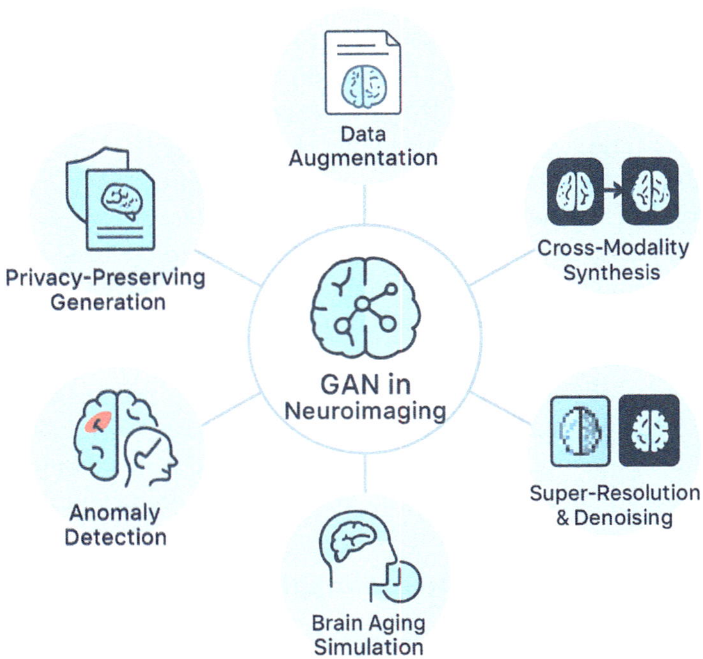

Figure 4.2 Overview of GAN applications in neuroimaging.

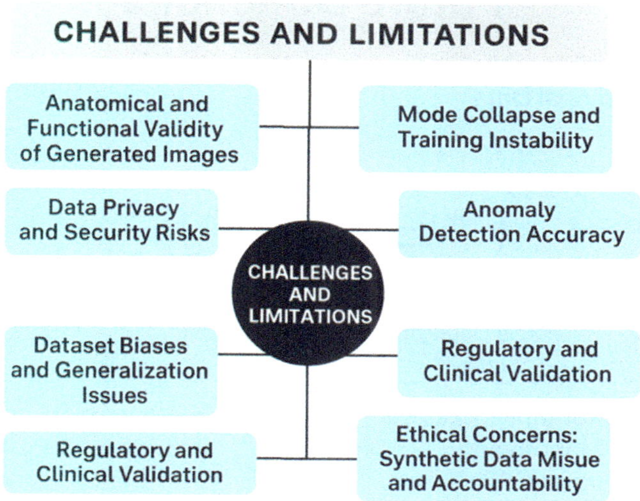

Figure 4.3 Challenges and limitations of GANs in neuroimaging.

3.1.1 Challenges with small and imbalanced datasets

The reasons why neuroimaging studies are conducted on small cohorts include their expense, time constraints, and ethical implications, which make it difficult to get labeled medical images. In addition, some conditions have very low data, for example, glioblastomas and early stage Alzheimer's, creating a class imbalance, hence affecting the performance of supervised learning models. The normal augmentation techniques, such as flipping, rotation, or contrast adjustment, do not consider the fact that brain structures are highly variable and complex at the same time. This is where GANs come in to change this whole concept; they learn what lies beneath by increasing available data through creating fake ones that closely resemble real ones.

3.1.2 GAN-based data augmentation strategies

The application of semisupervised GANs such as THS-GAN has been made in synthesizing T1-weighted MRI scans based on certain disease conditions, for example, Alzheimer's disease, MCI, and healthy state. By using labeled data and other data, these models can improve the classification with just a few training samples. In the same way, progressive growing generative adversarial networks (PGGANs) and MUNIT-based models can be employed for generating realistic tumor-infiltrated MRIs, which contain various morphological attributes in brain tumor imaging. Training for glioma segmentation and classification models using these augmented datasets yields superior results compared to those trained on genuine data alone.

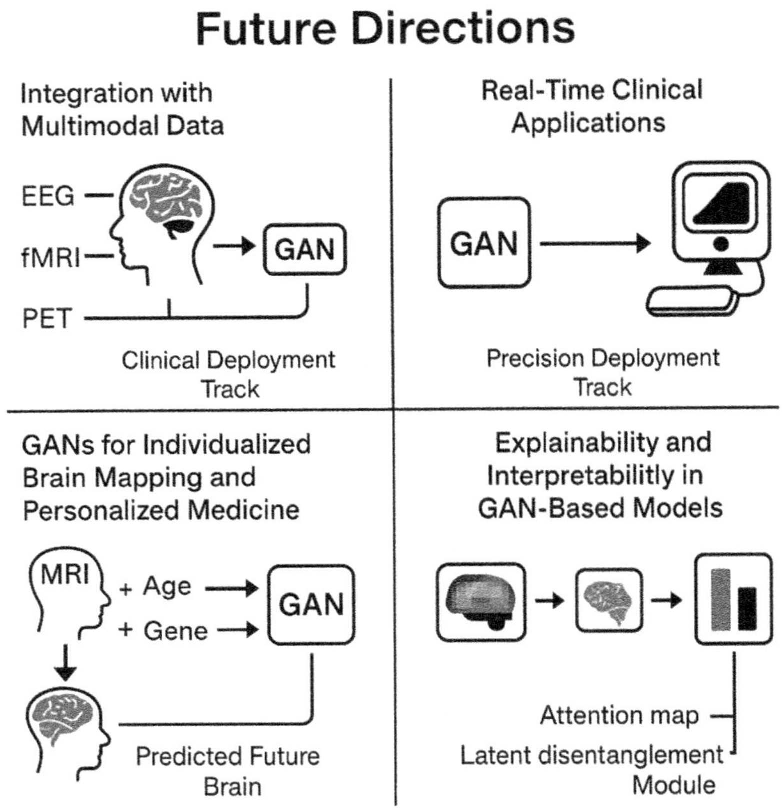

Figure 4.4 Future directions of GANs in neuroimaging.

3.2 Image synthesis and reconstruction

Image-to-image translation tasks have been revolutionized by GANs. For instance, it is possible to create a PET image from an MRI scan or improve the quality of low-resolution neuroimages using these models. The latter ones allow for high-res reconstruction; x-modality generation; as well as cleaning, which altogether increase clinical relevance and information content of the examinations. Such models also facilitate quick and cost-effective data collection by means such as MRI reconstruction from under-sampled data.

3.2.1 Cross-modality image synthesis

There is some unique biological information captured by various imaging modalities in neuroimaging, for example, MRI, PET, DTI, and fMRI. Nonetheless, it is not always possible to obtain complete multimodal data for every patient because of some factors, which may be either financial or logistical constraints. The solution lies in GANs, which can be employed for cross-modality synthesis with a focus on creating PET images out of MRIs. Conditional GANs (cGANs) have shown amazing results in this area. For example, a 3D cGAN architecture was employed to produce 18F-florbetapir PET images from T1-weighted MRI scans; these artificial PETs had very high SSIM (~0.95) scores and were nearly indistinguishable from real ones and could also predict progression from MCI to Alzheimer's (Nguyen et al., 2018). By allowing for bidirectional synthesis—MRI to PET and vice versa—even if there are no paired datasets, Rev-GANs have gone a step further from this. It demonstrates potential for producing realistic PET scans that maintain relevant anatomical anomalies, such as hippocampal atrophy.

3.2.2 Superresolution and denoising

It is not rare to find clinical images that are either low in clarity or corrupted in some way, especially when the scan time has been kept short to reduce patient stress. Some GANs, such as SRGAN (Superresolution generative adversarial network) and enhanced superresolution generative adversarial network (ESRGAN), have been employed to perform superresolution on brain MRIs by increasing their spatial resolutions without rescanning the patients (Yan et al., 2022). On top of that, GANs can be used for cleaning up images obtained from under low SNR scanners, such that recovery of fine anatomical structures can be achieved and interpretation of fMRI or DTI data may be enhanced. These techniques are particularly efficient with cycle-consistency losses during training, evident in CycleGAN-based pipelines.

3.2.3 MRI reconstruction from undersampled data

To speed up MRI scans, some data are missing in the k-space, which introduces errors in the final image. Nevertheless, there have been successful attempts to use GAN-based frameworks trained on fully sampled and undersampled pair images for producing reconstructions that closely resemble true fully sampled data and are of high quality. These methods reduce the scan time significantly without compromising the diagnosis.

3.3 Pathological feature simulation

GANs have the capacity to mimic patterns that are unique to a given disease, which enables them to create fake neuroimaging of pathological conditions, for example, atrophy, lesions, and tumors. Such images play a crucial role in training diagnostic models, testing algorithms under various environments, and monitoring the advancement of illnesses while still maintaining patient data confidentiality.

3.3.1 Generating disease-specific brain images

It is possible to train GANs so that they can create images with pathological features such as cortical atrophy, white matter hyperintensities, and neoplasms. Training these synthetic abnormal images is useful in various ways, including improving training datasets, mimicking the evolution of diseases, and helping radiologists see better certain signs. For instance, models such as CycleGAN and PGGAN have been employed to effectively synthesize atrophy patterns related to Alzheimer's, enabling the generation of brain pictures that are digitally set to have specific types of diseases at different stages (Zhang et al., 2020). In a similar fashion, GANs have also been applied to the modeling of temporal progression in malignant brain tumors from their incipient phase through to an advanced stage by employing progressive image generation techniques.

3.3.2 Applications for diagnostic model training

It has been proven that integrating artificial pathological images into training pipelines can enhance diagnostic performance. Classifiers who were trained with GAN-augmented datasets achieved notably higher AUC scores in binary and multiclass disease classification tasks. For instance, in certain research, the use of synthetic PET scans rather than traditional data augmentation techniques leads to higher classification accuracy for Alzheimer's prediction; this confirms the importance of GAN-generated data in practice.

3.4 Privacy-preserving data generation

GANs can be used to make synthetic neuroimaging data that still contains information about the group but does not show any details about an individual—thereby aiding in ethical data exchange and collaboration between various institutions. In addition, these artificial images are very effective for teaching AI models as they can also maintain their privacy.

3.4.1 Synthetic data for collaborative research

The sharing of medical imaging data among different facilities is limited by patient privacy concerns as well as legislation such as the General Data Protection Regulation (GDPR) and the Health Insurance Portability & Accountability Act (HIPAA). To handle this issue, GANs have been designed to create synthetic images that are statistically similar yet unrecognizable; hence, one can carry out studies in various centers without facing legal or moral issues. Some research has shown that it is feasible for the MedGAN model and other related approaches to produce artificial MR or PET brain scans that follow typical population statistics but remain nonidentifiable at the level of individual patients.

3.4.2 Deidentification and utility preservation

In addition to being anonymized, GANs may also be employed for removing identifying data from neuroimages, but still leaving behind clinically important data. By doing so, it becomes possible to share a dataset without compromising the participants' safety or even train models within such learning frameworks that operate across different institutions and involve decentralized AI without any fear of moving the data around. Research has indicated that it is possible for statistical models that are trained on synthetic data that maintain privacy to have comparable performance to those trained on real data, given high predictive.

4. Advanced techniques and enhancements

The initial version of GAN was ground-breaking for creating fake data, but when it was applied directly to neuroimaging, it showed that it had many shortcomings, such as training instability and poor spatial awareness, an inability to control certain output features, as well as problems in processing volumetric data. With time, a variety of architectural and methodological solutions have been proposed to combat these challenges. These improvements, on the other hand, tremendously improved the anatomical accuracy, clinical reliability, and performance of GAN outputs in neuroimaging. Four major advancements will be covered in this part, including conditional GANs, which are used to control synthesis; Wasserstein-based loss functions employed to address training instabilities within networks, attention mechanisms ensuring spatial coherence, and volumetric neuroimaging data handled through 3D GANs (Ali et al., 2022).

4.1 Conditional GANs for controlled synthesis

In comparison to the vanilla GAN architecture, conditional GANs (cGANs) make use of extra features, such as class labels and segmentation maps, at both the generator and discriminator. With this change, the model becomes capable of producing outputs that are highly focused and carry a specific meaning, especially in cases where attention should only be given to specific clinical properties.

These networks have proven to be very effective for cross-modality synthesis in neuroimaging. Cross-modality synthesis refers to generating one kind of data from another (e.g., PET images using an MRI scan). For example, researchers have employed 3D cGANs trained using paired PET and MRI scans for creating synthetic amyloid PET images that show a realistic pattern of distribution while also maintaining typical signs such as increased amyloid levels associated with Alzheimer's disease (Goodfellow et al., 2014). The increased realism provided by such models not only improves class-wise discrimination but also plays a key role in disease staging.

On top of that, cGANs have also shown success when employed for segmentation purposes by conditioning the generator on certain features, such as brain lesions or mask outlines. Integration of this nature has significantly enhanced the capacity to simulate MRIs with tumors & improve network training in low data regimes through a combination of these synthesized outputs.

Conditional learning further enables across-domain transformation, such as hypothetical clinical data generation by the models. For instance, with an input of an intact brain MRI image of a patient together with a label indicating disease presence, the cGAN may hypothesize on how the patient's brain would appear if affected by early stage Alzheimer's—hence aiding in early stage prediction as well as individualized diagnosis strategies.

4.2 Wasserstein GANs and improved training stability

Training WGANs is known to be difficult due to its instability. The problems, such as mode collapse—when the generator produces samples that are not diverse enough—and vanishing gradients, which stop the learning process, have been widely documented. Wasserstein GANs (WGANs) were introduced by Arjovsky et al. with an objective to combat two of the issues related to training GANs—mode collapse and vanishing gradients. WGAN does this by changing the loss function so that it measures how much two distributions differ in a more sensible way than through KL divergence: namely, by using the so-called Earth-Mover's distance or "Wasserstein metric."

The framework of WGAN is characterized by smoother gradients that lead to better convergence properties and make training more stable even when dealing with complex high-dimensional neuroimaging data. The introduction of WGAN-GP improves upon this further still since it refines the model through an additional Lipschitz constraint enforcement using gradients penalties approach, without requiring weight clipping, unlike in other GAN models, thereby ensuring consistent learning among various architectures.

WGANs have found application in producing synthetic T1-weighted and FLAIR MRI volumes that depict anatomic details such as gyri, sulci, and small focal lesions while still maintaining overall brain shape and without compromising either one too much, even if there were few examples available for training such models posing low risk on privacy concerns, especially among patients suffering from rare types of neurologic diseases. In addition, one may note that WGANs can also be trained for dataset harmonization so that they produce fake but very similar pictures, which will help reduce any differences between scans taken on different machines; thus, making it possible to integrate information from multiple locations without violating either side's confidentiality or compromising the overall accuracy of diagnostic algorithms at stake (Wang et al., 2023).

4.3 Attention mechanisms and spatial consistency

Standard convolutional architectures in GANs often have a difficult time when it comes to understanding spatial dependencies, which are long-range within medical images that have very high resolution. The outcome is that boundaries are not well defined, there is no relationship between adjacent regions of the brain, and anatomical details are missing. This problem was addressed by integrating attention mechanisms into the GAN architecture, enabling the model to adaptively attend to relevant portions while generating images.

The attention module introduced by Self-Attention GANs computes how distant pixels relate with each other in an image, thereby helping in keeping fine details and complex structures within generated samples (Islam & Zhang, 2020). Such capabilities are quite valuable when it comes to tasks such as lesion synthesis, which require maintaining both overall and local brain anomaly profiles.

Attention-based GANs have found application in neuroimaging for generating MRI scans with tumors whose characteristics are not only realistic but also vary widely and follow a particular pattern throughout an image. This kind of model, focusing on important areas, for example, the hippocampus or prefrontal cortex, gives rise to anatomically truthful data that can be employed to teach AI systems intended for early diagnostics and monitoring of illnesses.

On top of that, special dual-attention techniques that take into consideration spatial and channel attention were invented for even better performance enhancement. These mechanisms enable selective focusing on key feature maps as well as spatial regions so that across different imaging modalities, it becomes easier to handle superresolution or reconstruction tasks of all kinds.

4.4 Use of 3D GANs for volumetric neuroimaging data

Neuroimaging is basically about data, which is volumetric in 3D; this is especially so with MRI, PET, or fMRI. At the beginning of their development, GAN models could only process images one at a time (2D), leading to artifacts between neighboring slices and an overall lack of anatomical consistency. However, now there are some 3D GANs that can be used for this purpose. This was made possible by the introduction of 3-dimensional GANs (3D GANs) in the literature, which can process and generate volumetric data from any MR or CT scanner. The 3D GAN takes a 3D noise tensor as its input and generates 3D volumes in such a way that there is coherence not only within the generated slice but also across different slices. These are some examples where these models have been successfully used, such as generating volumetric PET images from MRI data, creating a full tissue cancer volume in silico, or adding extra details into brain scans, such as segmentation masks (Zhang et al., 2020). They have even found use in reconstructing undersampled 3D MRIs, hence facilitating quicker imaging procedures without reducing quality.

For instance, Lin et al. introduced a 3D Reversible GAN (Rev-GAN) which learns how to create high-quality volumetric images of both MRI and PET that look like each other while maintaining spatial information within them. It is capable of volumetric data synthesis as well as reconstruction, thereby improving interpretability and reducing memory consumption during training (Zhang et al., 2020). The use of 3D GANs is also very important in hippocampus modeling for predicting dementia because such models rely on fine-grained volumetric characteristics that can be employed in identifying even subtle anatomical changes. These provide an all-encompassing perspective on brain structure and therefore are more appropriate for studying the evolution of pathology, preparing for operations, and conducting radiomics analyses at hospitals.

5. GANs in neuroimaging: Roadblocks and realities

Even though GANs show a lot of potential in improving neuroimaging workflows, there are several challenges that prevent their use in hospitals and large research projects. Some of these challenges include, but are not limited to: how complex the algorithms are, how much data is needed, problems with evaluation, possible ethics issues, and whether they can be generalized. For this reason, it is important to recognize all these weaknesses of GANs with respect to neuroimaging and working on overcoming them (Zhang et al., 2020; Alrashedy et al., 2022).

5.1 Customized content production

The reason why training GANs can be said to be difficult lies in the fact that the two networks, the generator and the discriminator, work against each other in an adversarial manner. This may result in one network dominating the other, hence it fails to learn or vanish the gradient. One of such problems that can occur during training is mode collapse, which refers to a situation whereby the generator assigns similar or the same outputs for a range of different noise inputs. Such synthetic scans will not have various anatomical features, as well as pathological ones. For example, a GAN under mode collapse while trying to generate synthetic MRIs with tumor annotations might end up always producing similar-looking tumors at the same position and shape. Hence, compromising on diversity is needed for effective training of diagnostic models. On top of that, training becomes even more unstable due to the high dimensionality inherent in neuroimaging data, particularly volumetric data (e.g., 3D MRI/fMRI volumes). Although many remedies have been suggested, such as WGANs, spectral normalization, and minibatch discrimination, they are still not enough.

5.2 Limited dataset size and data quality

Acquiring a large, annotated neuroimaging dataset is not as easy as in the case of other natural image datasets; this is because it is very difficult to gain access to patients, the imaging protocols are very expensive, and people value their privacy. For deep models like GANs to learn effectively from data, they need thousands of samples. Without such information, one can observe overfitting as well as receive biased results comprising irrelevant anatomical structures. In most cases, neuroimaging data is gathered at multiple sites with different scanners that introduce diversity into the dataset. A GAN that has been taught using pictures from one place may not be able to adapt when taken to another place with a scanner set to various settings other than those used initially. This inconsistency reduces the reproducibility of models and prevents collaboration among different research locations. Even with data harmonization methods, it remains an open issue on how best to achieve this, that is, maintain anatomical truth while removing domain shifts.

5.3 Difficulty in the evaluation of generated images

It is much harder to determine the quality, as well as how realistic medically the neuroimages are that have been created by GANs, compared to those in other sectors. In computer vision, the IS or FID are some of the metrics used to evaluate this quality, but they cannot be directly applied to medical imaging because they measure how well particular statistical features of an image approximated by a deep learning architecture, matching those in natural pictures taken from publicly available databases (e.g., ImageNet). Also, it should be noted that SSIM and PSNR are not left out in neuroimaging literature; nevertheless, they focus on low-level image discrepancies while sidelining higher-order medical semantics. The reason behind this claim is that the structural similarity index (SSIM) and PSNR only consider low-level differences, which may not necessarily reflect the most important issue of medical images in question. For instance, a fake brain MRI could have high SSIM and still not show some small details like cortical thinning or lesions, making it inappropriate for clinical use. Although manual evaluation by radiologists is accurate, it is time-intensive, subjective, and not scalable.

5.4 Black-box nature and lack of interpretability

One of the characteristics of GANs and most other deep learning models is that they are black-box systems. It is hard to understand their internal data representations and what they produce, particularly attributing to any given generation to learn specifics on individual features or anatomical structures. The use of such models in hospitals does not create additional confidence for doctors because of this problem. To illustrate, if a GAN model is employed to mimic the evolution of a certain disease, it might anticipate decreasing volumes of gray matter at given locations; nevertheless, one cannot be sure whether such a prognosis follows from real data patterns or random correlations. The current efforts toward incorporating interpretability through means like attention-based GANs or gradient attribution into deep learning architectures are still at a primitive stage, with no backing from clinical data (Yan et al., 2022). If we cannot make sense of them, applying GANs in life-or-death situations will have many ethical and legal implications.

5.5 Generalization and domain shift

The performance of models trained on one population's or scanner's data and used in a different context is referred to as domain shift. Since they learn how to produce data under certain distribution assumptions, which might differ from one region to another, GANs may fail when encountering this problem. A good example is when you take a GAN trained on adult MRI scans and use it for kids; it will create images with incorrect anatomical features. It is common to observe such errors as well as changes in appearance profiles while trying to apply a model derived from 1.5T MRI upon 3T MRI images. Even though there have been some attempts to address this through domain adaptation techniques such as CycleGANs, these approaches could also eliminate certain abnormalities while adjusting for scanner noise.

5.6 Ethical concerns and regulatory uncertainty

The synthesis of medical data using GANs poses important issues both in terms of ethics and regulations. First, this kind of data can help in keeping data private while still being able to carry out some research, but on the other hand, it may also pose a danger. To be specific, fake scans may result in cases of insurance fraud or affect clinical decisions if put into practice within this context. The generation of realistic but nonauthentic patient information also creates confusion about who should take responsibility for different activities within hospital settings. At present, there are no clear guidelines on whether GAN-produced medical images meet regulatory standards set by bodies such as the filter deform attention (FDA) and expectation maximization algorithm (EMA). Because of this oversight, institutions find themselves at a crossroads on how best they can authenticate, approve, or even put these models into operation. Until such time as clear ethical boundaries and monitoring systems have been laid down, there will be little use of clinical applications based on GANs.

5.7 Overfitting to training data and hallucinated artifacts

Overfitting of GANs to the training dataset is possible, especially with a small sample trained. In this case, the generator learns to memorize individual scans, therefore violating patients' data privacy and preventing it from being used for other purposes. Even more terrible is the fact that GANs can produce fake images with certain anomalies, such as injuries or abnormalities; if one uses such data for analysis, it may cause certain errors of clinical judgment. Such "generated" indicators are hardly considered by standard quality measures. In case there is no proper validation, which should also include expert reviews and cross-sectional analysis, then it will be difficult to trust any kind of automation, increasing the risk associated with its clinical application.

Even though GANs could revolutionize neuroimaging research and clinical workflows, it is not easy for them to be integrated into the real world because of various technological, ethical, and legal issues. The application of these models may be unreliable due to problems such as training instability, small data sets, and interpretability, and also their nonscalability because of domain shifts and evaluation uncertainties. Besides, there are no clear rules on how these should be used within a healthcare setting, which makes their integration even more difficult. Overcoming these challenges will require collaboration among individuals with expertise spanning multiple fields, including but not limited to machine learning specialists, clinicians, ethicists, and policymakers, toward creating GANs that are transparent, strong, clinically reliable, and can form the basis for future data-driven neuroscience.

6. Evaluation and validation in the medical context

The integration of GANs in neuroimaging applications requires the use of thorough evaluation techniques to make sure that the results are medically correct and clinically relevant. When compared to other kinds of computer vision problems that may only need the output to look real, medical imaging has very strict requirements, including anatomical detail fidelity as well as being able to make sense out of it on a clinical level, and being able to use it for diagnosis. As a result, validation of synthetic neuroimages encompasses a combination of technical tests, professional opinions, and adherence to certain rules (Dinh et al., 2022).

6.1 Validation techniques

6.1.1 Expert radiologist assessment

GAN-generated neuroimages can be validated with a high level of confidence using human evaluation from either board-certified radiologists or neuroscientists. Such domain experts examine if the synthetic pictures retain important anatomical structures such as sulci, gyri, ventricles, as well as if they have been able to include any of the known disease features, such as tumors, lesions, and cortical thinning, in appropriate proportions or not.

- **Visual Turing Tests:** The ability of radiologists to tell if a scan is real or artificially created is tested under conditions where they do not know which is which. If many mistakes are made, it suggests that the synthetic images closely resemble real ones.
- **Clinical scoring:** Specialists use certain guidelines, such as how clear the outline is. Are there any artifacts? Does it match anatomical rules? In the end, they compare these scores for both synthetic and genuine pictures to determine the efficacy of GANs.

The approach may provide valuable information about the medical validity, but it is difficult to apply on a large scale, time-intensive, and subjective.

6.1.2 Impact on downstream model performance

An additional effective technique of validation involves evaluating the impact of synthetic images on the performance of the downstream model. Some of these tasks are included, but not limited to:

- Classification (e.g., Alzheimer's vs. control)
- Segmentation (e.g., tumor or lesion mapping)
- Disease progression modeling

When models trained on data sets augmented with GANs perform better or as good as those trained on real data only, then this can be taken to show that synthetic data may be helpful. To illustrate, there are studies that reveal that integrating GAN-generated MRIs for Alzheimer's disease in few-shot settings increases the classification accuracy up to 5%–10% compared to using natural data alone within models. In the same way, artificial scans were employed for U-Net training on glioma segmentation tasks, and it was observed that the network performs at most 3% worse than when trained with real annotated data (Alrashedy et al., 2022).

6.2 Regulatory considerations in synthetic data usage

Inasmuch as they can be useful in a technical sense, GANs also lead to queries in relation to medical regulatory authorities, for example, the FDA, EMA, and CDSCO, which demand that any technology intended for clinical use be proven safe, accurate, reproducible, and ethically compliant.

6.2.1 Data privacy and de-identification

Artificial data are usually recommended as the best way of maintaining the privacy of patients as opposed to using their real data. Nevertheless, when a GAN is trained on limited data, it might memorize individual examples and hence fail to generalize well; such an approach could lead to the creation of synthetic samples containing personal information. The latter has raised alarms with reverse engineering and data leakage being at the top of the list. Compliance with the law will require developers to prove:

- That synthetic data is statistically similar but not similar to any real patient scan.
- That through facial reconstruction, voxel patterns, or metadata, there are no risks for reidentification.

6.2.2 Clinical use of synthetic data

The use of GAN-generated images in making diagnostic decisions would mean that they are covered by the medical device regulation. For this, it is required that:

- **Prospective trials** validating synthetic scans in real-world diagnosis.
- **Audit trails** detailing how synthetic data was created, including architecture, training data, and evaluation metrics.

- **Risk management strategies** in case of misdiagnosis due to hallucinated or missing structures.

At present, there are no set rules by any regulatory body regarding the approval of synthetic images, although such guidelines should closely follow the software-as-a-medical-device (SaMD) frameworks.

7. The future of GANs in medicine: Unfolding possibilities in generative neurotechnology's

Despite the significant impact of GANs in some areas of clinical diagnostics and neuroimaging research, there is still more that can be done. The future development of GANs is predicted to progress into increased integration with multiple data modalities, real-time clinical applicability, personalized neuroscience, and model interpretability, which will extend current limits on precision medicine and computational neurodiagnostic (Ali et al., 2022).

7.1 Integration with multimodal data

The fusion of neuroimaging with other biosignals, such as electroencephalography (EEG), magnetoencephalography (MEG), and genetic or behavioral data is among the most viable research areas today. Integration of multimodal datasets has been very effective. For example, when we use fMRI plus EEG, it becomes easy to know which area of the brain activity occurred, and with the use of GANs. It may therefore be expected that high-resolution spatiotemporal brain activity mapping, which will provide a new level for understanding cognitive processes and identifying neurological disorders, can now be obtained through such integration. Currently, researchers apply cross-domain GANs to merge information across disparate sources, overcoming gaps between them in patient files, thus creating an effective framework for complex diagnostics.

7.2 Real-time clinical applications

The other significant development is the integration of GANs in live clinical workflows. The traditional neuroimaging pipelines require a lot of time and resources for analysis. These GANs may facilitate such issues as real-time denoising, accelerated transformation of an incomplete 2D object into a 3D shape, as well as fast conversion from one modality to another (e.g., CT to MRI). This feature would be very important in emergency settings like stroke units or trauma units because it would cut down on the time taken to make diagnoses, hence improving patient outcomes. On top of that, researchers are working on proof of concept (POC) devices that use less complex GAN models. Such applications are intended for use in portable communication or edge technologies within underprivileged areas, thereby increasing equity on a wider scale in relation to neuro healthcare services.

7.3 Personalized medicine and brain mapping

There is also an increasing interest in using GANs for personalized brain mapping and accurate neurology. It has been demonstrated by conditional and style-based GANs that we can generate individual-specific neuroimaging profiles given some information such as age, sex, genetic markers, or disease phenotypes. This creates an opportunity for customized progression models of diseases whereby clinicians can see the evolution of a particular patient's brain under various types of pathological issues or treatment courses offered (Islam & Zhang, 2020; Zhang et al., 2020). On top of that, through GANs, it would be possible to create fake twin data packs, which could be used in virtual experiments for tailoring treatment approaches without any high-risk operations done on real patients.

7.4 Explainability and interpretability

With the advancement of GANs toward their use in real clinical settings, it is vitally important that these systems can be explained and are transparent. The doctors should comprehend and have confidence in the results provided by the model. Most recent studies revolve around combining GANs with XAI frameworks, which can show what goes on inside them, such as attention maps or anomaly localization. Such techniques as saliency mapping and latent space interpolation are now being included in GAN training to increase model responsibility and eliminate "black box" doubt (Tavse et al., 2022). On top of that, scientists seek to integrate hybrid models that link GANs with conventional wisdom-driven or statistical approaches for sector-specific knowledge enhancement and better understanding. Such endeavors will be necessary to meet regulatory requirements and establish confidence among clinicians.

8. Conclusion: The transformative power of GANs in neuroimaging

GANs are now being used in the analysis of neuro images as they completely change how artificial brain data can be made, improved upon, and examined. The applications of these GANs span from increasing incompleteness within databases or simulating certain modalities to producing individualized brain images and determining possible progression of illnesses; thus it seems that in research as well as clinical practice, GANs have gone beyond what was believed to be feasible.

To recapitulate, this chapter has shown that GANs are capable of learning complex anatomical features through which they can overcome some current challenges in neuroimaging, such as a lack of available data, high costs related to their acquisition, or the need for personalized analysis. Anatomically realistic outputs refer to outputs that look like real human brains. This is important in the medical field because it will help doctors identify any abnormalities that might exist within patients' brains. Nonetheless, there are still many challenges that need to be overcome before these systems are widely accepted. For example, problems related to training, such as its unsteadiness, evaluation complexity, and not forgetting GANs black-box issues, require collective responsibility, particularly among researchers and developers. The other important things are ethics and regulations; they also need to consider if we are ready for any new development that may come up so that we make use of it rightly, especially concerning patients alone.

In conclusion, GANs signify a progressive development in computational neuroimaging with the potential to facilitate early stage diagnosis through personalized treatment and transform conventional brain mapping techniques. Nonetheless, realizing these potential calls for more than just technical prowess. The journey toward achieving this will only be possible through upholding moral standards, working across different fields, and having a common goal toward innovation within the healthcare sector.

References

Ali, H., et al. (2022). The role of generative adversarial networks in brain MRI: A scoping review. *Insights into Imaging, 13*(1). https://doi.org/10.1186/s13244-022-01237-0

Alrashedy, H. H. N., Almansour, A. F., Ibrahim, D. M., & Hammoudeh, M. A. A. (2022). BrainGAN: Brain MRI image generation and classification framework using GAN architectures and CNN models. *Sensors, 22*(11). https://doi.org/10.3390/s22114297

Dinh, T. Q., et al. (2022). Performing group difference testing on graph structured data from GANs: Analysis and applications in neuroimaging. *IEEE Transactions on Pattern Analysis and Machine Intelligence, 44*(2), 877–889. https://doi.org/10.1109/TPAMI.2020.3013433

Goodfellow, I., Pouget-Abadie, J., Mirza, M., Xu, B., Warde-Farley, D., Ozair, S., Courville, A., & Bengio, Y. (2014). Generative adversarial networks.

Islam, J., & Zhang, Y. (2020). GAN-based synthetic brain PET image generation. *Brain Informatics, 7*(1). https://doi.org/10.1186/s40708-020-00104-2

Nguyen, H., Morris, R. W., Harris, A. W., Korgoankar, M. S., & Ramos, F. (2018). Correcting differences in multi-site neuroimaging data using generative adversarial networks. Online. Available from: http://arxiv.org/abs/1803.09375.

Seeliger, K., Güçlü, U., Ambrogioni, L., Güçlütürk, Y., & van Gerven, M. A. J. (2017). Generative adversarial networks for reconstructing natural images from brain activity. https://doi.org/10.1101/226688.

Tavse, S., Varadarajan, V., Bachute, M., Gite, S., & Kotecha, K. (2022). A systematic literature review on applications of GAN-synthesized images for brain MRI. *Future Internet, 14*(12). https://doi.org/10.3390/fi14120351

Wang, R., et al. (2023). Applications of generative adversarial networks in neuroimaging and clinical neuroscience. *NeuroImage, 269*. https://doi.org/10.1016/j.neuroimage.2023.119898

Yan, W., et al. (2022). Deep learning in neuroimaging: Promises and challenges. *IEEE Signal Processing Magazine, 39*(2), 87–98. https://doi.org/10.1109/MSP.2021.3128348

Zhang, L., Wang, M., Liu, M., & Zhang, D. (2020). A survey on deep learning for neuroimaging-based brain disorder analysis. *Frontiers in Neuroscience, 14*. https://doi.org/10.3389/fnins.2020.00779

CHAPTER 5

Variational Autoencoders (VAEs) for neuroimaging dimensionality reduction, data augmentation and latent space analysis

Rosepreet Kaur Bhogal, Ajmer Singh and Asha Rani
Lovely Professional University, Phagwara, Punjab, India

1. Introduction

Neuroimaging is a non-invasive discipline that uses advanced imaging technologies to visualize, map, and quantify the structural and functional organization of the human brain. It serves as a foundational tool in both clinical diagnosis and neuroscientific research, enabling systematic investigation of brain morphology, connectivity, and metabolic activity (Kora et al., 2022). The field encompasses a range of complementary modalities; each designed to capture different dimensions of brain physiology and has transformed the understanding of neurological and psychiatric disorders.

The structural imaging modalities, such as Computed Tomography (CT) and Magnetic Resonance Imaging (MRI), capture static characteristics of the brain such as tissue continuity, volumetry, and morphological abnormalities. MRI, in particular, offers contrast resolution of tissues and is used to examine neurodegenerative changes and the integrity of the brain, including morphologic tumor patterns. Complementing these, functional modalities such as functional MRI (fMRI) and Positron Emission Tomography (PET) measure dynamic activity. The fMRI tracks blood-oxygen-level-dependent (BOLD) signals for neural activity and reveals patterns of functional connectivity across brain networks. Whereas, PET can provide information on metabolic processes and is paramount for investigations regarding neurotransmitter function and receptor dynamics (Kalafut et al., 2023; Rauch & Renshaw, 1995; Sander & Hesse, 2017). Altogether, these modalities form an integrated neuroimaging toolkit for investigating both the architecture and physiology of the brain.

Despite these evolutions, neuroimaging-based analysis remains challenging and statistically demanding. particularly when there is high-dimensional data. The key challenge is analyzing datasets that routinely contain millions of voxels (3D pixels) across time points. This complexity necessitates the use of advanced methods capable of identifying biologically meaningful patterns while preserving the subtleties intrinsic in the data (Bi et al., 2024).

Generative Artificial Intelligence for Neuroimaging
ISBN 978-0-443-45010-5
https://doi.org/10.1016/B978-0-443-45010-5.00011-3

To address these issues, deep generative methods, particularly Variational Autoencoders (VAEs), have emerged as powerful unsupervised learning models. They learn a low-dimensional latent space that encodes the statistical features of the high-dimensional input data and simultaneously generates the realistic synthetic patterns. By combining DL with probabilistic inference, VAEs can frame non-linear and complex relationships in data without using labeled annotations (Wei & Mahmood, 2020). Since VAEs can capture the variability in neuroimaging data, they have been used to synthesize new data with variations related to the original dataset, thus easing their interpretability and analysis (Saini & Dogra, 2025). Their capacity to learn feature representations in an unsupervised manner makes them suitable for extracting hidden patterns or associations from complex data sets (Cao et al., 2003). A study reports that among Principal Component Analysis (PCA), Independent Component Analysis (ICA), Kernel Principal Component Analysis (KPCA), and VAEs, VAEs perform best, followed by KPCA and then PCA in terms of accuracy, especially for small-sample-size problems. In general, integrating VAEs into neuroimaging data analysis addresses the high-dimensional problem by reducing data dimensionality while retaining important features.

This chapter presents a systematic analysis of VAE-based methods in neuroimaging. It is organized around three focused areas: data augmentation, dimensionality reduction, and latent representation. and the difficulty of handling high-dimensional data. Section 2 presents the mathematical foundations and architecture of VAEs. Sections 3, 4, and 5 address the focused areas in detail. ways to add more data, and Section 5 includes how to understand latent space. Section 6 discusses real-world applications with benchmark datasets, including ADNI and HCP. Section 7 discusses the limitations and future research directions.

2. Fundamentals of variational autoencoders

VAEs were first introduced by Kingma and Welling (2013) as a principled way to combine variational inference with deep generative modeling (Kingma & Welling, 2013). A generative model that learns the probabilistic mapping from its input to a latent representation, and back. It introduces a stochastic latent variable and optimizes a lower bound on the data likelihood. It can also be used to modify existing data through an encoding-modification-decoding process (Madathil et al., 2025). In neuroimaging, this framework is especially useful because brain data is noisy and complex (Pimpalkar et al., 2025).

2.1 Encoder-decoder architecture

The VAEs consist of two connected deep neural networks. Fig. 5.1 shows the encoder and decoder, where the input image is mapped from high-dimensional data to a lower-

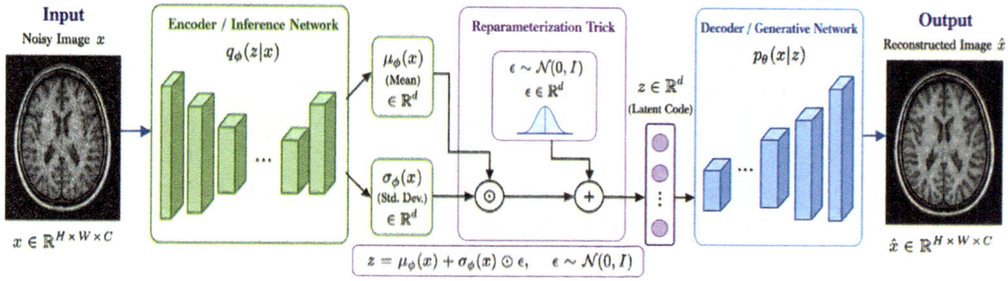

Figure 5.1 Schematic representation of the VAE architecture used for brain MRI denoising. The encoder $q_\phi(z|x)$ extracts latent features from the noisy input image, the reparameterization trick samples latent code z, and the decoder $p_\theta(x|z)$ reconstructs the clean image$\widehat{x}$. This framework effectively reduces noise while preserving anatomical details.

dimensional latent space. The architecture is handling the complexity of the neuroimaging dataset.

2.1.1 Mathematical expression for VAE

Let $x \in \mathbb{R}^n$ be high-dimensional observed data (e.g., an MRI).

Where $\mathbb{R}^n$ is an n-dimensional real-valued vector space

n: number of features/components used to represent the sample ŝ, for an MRI its number of pixels.

The encoder translates the original high-dimensional x input into a Gaussian latent low distribution parameter:

$$q_\varnothing(z|x) \tag{5.1}$$

Where x: input image

Output: latent variables usually

$z \in \mathbb{R}^d$: the latent variables of compressed dimension d and $d \ll n$, input size exceeding the output size

ϕ: Encoder weights and biases.

The input size exceeds the output size. A decoder network reconstructs the data

$$p_\theta(x|z) \tag{5.2}$$

It maps input data z to a Gaussian latent distribution over variables x and reconstructs the image

$$\widehat{x} \sim p_\theta(x|z) \tag{5.3}$$

Where θ: Decoder weights and biases.

Instead of directly forwarding the latent values to the decoder, VAEs use a latent vector to calculate a mean and a standard deviation. The input to the decoder is then sampled from the corresponding normal distribution. During training, VAEs encourage this normal distribution to be as close as possible to the standard normal distribution by including the Kullback-Leibler (KL) divergence in the loss function. KL divergence is a used to measure the difference one probability distribution is from another (Gomari et al., 2022). VAE will be altering or exploring variations in the input data, and not just randomly, but in a desired, specific direction. These allow one to reframe statistical inference problems (i.e., infer the value of one random variable from another) as statistical optimization problems. The latent variables, z, are drawn from a probability distribution depending on the input x, and the reconstruction is chosen probabilistically from z.

2.2 Latent space representation

The latent space in a VAE is a low-dimensional embedding that ideally captures the most informative aspects of the data (Vázquez-García et al., 2024). Fig. 5.2 illustrates the latent space representation in the VAE framework. It shows that high-dimensional noisy MRI input is encoded into a compact latent distribution that captures essential anatomical features. The learned latent representation enables robust reconstruction of denoised neuroimages. It also preserves clinically relevant information while suppressing noisy patterns. Latent representations can reveal insights into complex brain data as in fMRI. Instead of mapping to a single point z, the encoder outputs a distribution

$$q_{\varnothing}(z|x) = \mathcal{N}\left(z;\ \mu_{\phi}(x), \mathrm{diag}\left(\sigma_{\phi}^{2}\ (x)\right)\right) \tag{5.4}$$

Where N: Gaussian distribution,

$\mu_{\phi}(x)$: Mean vector produced by the encoder

$\sigma_{\phi}^{2}(x)$: Variance vector produced by the decoder

diag: Diagonal covariance matrix

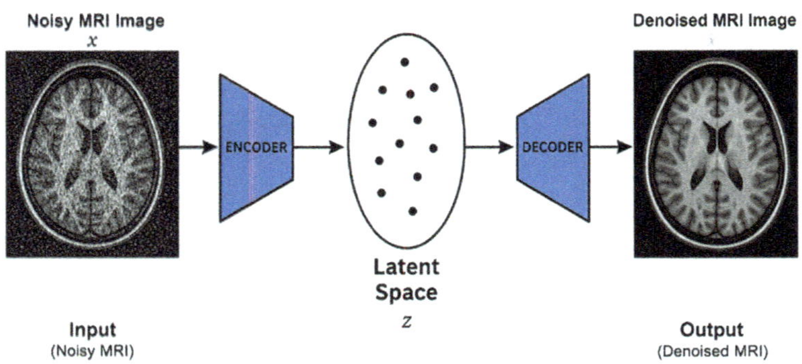

Figure 5.2 Latent space representation in the Variational Autoencoder (VAE) framework.

Encoder outputs a probability distribution over the latent space. μ_ϕ and σ_ϕ are learned using Convolutional Neural Networks (CNNs). So, the latent space is a distribution, not just points (Pinaya et al., 2022). This representation supports both regularization and robustness to input noise.

The input noisy MRI image x is encoded into a low-dimensional latent space z, where the underlying structural features of the neuroimage are compactly represented while suppressing noise variations. This latent representation serves as the intermediate feature embedding for reconstructing the denoised MRI image $\widehat{x}$ through the decoder network.

The objective is to maximize the marginal likelihood $p_\theta(x)$. To enable backpropagation through the stochastic process, the reparameterization trick is applied. Instead of sampling directly from the latent space, we rewrite sampling as

$$Z = \mu + \sigma \odot \epsilon, \epsilon \sim \mathcal{N}(0, \mathrm{I}) \tag{5.5}$$

Here

μ: mean vector from encoder

σ: standard deviation vector from encoder

ϵ: random noise sampled from the standard normal distribution

$\odot$: element-wise multiplication

I: identity matrix

The core training objective is to maximize the likelihood of the observed data. It is expressed as:

$$p_\theta(x) = \int p_\theta(x|z)p(z)dz \tag{5.6}$$

Taking logarithm:

$$\log p_\theta(x) = \log \int p_\theta(x|z)p(z)dz \tag{5.7}$$

This represents the marginal log-likelihood. However, this marginal likelihood is intractable because of a high-dimensional latent vector and a non-linear decoder. VAEs optimize the Evidence Lower Bound (ELBO).

2.3 Evidence lower bound

To optimize the encoder parameters θ and decoder parameters ϕ, the VAE maximizes a variational lower bound on the marginal log-likelihood of the observed data x. ELBO is a fundamental concept in variational inference, especially in latent-variable models like Variational Autoencoders (VAEs). Since the true posterior distribution $p(z|x)$ is intractable, it provides a tractable lower bound objective function to approximate inference via optimization.

The general form of ELBO is

$$\log p_\theta(x) \geq L_{\mathrm{ELBO}}(x) = \mathbb{E}_{q_\phi(z|x)}\left[\log p_\theta(x|z)\right] - D_{\mathrm{KL}}\left(q_\phi(z|x) // p(z)\right) \tag{5.8}$$

The first term measures how accurately the model reconstructs the input x from the latent variable z and encourages precise data prediction. The second term is the regularization term, which penalizes deviations from the prior and controls overfitting by minimizing the KL divergences to keep the reconstructed data close to the latent variable. Together, these encourage the model to produce reconstructions that are accurate and latent codes that follow a smooth, continuous distribution.

2.4 Loss functions and regularization strategies

The reconstruction loss measures the discrepancy between input image data x and reconstructed data $\widehat{x}$ given as follows:

For continuous-valued data

$$L_{\mathrm{rec}} = \|x - \widehat{x}\|^2 \tag{5.9}$$

Or for binary-valued data:

$$L_{\mathrm{rec}} = \sum \left[x \log\widehat{x} + (1 - x)\log(1 - \widehat{x})\right] \tag{5.10}$$

The choice depends on the type of input data. For neuroimaging data, Mean Squared Error (MSE) is typically preferred as pixel intensities are continuous.

The KL divergence regularizes the approximate posterior distribution by minimizing its divergence from the Gaussian distribution as follows:

$$L_{\mathrm{KL}} = \frac{1}{2}\sum\nolimits_{i=1}^{d}\left(\mu_i^2 + \sigma_i^2 - 1 - \log \sigma_i^2\right) \tag{5.11}$$

This encourages the latent variables to follow a standard normal distribution.

β-VAE Extension:

Conventional VAE can be extended by introducing a regularization coefficient β (Ouyang et al., 2022). The modified objective function is given as:

$$L_{\beta-\mathrm{VAE}} = L_{\mathrm{rec}} + \beta\, L_{\mathrm{KL}} \tag{5.12}$$

Where

L_{rec}: reconstruction loss

L_{KL}: the Kullback–Leibler divergence

It balances the latent-space regularization and reconstruction accuracy as follows:

B = 1: standard VAE

$B < 1$better reconstruction quality, weaker regularization

$B > 1$: stronger regularization, better disentanglement, higher compression

β-VAE helps learn more robust and structured latent representations

Eqs. (5.4)–(5.7) define the optimization framework for learning robust neuroimaging latent data. neuroimaging data.

3. Dimensionality reduction in neuroimaging using VAEs

Neuroimaging data consists of more spatial resolution than temporal. Each scan encodes many voxels in structural MRI (sMRI), fMRI, or diffusion weighted images. Each encoding information consists of thousands to millions of voxels per acquisition. This inherent high dimensionality presents a fundamental challenge for downstream analytical tasks, including classification, regression, and visualization. This necessitates the application of dimensionality reduction strategies that project data into lower-dimensional manifolds while preserving statistically meaningful structure. Dimension reduction will improve the computational complexity of the designed models. Conventional approaches to this problem have historically relied on linear transformations. Principal Component Analysis (PCA) identifies orthogonal directions of maximal variance. Independent Component Analysis (ICA) seeks statistically independent source signals underlying the observed measurements (Cao et al., 2003). However, their assumptions of linearity and statistical simplicity fail in the context of brain signals. The brain signals are nonlinear, hierarchical, and affected by diverse sources of noise (Ouyang et al., 2022). VAEs provide a non-linear alternative to this linear technique with compact latent representations. That is trained to balance reconstruction fidelity and regularization (Ménoret et al., 2017).

3.1 VAE-based projection of neuroimaging data

A trained VAE learns an encoder that maps the high-dimensional input to a lower-dimensional latent representation. This latent code captures the required features from the dataset. This is enabling three types of features, such as disease-relevant, grouping subjects based on similarity, and separating information from noise. In another way, the PCAs have fixed directions. VAEs can relate the nonlinear relationships within brain tissue and function (Ouyang et al., 2022).

3.2 Applications in feature extraction and disease modeling

The features from VAEs are used as input to classifiers. Examples of classifiers include Support Vector Machines, CNNs, and logistic regression. Feature extraction has shown better results compared to conventional techniques. It has been proven that feature estimation from VAEs is a more effective approach. For example, in Alzheimer's disease, embeddings from VAEs were used and gave better prediction of mild cognitive impairment conversion. In fMRI, the latent code was applied to provide visual support for learning tools and to separate disease subtypes. These projections also carried biological variables such as age, sex, and brain volume. In this way, the method was directly applied to solve both classification and regression problems.

3.3 Noise reduction in brain imaging

Neuroimaging data often contain noise due to scanner variability, subject motion, or other artifacts. VAEs are used as denoising models in such situations. The images are reconstructed from the latent code after removing irrelevant variance. Compared to traditional methods like median filters or Gaussian pyramids, VAE-based denoising preserves more anatomical details. It also adapts to the noise level and works across different scanners and datasets. The quality of reconstruction is usually measured with metrics such as Peak Signal-to-Noise Ratio (PSNR), Structural Similarity Index Measure (SSIM), and MSE.

3.4 Interpretation of latent dimensions

A trained VAE can learn a disentangled latent space. Each dimension captures a different anatomical variation. For example, one axis may show changes in brain volume with age, while another may reflect tissue contrast differences caused by disease. The level of interpretability depends on the model itself. Variants such as β-VAEs and Info-VAEs are often used to improve disentanglement through stronger KL regularization (Gomari et al., 2022).

4. Data augmentation for neuroimaging via VAEs

Data augmentation is a technique to create new training data from existing samples. It facilitates improving the performance of machine learning models, especially when datasets are small or imbalanced. Synthetic data generated by VAEs expands the training set by augmentation. This is valuable as patient data in medicine is often limited by cost, privacy concerns, and ethical restrictions.

VAEs generate new scans by learning a latent space from real data. This space captures the main brain features, from which samples are drawn and decoded into images. The generated images look realistic and preserve the statistical properties of the original dataset. They can be used for training, testing, and validation, reducing reliance on real data and improving efficiency.

Several applications of VAE-based augmentation have been reported. In fMRI, synthetic scans are used to enlarge datasets and improve prediction accuracy. In MRI, they support segmentation and classification tasks when real samples are limited. VAEs have even been applied in eye-tracking studies, showing that their use goes beyond brain imaging.

Validation is needed before synthetic data can be applied. Radiologists and neurologists often inspect the images visually. Quantitative measures such as SSIM, Dice coefficient, and image entropy are also used for evaluation. In some studies, adversarial validation is done by training a discriminator to separate real and synthetic scans. Hybrid models like VAE-Generative s and CVAEs generate sharper and more realistic images, but VAEs are still easier to train and interpret.

Data scarcity and class imbalance are common problems in neuroimaging. Rare diseases such as glioma or early dementia are often underrepresented in datasets. This imbalance reduces fairness and introduces bias in models. VAEs help by generating additional samples for minority classes. Latent space clustering can identify rare groups, and synthetic data from these groups are then added to the dataset. The quality of such data is checked using measures like SSIM or Fréchet Inception Distance. This balances the classes and makes classifiers more reliable.

Generalization also improves with VAE-based data. Models trained on augmented datasets show less overfitting and perform better across different scanners and populations. Latent regularization adds smoothness and stability, producing features that transfer more effectively. Studies report that models trained with synthetic data achieve higher PSNR, SSIM, and AUC in segmentation and classification tasks.

The latent space in VAEs also explains how features are stored. By adjusting individual dimensions in the latent vector, researchers can observe variations in the reconstructed images. For example, moving along an axis linked to hippocampal volume produces images that resemble Alzheimer's progression. Metrics such as MIG, Beta-VAE score, and DCI are used to measure disentanglement. Higher values indicate that latent factors are aligned with clear biological patterns.

Clustering in the latent space can reveal hidden groups. Methods like K-means or Gaussian Mixture Models form clusters that often match categories such as healthy controls, mild cognitive impairment, or Alzheimer's disease. Techniques like Uniform Manifold Approximation and Projection (UMAP) and t-Distributed Stochastic Neighbor Embedding (t-SNE) make these clusters easier to visualize. This approach helps in studying subtypes and supports targeted augmentation.

Interpretability is another strength of VAEs. The latent space is used not only to generate synthetic images but also to connect data with clinical markers. When combined with metadata, VAEs can highlight brain regions that drive predictions, simulate disease changes, and generate patient-specific data. This makes them valuable for both data generation and medical interpretation.

4.1 VAE-based synthetic neuroimage generation

After training, the decoder $p_\theta(x|z)$ can generate new images by sampling $z \sim N(0, I)$ or from conditioned distributions. The outputs are synthetic neuroimages that look like the training data but still show variation. These can be used to expand training sets for classifiers or segmentation models. They are also useful for simulation studies to test algorithm robustness. Another use is visual analytics, where researchers explore image variations across the latent space. In many cases, generated MRI slices look so real that even experts or similarity tests struggle to tell them apart from actual scans (Pinaya et al., 2022).

4.2 Addressing class imbalance with synthetic samples

Classification in neuroimaging suffers from imbalance. For example, public datasets may contain three times more healthy controls than patients with Alzheimer's or schizophrenia. This imbalance can distort the loss function and push the model toward majority predictions. VAEs help by:

- Generating minority-class samples from conditional latent distributions (e.g., $p\ (x|z, y = \text{AD})$).
- Equalizing sample counts during training, which improves recall and AUC scores.
- Enhancing cross-validation stability, especially in small cohorts.

For example, synthetic Alzheimer's images in ADNI can improve sensitivity to early symptoms.

4.3 Improving generalization in downstream tasks

Synthetic data from VAEs is used to balance datasets and improve generalization. In multi-site studies, scanner bias is reduced when synthetic samples are added. In domain adaptation, data from one population is generated to resemble another. In cross-modality tasks, synthetic T1 images are produced, and segmentation on T2 scans is improved (Chartsias et al., 2018). VAE-Generative Adversarial Network models are applied to create sharper images. Style transfer models are used to copy scanner artifacts or demographic variations. These methods increase the robustness and reliability of neuroimaging models (Yeoh et al., 2025).

4.4 Validation of synthetic neuroimages

Despite good results, validation is still necessary. Scans are examined by radiologists and neurologists through direct viewing (Yeoh et al., 2025). Quantitative evaluation is carried out using SSIM, Dice score, and image entropy. Adversarial testing is also applied, where a discriminator is trained to separate real scans from synthetic ones (Pol et al., 2019). Recent studies have shown that hybrid models such as VAE, GAN and Conditional VAE generate more realistic images. VAEs, however, remain simpler and easier to interpret.

5. Latent space analysis and interpretability

The latent space of a Variational Autoencoder is more than a compressed code. It is used as a map of the dataset, showing hidden variation and structure (Vázquez-García et al., 2024). In neuroimaging, this space is important because it captures anatomical and pathological details that are connected to clinical meaning.

5.1 Latent traversals and semantic variation

Latent traversal is done by changing one latent dimension while the others are kept fixed. The changes in the reconstructed image are then observed. In neuroimaging, these traversals can display gradual variation in features. Examples include expansion of ventricles or reduction of hippocampal volume with each step. These changes show that the latent space holds meaningful information about anatomy and disease (Huynh et al., 2025).

5.2 Disentanglement metrics in neuroimaging

The quality of a latent space is measured by disentanglement metrics. DCI, which stands for Disentanglement, Completeness, and Informativeness, is used to check how well each latent dimension explains a single factor age or diagnosis. MIG, or Mutual Information Gap, is applied to measure the separation between ground truth factors and latent encodings. In neuroimaging, a high level of disentanglement improves interpretability and diagnostic strength (Huynh et al., 2025). β VAEs perform better than vanilla VAEs in separating structure, and they support biomarker discovery (Ouyang et al., 2022).

5.3 Clustering and population stratification

Latent embeddings are grouped by applying clustering methods. K-means, Hierarchical clustering, or DBSCAN are used to reveal hidden patient subgroups. These clusters are linked with clinical variables that is, disease progression, cognitive score, or genetic markers. For example, analysis of fMRI latent codes has shown three subgroups of schizophrenia that differed in connectivity strength and symptom severity (Pinon et al., 2023; Huang et al., 2018).

5.4 Anatomical and clinical interpretability

The clinical value of latent dimensions is validated before use. Latent variables are correlated with biomarkers such as hippocampal volume or gray matter density. Regression and attribution methods, including SHAP and Integrated Gradients, are applied to connect latent features with clinical outcomes. Some studies have reported that a single latent node in a VAE trained on T1 MRIs was able to predict age with high accuracy ($r > 0.9$). This shows that VAEs can achieve biologically meaningful compression (Pinon et al., 2023).

5.5 Mechanisms of VAEs in neuroimaging

VAEs are trained to map high dimensional brain data into lower dimensional latent spaces, which approximate the underlying manifold (Pinon et al., 2023). A prior distribution is applied to create a structured latent space and to improve accuracy. Disentanglement methods are also used to increase interpretability, making it possible to separate distinct brain features.

5.5.1 Feature extraction, noise reduction, and visualization

Latent embeddings are used as low dimensional representations that support clustering, classification, and visualization. PCA, t SNE, and autoencoders are commonly applied to reduce dimensionality while important features are preserved (Pimpalkar et al., 2025). In case of noise reduction methods, the wavelet denoising and Gaussian filters are also applied to improve data quality. Autoencoders are considered powerful because they combine dimensionality reduction with denoising, which makes them suitable for extracting stable latent embeddings.

5.5.2 Visualization techniques

Visualization is important for understanding latent codes. PCA, t-SNE, and UMAP are applied to simplify the data and used as dimensionality reduction methods. PCA is used to capture overall variance, t SNE preserves local structure, and UMAP balances both local and global patterns (Pimpalkar et al., 2025). Interactive visualization methods are also used. Parallel Coordinate Plots allow correlations across dimensions to be examined (Itoh et al., 2017). Neural network-based visualization can reconstruct the original data from embeddings and allows controlled manipulation of latent codes (Yerebakan & Valadez, 2023). These methods provide useful insight, but they also face challenges in displaying complex relationships in high-dimensional spaces.

5.5.3 Semantic structure of latent space

Neuroimaging data are characterized by substantially greater spatial than temporal resolution, with structural MRI, functional MRI (fMRI), and diffusion-weighted imaging. These spaces are able to preserve biological and clinical information while also simplifying the data (Pinon et al., 2023). Deep learning methods apply latent spaces to model brain activity and connectivity, which reveals links with cognition and disorders (Pinon et al., 2023; Pol et al., 2019). Latent codes are also used to predict cognitive outcomes and to support anomaly detection (Afrasiyabi et al., 2024). Latent space cartography is applied to map dimensions to semantic properties, making it possible to interpret neuroimaging patterns (Pinon et al., 2023). Graph signal processing combines functional and structural data to expose higher-level organization of the brain (Huang et al., 2018).

6. Case studies and real-world applications of VAEs in neuroimaging

Variational Autoencoders (VAEs) have shown significant promise across diverse neuroimaging. They learn compact and meaningful representations of brain data. This section presents real-world applications, including disease classification, anomaly detection, and connectome compression, supported by publicly available datasets (Pinon et al., 2023; Molnár & Tamás, 2024).

6.1 Disease classification

VAEs are used to separate neurodegenerative and neuropsychiatric disorders by extracting latent features that are linked with disease. The Alzheimer's Disease Neuroimaging Initiative (ADNI) dataset has been widely applied for this purpose (Ouyang et al., 2022). In this approach, VAEs are trained on structural MRI scans. Latent embeddings are then extracted and used as input for SVM or Random Forest classifiers. Studies have reported AUC scores above 0.90 when distinguishing Alzheimer's disease from healthy controls with the use of VAE-based features (Chartsias et al., 2018). These embeddings also capture continuous disease progression more effectively than raw voxel intensities.

6.2 Anomaly detection

VAEs are well suited for unsupervised anomaly detection, and this is carried out through reconstruction error. Local hospital fMRI scans and schizophrenia cohort data have been used for such studies. In this method, VAEs are trained only on healthy control data. When patient data are tested, high reconstruction errors appear, and these errors reveal abnormal patterns. The Mean Squared Error is applied as the metric, comparing the input with the reconstructed output. Reported results show sensitivity and specificity above 85% in detecting subtle brain deviations without the need for explicit labels (Yeoh et al., 2025).

6.3 Connectome compression

VAEs are used to provide compact embeddings for functional connectivity matrices. These embeddings reduce data size while keeping the main patterns of brain networks. The Human Connectome Project (HCP) dataset is applied in this area. Correlation matrices, 360 by 360 connectomes, are given as input to a convolutional VAE. The latent codes preserve the topology of the network while lowering dimensionality. This allows clustering of brain states, tracking of changes over time, and alignment with genetic or behavioral data. The results show that VAE compressed representations retain a substantial proportion of the functional connectivity variance and perform better than PCA in predicting cognitive traits (Vázquez-García et al., 2024).

6.4 Some referenced datasets

Some datasets are named with the modality and uses, as in Table 5.1.

Table 5.1 Comparative analysis of neuroimaging datasets, imaging modalities, deep learning models, tasks, evaluation metrics, and key findings in VAE-driven neuroimaging applications.

Dataset	Modality	Model	Task	Metric	Key finding
COBRE (Center for Biomedical Research Excellence) (Avberšek & Repovš, 2022; Pinon et al., 2023)	fMRI	CNN, VAE, GAN (Pol et al., 2019; Wei & Mahmood, 2020)	Schizophrenia anomaly detection	Specificity, accuracy, sensitivity	Deep latent-space modeling enhances abnormal brain pattern identification in schizophrenia (Pinon et al., 2023; Pol et al., 2019)
Alzheimer's Disease Neuroimaging Initiative (ADNI) (Yeoh et al., 2025; Ouyang et al., 2022)	PET, sMRI	CNN, VAE, transformer (Bi et al., 2024; Liu et al., 2019; Wei & Mahmood, 2020)	Alzheimer's staging and disease progression modeling	PSNR, SSIM, accuracy, AUC,	Latent representations effectively improve diagnostic classification performance and capture disease progression patterns
IXI dataset (Chartsias et al., 2018)	MRI	VAE, DnCNN, Autoencoder (Molnár & Tamás, 2024; Chartsias et al., 2018)	MRI dataset denoising and reconstruction	PSNR, SSIM, RMSE	VAE-based denoising suppresses Gaussian and Rician noise while preserving structural information (Chartsias et al., 2018)
HCP (Human Connectome Project) (Huang et al., 2018; Afrasiyabi et al., 2024)	rs-fMRI	Autoencoder, VAE, Graph Neural Network (GNN), (Huang et al., 2018; Wei & Mahmood, 2020)	Functional connectome modeling and clustering	Silhouette score, accuracy, reconstruction error	VAE-based latent embeddings improve brain connectivity analysis and subject clustering (Huang et al., 2018)

Brain tumor segmentation challenge (Kalafut et al., 2023)	MRI (T1, T2, FLAIR)	VAE, GAN U-Net (Hattingen et al., 2024; Molnár & Tamás, 2024)	Tumor segmentation and denoising	PSNR, SSIM, Dice ScoMayre,	Multi-modal latent learning improves denoising quality and tumor boundary preservation (Hattingen et al., 2024)
UK Biobank (Pinaya et al., 2022)	sMRI	Diffusion model, VAE, GAN (Pinaya et al., 2022)	Population-level augmentation and biomarker discovery	PSNR, SSIM, FID,	Large-scale generative modeling improves structural biomarker extraction and data augmentation (Pinaya et al., 2022)
OpenNeuro (Vázquez-García et al., 2024)	MRI, fMRI	VAE, transformer (Bi et al., 2024; Wei & Mahmood, 2020)	Neuroimage feature learning and reconstruction	Accuracy, reconstruction loss	Multi-modal latent learning improves representation robustness across imaging modalities (Pol et al., 2019; Afrasiyabi et al., 2024)

7. Challenges, limitations, and future directions

VAEs in neuroimaging show strong promise, but several challenges are still present. Technical and clinical issues must be resolved before these models can be deployed safely in practice.

7.1 Limitations

One limitation is over smoothing in reconstructions. The Gaussian assumptions in the latent space produce blurry images. This lowers anatomical sharpness and makes it difficult to examine small but important structures for microbleeds or tumors, where detail is essential.

Posterior collapse is another challenge. During training, the decoder may ignore the latent code and rely only on learned biases. This reduces the usefulness of the latent representation. Methods including KL annealing, skip connections, and alternative priors are proposed to reduce this problem.

Interpretability is also restricted. Latent dimensions are abstract and do not map directly to biological meaning. Correlation with biomarkers or the use of semi-supervised training is required to make latent features clinically relevant.

Generalization is a further concern. Models trained on one scanner or one population fail when tested on another. Differences in protocols, hardware, and demographics create this gap, which makes cross-site deployment difficult.

GAN-based models produce sharper, more visually realistic neuroimages than VAEs due to adversarial learning. However, GANs are limited by training instability, low sample quality, and mode collapse, restricting their clinical reliability. VAEs, on the other hand, provide structured probabilistic latent representation, but are limited by over-smoothing in reconstructions. So, a trade-off between reconstruction fidelity, latent space representation, and space interpretability is a key limitation in neuroimaging applications.

7.2 Future research directions

Future work is focusing on several solutions. One approach is multimodal integration. VAEs can combine imaging data with genetics, cognition, or clinical features through joint encoders. For example, neuroimage genome fusion VAEs are being studied to identify polygenic risk patterns.

Another direction is the use of semi-supervised and conditional VAEs. By conditioning on demographic or disease labels, the latent space becomes more structured. M2 VAEs and CVAEs variants improve classification and make latent codes easier to interpret. Federated learning is also being developed. In this method, VAEs are trained across several institutions without sharing raw data. This protects patient privacy while improving generalization across sites.

Hybrid models are another area of progress. VAEs combined with GANs, VAE GAN and InfoVAE, are able to generate sharper and more realistic images. These models aim to balance visual quality with interpretability and clinical reliability.

8. Future outlook

The role of Variational Autoencoders in neuroimaging continues to develop. New variants, fusion of multiple data types, and advanced training methods are shaping how brain data are analyzed, interpreted, and applied in both research and clinical practice. Fig. 5.3 summarizes major applications of VAE frameworks in neuroimaging, including Multimodal Integration, Federated Learning, Semi-Supervised, and Hybrid Models

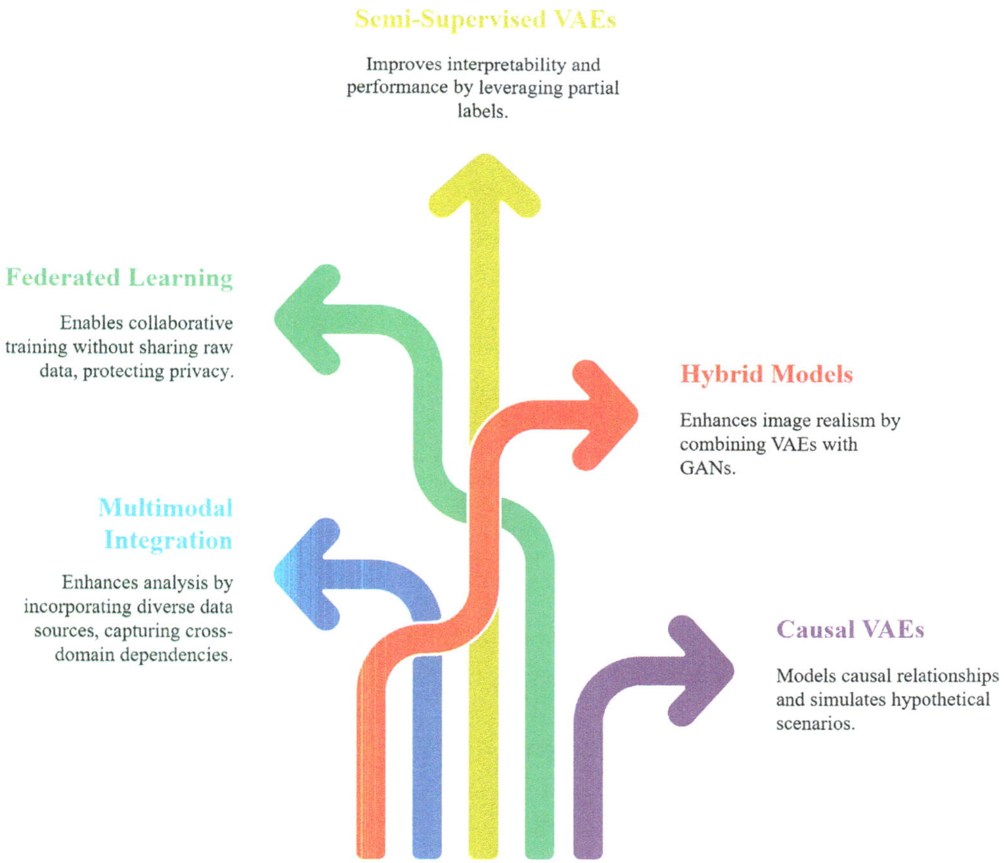

Figure 5.3 VAE approaches use for neuroimaging analysis.

8.1 Multimodal integration

VAEs excel in modeling complex distributions, but brain health extends beyond imaging. When genomics, cognitive profiles, demographics, and behavioral metrics are included, the analysis becomes richer and more multidimensional. Multimodal VAEs (MVAEs) learn joint representations across these different sources of data, capturing cross-domain dependencies (Kalafut et al., 2023).

Applications include:

- Predicting polygenic risk from combined neuroimage-genomic encodings.
- Identifying behavioral phenotypes through shared latent spaces.

Future pipelines may use shared encoders with modality-specific decoders. This design allows both unified analysis and accurate reconstruction.

8.2 Federated learning and privacy-aware training

Data silos across hospitals and research institutes limit how well models generalize. Federated learning makes it possible for multiple institutions to train VAEs together without sharing raw scans (Huang et al., 2018). Instead, only model weights are exchanged, which protects patient privacy while still drawing on diverse datasets. This approach is especially valuable for rare disease studies and cross-cultural neuroimaging. Federated VAEs can bring together distributed resources like ADNI, HCP, and the UK Biobank into a single learning framework.

8.3 Semi-supervised and conditional VAEs

For diagnosis or age, many neuroimaging datasets contain partial labels. Leveraging these labels in VAE training improves interpretability and downstream performance. Semi-Supervised VAEs (SS-VAEs) augment the ELBO with classification objectives, guiding latent space separation (Madathil et al., 2025). Similarly, Conditional VAEs (CVAE) include side information in both encoder and decoder paths to condition image synthesis and latent structure.

Potential applications:

- Disease progression modeling across time points.
- Stratified anomaly detection based on demographic factors.

8.4 Hybrid models for improved generative fidelity

VAEs generate reconstructions that look smooth but slightly blurry. To improve sharpness, researchers have developed GAN-based hybrids that add adversarial loss during training (Pinaya et al., 2022). Some notable examples include:

- VAE-GAN: Combines probabilistic encoding with GAN-style decoding.

- InfoVAE: Balances mutual information with reconstruction quality (Huynh et al., 2025).
- Hierarchical VAEs: Use multilevel latent structures to represent anatomy across scales.

These approaches mark an important step forward in image realism. By producing sharper and more natural-looking synthetic scans, they make VAE-based data better suited for clinical decision support and radiologist evaluation.

8.5 Causal and interventional VAEs

A critical direction is embedding causal reasoning into VAEs:

- Disentangling structural changes due to disease versus aging.
- Modeling interventional shifts (e.g., effect of treatment on scan appearance).

Future VAEs may simulate outcomes under hypothetical scenarios for example, predicting post-surgical brain morphology or estimating impact of drug interventions on functional connectivity.

9. Conclusion

Variational Autoencoders (VAEs) have introduced new possibilities for analyzing neuroimaging data. By learning compressed, interpretable, and generative representations, they have become valuable tools in both research and clinical contexts. This chapter examined VAEs in three key areas. First, in dimensionality reduction, where nonlinear embeddings outperform classical methods. For an example, PCA preserving brain structure and supporting downstream tasks. Second, in data augmentation, where synthetic images address data scarcity, enhance classifier accuracy, and balance class distributions. Third, in latent space analysis, where traversals, clustering, and disentanglement reveal patterns linked to diagnosis, disease progression, and cognitive profiling.

VAEs are highly adaptable across modalities, including structural MRI, fMRI, PET, and multimodal fusion with genomic and clinical data. Extensions of their architecture also enable application to both 2D and 3D imaging. These strengths position VAEs as versatile solutions for contemporary challenges in neuroimaging.

Nevertheless, several limitations remain. Over-smoothing reduces fine anatomical detail, affecting small but clinically important structures (Wei & Mahmood, 2020). Posterior collapse can weaken the latent code, and interpretability remains a barrier since latent variables lack direct biological meaning (Goyal et al., 2017). In addition, tailored architectures are needed to handle the complexity of neuroimaging data (Doersch, 2016).

Future research is actively addressing these issues. Integration with multimodal neuroimaging, combining fMRI and EEG, can enrich latent representations and strengthen model power (Wei & Mahmood, 2020). Semi-supervised frameworks allow VAEs to leverage both labeled and unlabeled data, improving performance where annotated data are scarce (Avberšek & Repovš, 2022). Federated learning further enables decentralized training across institutions, preserving patient privacy while enhancing generalization.

In sum, VAEs act as a bridge between complex neuroimaging data and actionable insights. For researchers, they uncover hidden structures; for clinicians, they provide decision support; and for developers, they offer flexible frameworks for brain modeling. As methods continue to evolve, VAEs are poised to play an increasingly central role in building robust, interpretable, and clinically relevant neuroimaging systems.

References

Afrasiyabi, A., Bhaskar, D., Busch, E. L., Caplette, L., Singh, R., Lajoie, G., Turk-Browne, N. B., & Krishnaswamy, S. (2024). *Latent representation learning for multimodal brain activity translation.* http://arxiv.org/abs/2409.18462.

Avberšek, L. K., & Repovš, G. (2022). Deep learning in neuroimaging data analysis: Applications, challenges, and solutions. *Frontiers in Neuroimaging, 1.* https://doi.org/10.3389/fnimg.2022.981642

Bi, Y., Abrol, A., Fu, Z., & Calhoun, V. D. (2024). A multimodal vision transformer for interpretable fusion of functional and structural neuroimaging data. *Human Brain Mapping, 45*(17). https://doi.org/10.1002/hbm.26783

Cao, L. J., Chua, K. S., Chong, W. K., Lee, H. P., & Gu, Q. M. (2003). A comparison of PCA, KPCA and ICA for dimensionality reduction in support vector machine. *Neurocomputing, 55*(1–2), 321–336. https://doi.org/10.1016/s0925-2312(03)00433-8

Chartsias, A., Joyce, T., Giuffrida, M. V., & Tsaftaris, S. A. (2018). Multimodal MR synthesis via modality-invariant latent representation. *Medical Imaging, 37*(3), 803–814. https://doi.org/10.1109/TMI.2017.2764326

Doersch, C. (2016). Tutorial on variational autoencoders. https://doi.org/10.48550/arXiv.1606.05908.

Gomari, D. P., Schweickart, A., Cerchietti, L., Paietta, E., Fernandez, H., Al-Amin, H., Suhre, K., & Krumsiek, J. (2022). Variational autoencoders learn transferrable representations of metabolomics data. *Biology, 5*(1). https://doi.org/10.1038/s42003-022-03579-3

Goyal, P., Liang, X., Mellon, C., Xing, E. P., Wang, C., & Hu, Z. (2017). *Nonparametric variational autoencoders for hierarchical representation learning* (pp. 5104–5112). https://doi.org/10.1109/iccv.2017.545

Hattingen, E., et al. (2024). Challenges, limitations, and pitfalls of PET and advanced MRI in patients with brain tumors: A report of the PET/RANO group. *Neuro-Oncology, 26*(7), 1181–1194. https://doi.org/10.1093/neuonc/noae049

Huang, W., Bolton, T. A. W., Medaglia, J. D., Bassett, D. S., Ribeiro, A., & Van De Ville, D. (2018). A graph signal processing perspective on functional brain imaging. *Proceedings of the IEEE, 106*(5), 868–885. https://doi.org/10.1109/JPROC.2018.2798928

Huynh, T. N., et al. (2025). *Latent representation learning from 3D brain MRI for interpretable prediction in multiple sclerosis.* http://arxiv.org/abs/2510.00051.

Itoh, T., Kumar, A., Klein, K., & Kim, J. (2017). High-dimensional data visualization by interactive construction of low-dimensional parallel coordinate plots. *Journal of Visual Languages & Computing, 43*, 1–13. https://doi.org/10.1016/j.jvlc.2017.03.001

Kalafut, N. C., Huang, X., & Wang, D. (2023). Joint variational autoencoders for multimodal imputation and embedding. *Nature Machine Intelligence, 5*(6), 631–642. https://doi.org/10.1038/s42256-023-00663-z

Kingma, D.P., & Welling, M. (2013). Auto-encoding variational Bayes. arXiv preprint arXiv:1312.6114. http://arxiv.org/abs/1312.6114.

Kora, P., Ooi, C. P., Faust, O., Raghavendra, U., Gudigar, A., Chan, W. Y., Meenakshi, K., Swaraja, K., Plawiak, P., & Acharya, U. R. (2022). Transfer learning for medical image analysis: A review. *Computer*

Methods and Programs in Biomedicine, 216, Article 106624. https://doi.org/10.1016/j.cmpb.2022.106624

Liu, Y., Heer, J., Li, Q., & Jun, E. (2019). Latent space cartography: Visual analysis of vector space embeddings. *Computer Graphics Forum, 38*(3), 67–78. https://doi.org/10.1111/cgf.13672

Madathil, N. T., Dankar, F. K., Gergely, M., Belkacem, A. N., & Alrabaee, S. (2025). Revolutionizing healthcare data analytics with federated learning: A comprehensive survey of applications, systems, and future directions. *Computational and Structural Biotechnology Journal, 28*, 217–238. https://doi.org/10.1016/j.csbj.2025.06.009

Ménoret, M., Farrugia, N., Pasdeloup, B., & Gripon, V. (2017). Evaluating graph signal processing for neuroimaging through classification and dimensionality reduction. arXiv preprint arXiv:1703.01842.

Molnár, S., & Tamás, L. (2024). Variational autoencoders for 3D data processing. *Artificial Intelligence Review, 57*(2). https://doi.org/10.1007/s10462-023-10687-x

Ouyang, J., Zhao, Q., Adeli, E., Zaharchuk, G., & Pohl, K. M. (2022). Disentangling normal aging from severity of disease via weak supervision on longitudinal MRI. *Medical Imaging, 41*(10), 2558–2569. https://doi.org/10.1109/TMI.2022.3166131

Pimpalkar, A. P., Patil, V. S., Thorat, N. N., Gulame, M., Palkar, J. D., & Gham, P. S. (2025). Generative models in medical imaging. In *Recent advances in artificial intelligence and healthcare systems*. https://doi.org/10.1201/9781003539483-3

Pinaya, W. H. L., Tudosiu, P.-D., Dafflon, J., da Costa, P. F., Fernandez, V., Nachev, P., Ourselin, S., & Cardoso, M. J. (2022). Brain imaging generation with latent diffusion models. arXiv preprint arXiv: 2209.07162.

Pinon, N., Oudoumanessah, G., Trombetta, R., Dojat, M., Forbes, F., & Lartizien, C. (2023). *Brain subtle anomaly detection based on auto-encoders latent space analysis: application to de novo Parkinson patients*. http://arxiv.org/abs/2302.13593.

Pol, A. A., Cerminara, G., Berger, V., Pierini, M., & Germain, C. (2019). *Anomaly detection with conditional variational autoencoders. 1651–1657*. https://doi.org/10.1109/icmla.2019.00270

Rauch, S. L., & Renshaw, P. F. (1995). Clinical neuroimaging in psychiatry. *Harvard Review of Psychiatry, 2* (6), 297–312. https://doi.org/10.3109/10673229509017151

Saini, A., & Dogra, A. (2025). From spatial domain to learning based methods: A survey on MRI and HRCT image denoising methods. *Biomedical and Pharmacology Journal, 18*(3). https://doi.org/10.13005/bpj/3547

Sander, C. Y., & Hesse, S. (2017). News and views on in-vivo imaging of neurotransmission using PET and MRI. *The Quarterly Journal of Nuclear Medicine and Molecular Imaging, 61*(4), 414–428. https://doi.org/10.23736/S1824-4785.17.03019-9

Vázquez-García, C., Martínez-Murcia, F. J., Román, F. S., & Górriz, J. M. (2024). A review of latent representation models in neuroimaging. arXiv preprint arXiv:2412.19844.

Wei, R., & Mahmood, A. (2020). Recent advances in variational autoencoders with representation learning for biomedical informatics: A survey. *IEEE Access, 9*, 4939–4956. https://doi.org/10.1109/access.2020.3048309

Yeoh, P. S. Q., Hasikin, K., Wu, X., Goh, S. L., & Lai, K. W. (2025). Trends and applications of variational autoencoders in medical imaging analysis. *Computerized Medical Imaging and Graphics, 126*, Article 102647. https://doi.org/10.1016/j.compmedimag.2025.102647

Yerebakan, H. Z., & Valadez, G. H. (2023). *Deep generative neural embeddings for high dimensional data visualization*. http://arxiv.org/abs/2302.10801.

CHAPTER 6

Diffusion models for high-fidelity neuroimage generation and enhancement

Shiveta Bhat[1], Gurpreet Kour[1] and Kanika Bhalla[2]
[1]Department of Electronics and Communication Engineering, Model Institute of Engineering and Technology, Jammu, Jammu and Kashmir, India; [2]Washington University School of Medicine, St. Louis, MO, United States

1. Introduction

Neuroimaging has emerged as a boon in understanding the structure of the human brain by providing noninvasive, high-resolution insight into its composition, function, and biochemical processes. Techniques such as magnetic resonance imaging (MRI), computed tomography (CT), positron emission tomography (PET), and functional MRI (fMRI) play a key role in the diagnosis and surveillance of conditions such as Alzheimer's disease (AD), brain tumors, epilepsy, and many sclerosis (Sultana et al., 2024). These imaging systems support basic research in cognitive neurological science by enabling the view of brain connections, neuroplasticity, and activity patterns. Despite the significant progress, neuroimaging continues to meet major technical and practical challenges. High-resolution scans demand long procurement time, the sensitivity to speed and physical noise, which can impair the quality of the image and limit diagnostic utility (Padmanabhan et al., 2017). Various factors, such as a resource-constrained environment and noiseless scans, are essential for visual assessment and algorithm analysis. This can lead to a wrong dataset, which can affect the machine learning procedure. Thus obtaining multimodal data (e.g., both T1-weighted and T2-weighted MRI) for patients turns out to be time-consuming and an expensive process (Rudie et al., 2022).

Generative modeling, particularly using probabilistic models, has emerged as a very promising method to overcome these limitations. To create data, diffusion models simulate an additional process that progressively introduces noise into a sample. This is followed by a learned reverse process that eliminates the noise and restores the original data division (Chen et al., 2025). It has been demonstrated that this denoising diffusion framework, which is mathematically based on stochastic differential equations (SDEs), generates images with higher fidelity than earlier generative techniques, such as variational autoencoders (VAEs) and generative adversarial networks (GANs) (Bengesi et al., 2024).

Diffusion models have a stable, structured, and systematic approach, which is advantageous in medical imaging domains where both diversity and correctness of generated

Generative Artificial Intelligence for Neuroimaging
ISBN 978-0-443-45010-5
https://doi.org/10.1016/B978-0-443-45010-5.00009-5

images are crucial (Kazerouni et al., 2023), in contrast to GANs, which frequently experience training instability and mode collapse. Furthermore, diffusion models' sampling method works in several fine-grained phases rather than in a single pass, making them ideal for iterative refinement and feature preservation, which are two characteristics of medical image improvement.

Diffusion models have proven useful in a variety of applications, particularly in neuroimaging. The synthesis of missing modalities (e.g., generating T2 images from T1 scans), inpainting of missing or occluded brain regions, denoising of noisy or motion-corrupted images, superresolution of low-quality scans (Alexander et al., 2017), and the creation of synthetic data for rare disease augmentation or anonymization purposes are a few examples. These developments not only increase neuroimages' clinical usefulness but also aid in resolving the data scarcity problem that plagues medical imaging. In this chapter, a thorough analysis of diffusion models for the creation and improvement of high-fidelity neuroimages is covered.

The main focus of the chapter is on a comprehensive examination of diffusion models for high-fidelity neuroimage generation and enhancement. With the help of benchmark assessments utilizing datasets including the Human Connectome Project (HCP), BraTS, and the Alzheimer's Disease Neuroimaging Initiative (ADNI), the conversation then shifts to real-world applications across a range of neuroimaging tasks (Pinaya et al., 2022). In addition, existing constraints such as computing overhead and generalizability are addressed, highlighting ethical issues and unanswered research problems that are essential for safe implementation in clinical processes.

In addition, this chapter bridges the gap between algorithmic innovation and clinical application by offering a road map for developers, researchers, and clinicians who wish to implement or enhance diffusion-based neuroimaging techniques.

2. Fundamentals of diffusion models

2.1 Basic principles

Diffusion models are generative probabilistic models that convert noise into structured data (e.g., images and medical scans) by reversing a gradual noising process. This transformation happens in two key phases:

2.1.1 Forward process (diffusion)

A sample x_0 from the real data distribution is progressively corrupted over T time steps by adding Gaussian noise. As t → T, the data becomes indistinguishable from noise:

$$x_T \sim N(0, I) \tag{6.1}$$

The process is Markovian and modeled as:

$$q(x_{t-1}) = N\Big(x_t; \sqrt{1-\beta_t x_{t-1}}, \beta_t I \qquad (6.2)$$

where

β_t: Variance schedule at time t

$$\alpha_t = 1 - \beta_t$$

$\overline{\alpha}_t = \prod_{s=1}^{t} \alpha_s$: Cumulative product for sampling equations

This forward process $q(x_0)$ is analytically tractable and forms the basis of the training framework (Ho et al., 2020).

2.1.2 Reverse process (denoising)

Since the exact reverse distribution $p(x_t)$ is intractable, a neural network $\in_0(x_t, t)$ is trained to predict the noise:

$$L_{\text{simple}} = E_{x_0,t,\in}\Big[\|\in - \in_0\Big(x_t, t\|^2\Big] \qquad (6.3)$$

where

$\in = N(0, I)$: Gaussian noise

$\in_0$: Network's prediction of noise

This objective encourages the model to accurately reverse the noising process and recover x_0. Through repeated denoising, high-quality data can be synthesized from pure noise.

2.2 Key architectures

2.2.1 Denoising diffusion probabilistic models (DDPMs)

DDPMs were introduced (Ho et al., 2020) as the first practical implementation of diffusion models for image synthesis. They use a U-Net with skip connections and sinusoidal time embeddings for timestep conditioning. DDPMs generate sharp, high-fidelity images but are computationally slow as their generation requires hundreds of iterative steps.

2.2.2 Score-based generative models

Score-based models use SDEs instead of discrete Markov chains (Song et al., 2021). These models estimate the score function:

$$\nabla_x \text{loglog} p_t(x) \qquad (6.4)$$

which represents the gradient of the data distribution at various noise levels. Sampling is performed using Langevin dynamics or by solving a reverse-time SDE.

Advantages:

- Smooth noise control (continuous time)
- Stable training

- Flexible for various data types

2.2.3 *Latent diffusion models (LDMs)*

To reduce high-dimensional compute demands, latent diffusion models (LDMs) were proposed (Rombach et al., 2022). A VAE first compresses the input into the latent space, and diffusion is applied there. This dramatically cuts memory usage while preserving semantic structure. The comparison of different models for diffusion is shown in Table 6.1.

Strengths:

- Computational efficiency
- Feasible for large images or 3D medical volumes
- Effective for conditional tasks (e.g., T1 to fluid-attenuated inversion recovery (FLAIR) MRI translation)

2.3 Diffusion versus GANs and VAEs

Before diffusion models gained prominence, GANs and VAEs were the two primary families of generative models. Each of them offered unique capabilities and innovations, but they also had significant drawbacks, particularly when used in sensitive domains such as medical imaging.

2.3.1 *Generative adversarial networks*

GANs, proposed by Goodfellow et al. (2014) revolutionized the area of image generation using a two-player approach, which involves a discriminator and a generator. The discriminator tries to distinguish the synthetic samples from the real samples, while the generator aims at creating samples similar to the real samples.

Strengths:

- Capable of producing visually realistic and sharp images.

Table 6.1 Comparison of diffusion model variants.

Model type	Key mechanism	Pros	Cons	Use case in neuroimaging
DDPM	Discrete timestep denoising	High fidelity, flexible	Slow inference	Superresolution, denoising
Score-based models	SDE-based score estimation	Continuous noise modeling	Complex estimation	Lesion inpainting, MRI synthesis
Latent diffusion	Diffusion in latent representation	Fast, memory-efficient	Some detail loss	Modality translation, fast generation

- Fast inference because of one-step generation.

Limitations:

- Unstable training dynamics because of adversarial optimization.
- Mode collapse, where the generator produces limited types of outputs.
- No explicit likelihood estimation, making it hard to evaluate or compare samples probabilistically.
- Challenging conditioning mechanisms, particularly when dealing with complex input types (e.g., multimodal medical data).

GANs have been extensively explored in image synthesis, superresolution, and image-to-image translation, but their instability and lack of control of GANs have prevented their widespread use in clinical settings.

2.3.2 Variational autoencoders

VAEs take a probabilistic approach by learning a mapping from input data to a latent distribution, and then decoding samples back to the original data space (Kingma & Welling, 2014).

Strengths:

- Stable and efficient training via variational inference.
- Supports uncertainty modeling, valuable in medical applications.
- Enables explicit likelihood estimation through the evidence lower bound (ELBO) framework.
- Well-suited for latent space exploration and conditional generation.

Limitations:

- Outputs often suffer from blurriness because of reconstruction-based loss (such as MSE).
- Can struggle with capturing fine-grained visual details, critical in medical imaging.
- May underfit complex distributions, especially in high-resolution data.

Although VAEs have shown promise in semisupervised learning, anomaly detection, and modality fusion, the trade-off between tractability and fidelity is still difficult to achieve.

2.3.3 Diffusion models: A new generation of generative learning

Diffusion models emerged as a powerful alternative, addressing the key weaknesses of both GANs and VAEs as shown in Table 6.2. By avoiding adversarial training and preserving probabilistic interpretability, they offer a compelling balance between image quality, stability, and diversity.

Table 6.2 Comparative overview of GANs, VAEs, and diffusion models.

Aspect	GANs	VAEs	Diffusion models
Training stability	Unstable because of adversarial training	Stable via variational inference	Very stable, no adversarial dynamics
Mode collapse	Common issue; poor coverage of data modes	Less prone to collapse	Rare; models full data distribution
Image sharpness	High-quality images, but may hallucinate artifacts	Often blurry because of reconstruction loss	High-fidelity outputs, especially with DDPMs and LDMs
Diversity of outputs	Limited by mode collapse	High diversity	High and controllable via conditioning
Interpretability	Low; lacks explicit probabilistic framework	Moderate; provides latent representations	High; based on probabilistic noise estimation
Likelihood estimation	Not possible	Yes (through ELBO)	Yes (via variational bounds and score estimation)
Ease of conditioning	Moderate; requires architectural tuning	High; supports various conditional setups	Very high; conditioning is flexible and effective.
Clinical suitability	Limited reliability in high-risk domains	Acceptable, but may lack output fidelity	Excellent; reliable, interpretable, and robust for clinical use

3. Applications of diffusion models in neuroimage generation

Diffusion models have emerged as a transformative technology in neuroimaging, offering the ability to generate high-resolution, anatomically accurate, and clinically useful brain images. Their iterative denoising process and probabilistic nature make them exceptionally well-suited for tasks that demand fidelity, diversity, and interpretability, which are the key requirements in medical diagnostics and neuroscience research. Some of the key insights on the use of diffusion models in neuroimaging are depicted in Table 6.3.

3.1 MRI synthesis

MRI plays a crucial role in the field of neuroimaging as it has the ability to capture and provide detailed views of brain structure and function. However, because of challenges in the clinical settings, there are challenges such as missing sequences, low-field scanner limitations, and scanner access disparities that pose significant barriers in both clinical and research contexts. For this, diffusion models have emerged as a promising solution to

Table 6.3 Applications of diffusion models in neuroimaging.

Key references	Application area	Objective	Output type	Key benefits
Pinaya et al. (2022)	MRI modality synthesis	Generate missing modalities (e.g., T2 from T1) using latent diffusion models	Full MRI volumes	Completeness, reduced scan time
Dar et al. (2019)	MRI modality synthesis (GAN-based baseline for comparison)	Multi-contrast MRI synthesis using conditional GANs	Full MRI volumes	Provides baseline comparison for diffusion performance
Wolleb et al. (2022)	Medical anomaly detection	Anomaly detection using diffusion models in medical images	Anomaly probability maps	Improved unsupervised detection of anomalies
Chung and Ye (2022)	MRI acceleration and reconstruction	Accelerated MRI reconstruction via score-based diffusion models	Reconstructed MR images	Faster acquisition with high quality
Akbar et al. (2023)	Lesion simulation and segmentation	Brain tumor segmentation using synthetic MRIs generated by diffusion models versus GANs	Annotated MR images	Enhanced segmentation robustness with synthetic data
Zhang et al. (2023)	Lesion simulation (stroke)	Stroke lesion segmentation with synchronous image-label diffusion models	Annotated CT lesion masks	Accurate segmentation and data-efficient learning

these issues by generating missing scan types, sharpening low-resolution images, and even creating high-quality synthetic brain scans. As a result, they offer a powerful way to support diagnosis, research, and treatment planning, especially in settings where resources are limited or full MRI scans are not always possible.

3.1.1 Cross-modality MRI generation

One significant application is the generation of missing MRI modalities. In clinical practice, certain MRI modalities are often missing because of motion artifacts, time constraints, or cost. Diffusion models, particularly LDMs and score-based generative models, enable cross-modality synthesis, generating missing modalities such as T2-weighted scans from T1-weighted inputs, preserving anatomical accuracy. Pinaya et al. demonstrated the effectiveness of using LDMs in synthesizing missing MRI modalities in the HCP (Pinaya et al., 2022) dataset, achieving high fidelity in cross-modality reconstructions. Various studies confirm that synthetic MRI modalities via generative models can enhance multimodal fusion and reduce patient burden in repeated scanning (Dar et al., 2019; Wolterink et al., 2017).

3.1.2 Clinical implications

- Reduces patient scan time by synthetically generating sequences,
- Supports diagnostic completeness for AI systems requiring full modality input,
- Improves downstream tasks such as tumor segmentation or tissue classification,
- Lowers costs in low-resource settings by reducing dependency on full protocols.

3.1.3 Enhancing low-field MRI quality

Low-field MRI scanners are economically viable, particularly in low-resource environments. However, these scanners typically produce images with higher noise, lower resolution, and less contrast than high-field MRIs. Recent advancements in the field of diffusion models have shown significant promise in bridging this quality gap (Kim et al., 2020). A study revealed that applying score-based diffusion to enhance 0.5T MRIs to near-3T quality outperforms traditional superresolution algorithms in terms of resolution and contrast. Compared with conventional CNN-based superresolution, diffusion-enhanced images had a higher signal-to-noise ratio (SNR), improved clinical visibility of lesions, and superior perceptual quality, as rated by radiologists.

3.1.4 Clinical implications

- Enables deployment of low-cost scanners in underresourced hospitals or mobile clinics,
- Improves image utility for diagnosis and AI analysis in telemedicine setups,
- Supports longitudinal studies where access to high-field imaging is not guaranteed.

3.2 Data augmentation

In medical imaging, deep learning models often encounter a significant challenge that includes limited and imbalanced datasets. This issue is particularly pronounced in neuroimaging, where certain conditions such as glioblastomas, cerebral microbleeds, or congenital brain anomalies are inherently rare. Moreover, the availability of data is often constrained by patient privacy concerns, ethical regulations, and the logistical complexity of large-scale data collection. Diffusion models, owing to their ability to generate high-resolution, anatomically realistic synthetic brain images, provide a powerful solution for augmenting neuroimaging datasets in a targeted and controlled manner. By simulating underrepresented pathologies or demographic groups, diffusion-based augmentation improves both dataset diversity and model generalizability, ultimately enhancing the performance and reliability of AI-driven diagnostic tools.

3.2.1 Addressing class imbalance in training sets

Training deep learning models on medical data often faces the challenge of class imbalance, especially in detecting rare pathologies such as glioblastomas, hemorrhages, or rare genetic syndromes. Diffusion models can be trained to generate synthetic brain MRIs containing such rare findings, thus balancing datasets and improving generalization.

Diffusion models, such as denoising diffusion probabilistic models (DDPMs), offer a data-driven approach by learning the full data distribution and generating synthetic brain MRIs that include rare or underrepresented pathological features. A study utilized DDPMs to augment datasets with synthetic tumor and nontumor images for brain segmentation (Wolleb et al., 2022). The inclusion of generated data improved model robustness and classification accuracy.

Benefits:

- Controlled pathology injection: Specific lesion types or sizes can be targeted.
- Balanced training sets: Equal representation of healthy and diseased brains.
- Improved generalization: Models trained on synthetic-augmented data perform better in real-world clinical cases.

3.2.2 Augmenting pediatric and aging datasets

Datasets in pediatric and aging populations are particularly limited because of ethical and practical constraints. Using diffusion models, synthetic samples of underrepresented age groups can be generated to ensure more inclusive training. Also, score-based models could simulate anatomically valid age-progressed or regressed brain images for age prediction and dementia research (Chung & Ye, 2022).

Benefits:

- Data inclusivity: Enabling equitable AI development across all age groups.
- Longitudinal modeling: Simulating disease progression or brain maturation over time.
- Bias mitigation: Reducing overfitting to midlife adult brain data.

3.3 Brain lesion simulation

For training, validating, and testing an AI-driven system designed for neurodiagnostics, the simulation of brain lesions is a crucial step. Pathological diversity is frequently insufficient in traditional datasets, particularly when it comes to especially in rare or complex lesion types. The diffusion models provide a powerful tool for this purpose as they allow for the creation of pathologically and anatomically accurate lesions in brain images. Under controlled conditions, these models can simulate a wide spectrum of lesion characteristics, including shape, size, location, and intensity. Their capacity to learn and reproduce complex data distributions makes them particularly valuable in scenarios with limited annotated data or rare pathologies. Consequently, diffusion-based lesion simulation enhances model reliability and supports the creation of comprehensive, stress-tested datasets suitable for both clinical and research applications.

3.3.1 Synthetic lesions for robustness testing

Conventional datasets often lack sufficient diversity in lesion characteristics, leading to poor model generalization. Diffusion models can synthesize lesions such as gliomas, infarcts, or white matter hyperintensities within specific brain regions, allowing researchers to rigorously test segmentation and detection algorithms. LDM trained to generate synthetic glioma-injected MRIs with precise lesion control. This approach allows for fine-grained control over lesion attributes, including shape, location, and intensity, enabling the creation of custom test cases to evaluate model robustness under varying pathological conditions (Akbar et al., 2023).

This controlled simulation is especially useful for:

- Validating the generalizability of models trained on limited real lesion data,
- Evaluating edge-case performance in clinical AI tools,
- Generating large-scale synthetic datasets for training when real data is sparse or ethically restricted.

3.3.2 Stroke and tumor scenario simulation

Beyond isolated lesion injection, diffusion models can simulate complex pathological processes such as stroke onset or tumor evolution, which are valuable for temporal prognosis modeling and early detection algorithms.

For stroke analysis, a diffusion model can be used to simulate ischemic stroke lesions on FLAIR sequences, enabling better training of automated stroke detection models (Zhang et al., 2023). Similarly, researchers have used such models to emulate tumor growth and progression patterns over time, aiding in prognosis modeling. This controlled lesion simulation capability is particularly important for developing AI models for low-data or rare-event scenarios, where collecting real data is infeasible.

4. Neuroimage enhancement with diffusion models

Diffusion models provide a reliable and theoretically supported method for improving neuroimaging data, especially when image quality is compromised because of low resolution, noise, or missing regions. Because of better iterative refining processes to acquire rich, data-driven priors, these models perform better than conventional

Table 6.4 Enhancement techniques comparison.

Procedure	Conventional methodology	Diffusion model approach	Advantages
Superresolution	Interpolation/U-Nets	Multiscale denoising with DDPMs	High anatomical fidelity
Denoising	Nonlocal means/BM3D	Score-based denoising	Adaptive, model-driven
Inpainting	Patch-based/GANs	Conditional denoising via masking	Realistic and anatomically valid

denoising and reconstruction algorithms, making them ideal for use in research and clinical applications. A comparative approach between the conventional and diffusion models is illustrated in Table 6.4.

4.1 Superresolution

4.1.1 Enhancing low-resolution scans

In order to improve the visualization of brain components that are essential for diagnosis, such as cortical thickness and lesion boundaries, diffusion models have been used to resolve such issues. Low-resolution MRIs, especially those from rapid acquisitions or low-field scanners, can obscure fine anatomical details, particularly those obtained from quick acquisitions or low-field scanners. A plug-and-play framework can be used for MRI superresolution utilizing score-based diffusion models (Mármol-Rivera et al., 2025). By maintaining anatomical accuracy and improving resolution, the model outperformed conventional CNN-based and GAN-based superresolution approaches.

4.1.2 Multi-scale learning strategies

Using LDMs, a cascaded superresolution technique for medical pictures was developed in which various resolution levels were gradually improved. Both coarse-grained and fine-grained reconstruction are made possible by multiscale diffusion processes, which enable learning characteristics across different spatial resolutions. This hierarchical approach offers good alignment with multiresolution structures of the brain. These techniques greatly help neuroradiologists comprehend structures that are sometimes difficult to evaluate in low-resolution pictures, such as brainstem nuclei or hippocampal subfields (Muhammad et al., 2024).

4.2 Denoising

4.2.1 Removal of Gaussian, Rician, and motion artifacts

Gaussian (thermal), Rician (magnitude reconstruction), and motion-induced noise frequently contaminate neuroimages, especially in fMRI and pediatric scans. Diffusion models have demonstrated an impressive capacity to eliminate such intricate noise patterns while preserving fine features and tissue contrast. For Rician noise and mixed-noise removal in MRI, it was presented that score-based diffusion models perform better than cutting-edge techniques, such as block-matching and 3D filtering (BM3D) and nonlocal means (Dolui et al., 2013). By learning conditional score functions, these models adjust to the noise distribution, enabling precise noise estimation and elimination.

4.2.2 Adaptive noise-aware denoising

Modern diffusion models can be trained to be noise-aware by conditioning the reverse process on noise levels estimated from corrupted images. Adaptive noise-aware denoising (ANAD) GAN model that uses self-updating feature vectors (Jin et al., 2024). This noise-aware memory module adaptively recalls and decodes high-level hidden aspects of noise of varying intensities. An elaborately constructed block called BLOGS is integrated into the system to investigate a more realistic sense of noise distribution. To further improve the denoised images and visual quality, variations in tube voltage, slice thickness, and scanning locations result in reconstructed pictures with varying noise levels in the low-dose CT domain. For large noise images, this tendency leads to inadequate denoising, whereas for weak noise images, the results are excessively smooth. The novel network produces outstanding denoising outcomes by dynamically adapting to changing noise levels.

4.3 Inpainting and completion

4.3.1 Reconstruction of missing slices or voxels

Inpainting refers to the task of recreating the missing or contaminated parts of an image. During neuroimaging, it becomes necessary in cases where slices are lost because of motion or technical failure. Diffusion models can be used to reconstruct the lack of brain slices by learning to fill physically admirable structures from the surrounding reference. A solution to restore unusable voxels by evaluating a picture restoration technique termed inpainting and contrasting it with conventional interpolation techniques was also proposed (Torrado-Carvajal et al., 2021). For voxel-wise studies, inpainting is a promising method for recovering missing or unusable voxels, especially in imaging modalities with low SNR, such as chemical shift imaging (CSI).

4.3.2 Application to fMRI sequence correction

Head movement in fMRI can introduce dropouts in fMRI data with the aid of Temporal-aware diffusion processes. A Temporal LDM (TempDiff) diffusion-based framework created especially for real-world video superresolution, where degradations are varied. TempDiff improves temporal awareness by using a pretrained flow net for latent optimization and propagation across video sequences (Jiang et al., 2024).

5. Case studies and benchmarks

Benchmarking diffusion models in neuroimaging is essential to assess their performance, clinical utility, and potential for real-world deployment. This section reviews commonly used public datasets, evaluation metrics, and significant case studies that illustrate the application of diffusion models in various neuroimaging tasks, including modality synthesis, resolution enhancement, data augmentation, and lesion simulation.

5.1 Public datasets utilized for benchmarking

A variety of publicly available neuroimaging datasets have been leveraged to train and evaluate diffusion models. These datasets provide well-annotated, high-quality scans across different imaging modalities and disease states, supporting robust comparative analysis.

5.1.1 Human connectome project

For healthy adult brains, the HCP provides high-resolution multimodal MRI scans, such as resting-state fMRI, T1-weighted, T2-weighted, and diffusion-weighted imaging (DWI). With standardized preprocessing and demographic diversity, HCP serves as a benchmark for modality synthesis and reconstruction tasks (Schilling et al., 2022). It is ideal for training diffusion models aimed at cross-modal generation, such as synthesizing missing MRI sequences.

A LDM was trained on HCP data to create T2-weighted scans from T1-weighted images (Pinaya et al., 2022). With structural similarity indices (SSIM) above 0.90, the model accomplished high fidelity, showing that the generated and real images were closely aligned. It significantly demonstrated the feasibility of scan time reduction, maintaining the clinical utility even in resource-limited environments.

5.1.2 Alzheimer's disease neuroimaging initiative (ADNI)

The ADNI dataset comprises longitudinal imaging data from individuals across the spectrum of cognitive decline, normal aging, mild cognitive impairment (MCI), and AD. The dataset includes T1-weighted MRI, PET imaging, and cognitive scores. It enables the assessment of diffusion models in aging-related MRI synthesis, superresolution, and disease progression modeling.

Score-based diffusion models have been implemented to simulate age-conditioned MRI scans. The generated images mimicked the structural patterns of aging and early AD, supporting their use in training diagnostic AI models for early detection. It facilitated the synthetic augmentation of underrepresented age groups and disease stages, improving inclusivity and diagnostic sensitivity in predictive modeling (Song, Sohl-Dickstein, et al., 2020).

5.1.3 Brain tumor segmentation (BraTS) challenge

The BraTS dataset offers preoperative multimodal MRI scans (T1, T1c, T2, FLAIR) of patients with gliomas, along with voxel-wise tumor annotations. It is a benchmark for tumor segmentation, synthesis, and simulation tasks (Han et al., 2019). It is critical for evaluating lesion simulation and enhancement models because of its richly annotated tumor data.

Brain tumor segmentation using synthetic MR images generated by diffusion models significantly boosts segmentation model generalization by introducing controlled synthetic variance in training data, especially for rare morphologies (Akbar et al., 2023).

5.2 Evaluation metrics

The evaluation of diffusion models in neuroimaging requires a multi-dimensional assessment to ensure that generated images are not only visually realistic but also clinically useful. Metrics are broadly categorized into image quality metrics and segmentation or clinical performance metrics, as discussed in the following sections.

5.2.1 Image quality metrics

Peak Signal-to-Noise Ratio (PSNR)

PSNR quantifies the pixel-level fidelity of a reconstructed or generated image by measuring the ratio between the maximum possible pixel intensity and the mean squared error (MSE) between the generated and ground truth images. The higher value of PSNR indicated a better quality of image (Bhat & Koundal, 2020). PSNR is calculated using Eq. (6.4) as given below:

$$\text{PSNR} = 10\log_{10}\left(\frac{\text{Max}_I^2}{\text{MSE}}\right) \tag{6.5}$$

where MAX_I is the maximum pixel value (e.g., 255 for 8-bit images).

A PSNR of 30 dB and higher is considered acceptable in medical imaging, with higher values correlating with improved clinical interpretability. It is used in neuroimaging to evaluate superresolution and denoising performance. For example, enhancing low-field MRI images to near high-field quality requires PSNR improvement while preserving anatomical accuracy (Chung & Ye, 2022).

Structural Similarity Index Measure (SSIM)

SSIM helps to correlate the local patterns of intensities of pixels of two images. Eq. (6.5) depicts the equation used for calculating SSIM.

$$\text{SSIM} = \frac{(2\mu_R\mu_F + C_1)(2\sigma_{RF} + C_2)}{\left(\mu_R^2 + \mu_F^2 + C_1\right)\left(\sigma_R^2 + \sigma_F^2 + C_2\right)} \tag{6.6}$$

where the constants C_1 and C_2 are appended to maintain stability whenever the values of $\mu_R^2 + \mu_F^2$ and $\sigma_R^2 + \sigma_F^2$ approaches zero. Ranges from -1 to 1, where 1 indicates identical images (Bhat & Koundal, 2021; Shiveta Bhat & Koundal, 2021). SSIM is particularly suited for neuroimaging because it accounts for anatomical structure preservation, crucial for diagnostic validity. For cross-modality MRI synthesis, SSIM scores above 0.90 are indicative of clinically viable synthetic images (Pinaya et al., 2022).

Fréchet Inception Distance (FID)

FID calculates the distance between feature distributions of real and generated images, where features are extracted from an intermediate layer of a pretrained deep network (typically InceptionV3 for natural images) (Heusel et al., 2017). The FID calculator is shown in Eq. (6.6):

$$\mathrm{FID} = \left\| \mu_t - \mu_g \right\|^2 + T_r\left(\Sigma_r + \Sigma_g - 2\left(\Sigma_r \Sigma_g\right)^{\frac{1}{2}} \right) \tag{6.7}$$

Lower FID indicates greater realism and diversity of synthetic images. However, for neuroimaging, domain-adapted encoders are often used because natural image features are suboptimal.

5.2.2 Segmentation and clinical performance metrics

Dice Similarity Coefficient (DSC)

DSC measures the overlap between predicted segmentation masks and ground truth annotations, crucial for assessing lesion segmentation or tumor delineation (Wolleb et al., 2022). Eq. (6.7) is used to calculate the DSC.

$$\mathrm{DSC} = \frac{2|A \cap B|}{|A| + |B|} \tag{6.8}$$

It ranges from 0 (no overlap) to 1 (perfect overlap). Clinical applications generally require dice scores above 0.85 for segmentation reliability.

In a study, it was reported that tumor segmentation models trained with diffusion-augmented data achieved ~7% improvement in dice scores, validating the utility of synthetic data in enhancing model performance (Wolleb et al., 2022).

Hausdorff Distance (HD)

HD measures the maximum distance of a set to the nearest point in another set, capturing boundary discrepancies between predicted and true segmentation (Schilling et al., 2022). Eq. (6.8) illustrates the equation for the calculation of HD.

$$\mathrm{HD}(A, B) = \max\left\{ \sup_{\{a \in A\}} \inf_{\{b \in B\}} d(a, b), \sup_{\{b \in B\}} \inf_{\{a \in A\}} d(a, b) \right\} \tag{6.9}$$

Lower HD values indicate better alignment of structural boundaries, crucial in tasks such as tumor margin delineation for surgical planning. It is used to assess the precision of synthetic lesion simulations or postenhancement segmentations.

Voxel-wise Accuracy and Sensitivity

Voxel-wise accuracy and sensitivity quantify classification performance at the voxel level, especially important in lesion detection, where missing pathological voxels can have severe clinical consequences. Higher sensitivity ensures that true pathological voxels are identified, while high specificity minimizes false positives (Kazerouni et al., 2023). It is employed in evaluating models trained on diffusion-augmented datasets for tasks such as microbleed or infarct detection.

An overview of metrics and their corresponding application in diffusion model-based neuroimaging studies is shown in Table 6.5.

6. Clinical and research implications

Diffusion models are used in research to simulate uncommon conditions, create synthetic data, and test hypotheses. The use of diffusion models in neuroimaging workflows has created new ways to improve clinical diagnosis, speed up imaging techniques, and enhance personalized medicine. These models help in the early and precise diagnosis of neurological illnesses by improving the visual quality and interpretability of scans.

6.1 Diagnostic improvement

6.1.1 Enhanced lesion visibility and delineation

Diffusion models are significant as they have the ability to make visible any tiny pathological features, such as microbleeds, demyelinating lesions, or cancers in their early stages that might be hard to locate in noisy and low-resolution scans.

Table 6.5 Summary of evaluation metrics for diffusion model-based neuroimaging.

Metric	Purpose	Typical thresholds	Example use case
PSNR	Pixel-level reconstruction fidelity	>30 dB	Low-field MRI enhancement
SSIM	Structural and perceptual similarity	>0.90	Cross-modality MRI synthesis
FID	Realism and distribution similarity	Lower is better	Evaluating synthetic data realism
Dice	Segmentation overlap accuracy	>0.85	Tumor/lesion segmentation postenhancement
HD	Boundary accuracy	Lower is better	Tumor margin or infarct boundary precision
Voxel sensitivity	Lesion detection accuracy	Higher is better	Microbleed or infarct identification performance

An application of deep learning-based reconstruction (DLR) to improve the image quality and lesion detection of white matter hyperintensities (WMH) in FLAIR MRI, in contrast to conventional reconstruction (CR) (Sun et al., 2025).

By maintaining the best structural details and minimizing noise, diffusion more suitable pics can serve as advanced inputs for both manual interpretation and AI-based diagnostic pipelines.

6.1.2 Multimodal completion for comprehensive diagnosis

In actual situations, a few imaging modalities may be unavailable because of acquisition constraints. To synthesize lacking modalities (e.g., T2 from T1 or PET from MRI), enabling clinicians to make knowledgeable selections even in statistics-scarce conditions, diffusion models can be used.

A latent diffusion version (LDM) framework used to generate realistic 3D brain MRI scans, training T1-weighted snapshots produced incredibly sensible brain images (Pinaya et al., 2022).

6.2 Reduced scan time and cost

6.2.1 Acceleration of imaging protocols

Prolonged MRI procedures may result in motion artifacts, discomfort for the patient, and higher expenses. By reconstructing high-quality pictures from low-quality or undersampled acquisitions, diffusion models can shorten the necessary scan time.

In a study, the diagnostic quality of MRI acquired using compressed SENSE (CS) acceleration in standard neuroimaging protocols was evaluated. It was analyzed that employing CS acceleration allowed for a significant reduction in MRI scan times without compromising diagnostic image quality (Molnar et al., 2021). In particular, TOF and SPGR sequences maintain diagnostic reliability up to 3×, whereas T2-FLAIR and FIESTA-C sequences maintain good picture quality and lesion detectability with acceleration factors up to 4×. CS performs better than low-resolution approaches and conventional methods. The results highlight the need for sequence-specific optimum acceleration, which enables physicians to improve patient comfort and workflow effectiveness without sacrificing diagnostic accuracy.

6.2.2 Retrospective quality improvement

Diffusion models can be used to retrospectively improve older scans that have poor quality because of patient movements or obsolete hardware, supporting clinical reevaluation or longitudinal investigations.

By efficiently performing retrospective correction of real-world motion distortions, a study presented that a 3D convolutional neural network trained on simulated motion-corrupted T1-weighted MRIs may greatly increase the success rate of cortical surface reconstructions. In Parkinson's patients, motion-corrected scans showed more extensive and statistically significant cortical thinning, suggesting increased sensitivity for morphometric analysis. These results imply that structural MRI data that would otherwise be worthless can be saved by deep learning-based retrospective motion correction, improving picture quality and cortical surface analysis later on in neuroimaging studies (Duffy et al., 2021).

By reducing low-value imaging in first-episode psychosis (FEP) significantly without omitting clinically relevant pathology, the study shows that focused multidimensional techniques can improve resource usage and bring practice into line with evidence-based guidelines.

Another study aimed to decrease the number of needless brain scans (such as CT scans) performed on patients going through their first psychotic episode. Approximately 50% of the patients were receiving scans that weren't necessary for their medical conditions prior to the intervention. The frequency of needless scans decreased to just 16% after adjustments were made, such as training physicians, incorporating warnings into the system, and providing feedback. No significant medical reasons for psychosis were overlooked, even after the scans were reduced. The findings demonstrate that hospitals may safely cut back on unnecessary imaging, save money, and yet deliver high-quality care by implementing straightforward measures (Srivastava et al., 2021).

6.3 Personalized medicine and predictive simulation

6.3.1 Simulation of patient-specific progression

Diffusion models can be conditioned on patient-specific variables (age, sex, and diagnosis stage) to generate future brain states, offering a data-driven method for disease progression modeling.

The multiscale brain network modeling provides a potent framework to better understand and potentially treat neurodegenerative diseases (NDDs) such as AD, Parkinson's disease (PD), and cortical spreading depression (CSD), as emphasized (Shaheen & Melnik, 2025). Multiscale brain network modeling integrates electrophysiological dynamics, detailed cellular events, and large-scale network interactions. This review also discusses how AI, including artificial neural networks and data-driven methodologies, might improve these models' precision, validation, and personalization.

Overall, the research comes to the conclusion that coupled multiscale models—-which span molecular, cellular, and systems-level scales are essential for understanding disease mechanisms, connecting symptoms at the macro level to microlevel pathology, and directing the creation of therapeutic strategies.

6.3.2 Synthetic trials and rare disease simulation

Generating synthetic cohorts that mimic rare diseases or extreme presentations allows for robust model training and validation, improving generalization across clinical populations. Creating realistic synthetic brain PET pictures for various phases of AD (normal control, MCI, and AD) presented a GAN-based synthetic brain PET image generation technique using Deep Convolutional GANs (DCGANs) (Islam & Zhang, 2020). Visual inspection and similarity metrics such as PSNR and SSIM verified that the model, which was trained using ADNI PET datasets, successfully generated high-fidelity slices that closely matched actual scans. This method protects patient privacy while addressing the issues of sparse and unbalanced neuroimaging datasets. According to the study's findings, artificial PET pictures of this kind can improve the training and classification accuracy of AI models in neurodegenerative research. It also suggests that future research focuses on creating 3D volumes with greater resolution and using conditional GANs for more regulated results.

7. Challenges and future directions

Instead of remarkable potential, diffusion models in neuroimaging also have a number of drawbacks that need to be resolved for their safe, efficient, and moral application in authentic clinical environments. These difficulties include interpretability, generalizability, regulatory concerns, and integration within clinical workflow. Table 6.6 lists some common challenges with possible solutions.

7.1 Generalizability across populations and devices

7.1.1 Dataset-specific biases

Ensuring generalizability across various institutions, scanners, and patient populations is one of the most crucial and challenging scenarios when implementing diffusion models in clinical imaging. When applied to data obtained from other hospitals or devices, models trained on curated datasets such as HCP, ADNI, or the Brain Tumor Segmentation Challenge (BraTS) frequently reflect a limited demographic or imaging protocol, which results in a notable decline in performance. The therapeutic value of such models is limited by this dataset-specific bias, particularly in rural or low-resource settings where imaging circumstances can vary significantly.

Diffusion models trained on high-field scans, such as HCP, are unable to provide a targeted diffusion-based correction and generalize to low-field images (Kim et al., 2020). This study directly addresses the boundaries of dataset-specific biases in the medical imaging model. By transferring knowledge from high-quality datasets to low-quality clinical images, it allows the distribution of high-performance AI models in subpaired settings. Existing diffusion models are trained on high-field MRI datasets, which leads to biased learning.

Low-field MRI data, which is more prevalent in rural and resource-constrained environments, shows poor performance from these models. A 3D conditional diffusion model that uses image quality transfer (IQT) to learn how to convert low-field MRI pictures into high-field quality images has been used. By learning mappings across device-

Table 6.6 Challenges and mitigation strategies.

References	Challenge	Description	Proposed solution
Kushibar et al. (2021)	Generalizability	Dataset shift across hospitals/scanners (e. g., different manufacturers, field strengths, and protocols)	Domain adaptation via unsupervised or adversarial strategies
Zhang et al. (2022)		Dataset shift across hospitals/scanners	Federated learning with heterogeneity-aware aggregation (SplitAVG)
Yang et al. (2023)		Dataset shift across hospitals/scanners	Federated learning with dynamic aggregation/ knowledge distillation
Song et al. (2020)	Computational demand	Slow generation times with high sampling steps and resource demands	DDIM—faster deterministic sampling with fewer steps
Jiang et al. (2025)		Slow generation for medical image tasks	Fast-DDPM—reduces sampling to 10 steps using custom schedulers
Babu et al. (2025)	Ethical/ Regulatory issues	Clinicians and patients distrust synthetic medical images; legal frameworks are unclear	Build trust via transparency, empirical validation frameworks
Lekadir et al. (2023)		Opaque AI systems, insufficient transparency, and clinical validation in healthcare tools	Develop consensus-driven frameworks: FUTURE-AI principles

dependent quality gaps, it enhances generalizability and lessens bias brought on by scanner resolution or field strength.

7.1.2 Domain adaptation needs

Future studies should use domain adaptation strategies or federated learning frameworks, which enable models to learn from diverse, decentralized datasets without compromising patient privacy, to improve generalizability.

Domain adaptation techniques allow models to align feature distributions between the source and target domains without the need for labeled data. Furthermore, federated learning provides a collaborative learning paradigm that protects privacy by training models across several decentralized datasets without disclosing patient information. Methods such as dynamic aggregation and SplitAVG have shown promise in managing data from many institutions. When combined, these strategies increase the diffusion models' resilience and flexibility in actual clinical settings. Moreover, training on multi-institutional and multivendor datasets will help reduce bias and promote general use in global healthcare systems.

7.1.3 Computational demand

Despite the ability to provide high-quality medical images, diffusion models frequently have large processing costs because of their iterative sampling procedure. Conventional DDPMs are not feasible for real-time or resource-constrained applications since they need hundreds to thousands of steps to generate a single image. Researchers have implemented a number of optimizations to overcome this issue.

By using deterministic reverse diffusion processes, denoising diffusion implicit models (DDIM) drastically cut down on sampling time, allowing for similar image quality with a substantially smaller number of iterations. In generative tasks, diffusion models have shown impressive performance, especially in medical imaging. However, the tremendous computational demand brought on by their sluggish sampling speeds and significant memory utilization frequently prevents their practical application. The following research initiatives listed below offer practical solutions to these problems without appreciably sacrificing image quality.

By using a deterministic alternative to the stochastic reverse process of conventional DDPMs, DDIM allows for fewer sampling steps. This makes diffusion models more practical for time-sensitive medical tasks by dramatically speeding up inference while maintaining sample quality (Song et al., 2020).

For medical image synthesis tasks such as denoising and superresolution, this study adapts the noise scheduling and learning rate adaptation to cut the number of sample steps from 1000 to just 10. Diffusion methods are now feasible for clinical applications where computing efficiency is crucial (Jiang et al., 2025).

7.2 Regulatory and ethical considerations

7.2.1 Trust in synthetic data

There are serious concerns regarding trust, authenticity, and validation when synthetic neuroimages are used for diagnosis, surgical planning, or disease progression tracking.

- Can doctors make a diagnosis using generated images?
- Do these photos need to be classified as synthetic?
- Who is accountable if a synthetic image leads to an inaccurate diagnosis?

There are serious ethical and legal issues with the growing use of artificial imagery in medical AI, especially with regard to legal accountability, transparency, and trust. The dependability and interpretability of synthetic images are frequently questioned by clinicians and patients, particularly when models are trained on opaque datasets or employ black-box generation approaches. In addition to impeding clinical adoption, this lack of openness may give rise to concerns regarding culpability in misdiagnosed situations. Validation frameworks have been put out to evaluate the clinical value and realism of synthetic data in order to overcome these problems.

7.2.2 Real-time and resource constraints

To produce a single image, diffusion models need hundreds to thousands of iterations, which is computationally demanding. This restricts their application in therapeutic settings where time is of the essence. In order to solve this, scientists have suggested information distillation techniques and accelerated diffusion models, for example, DDIMs (Song et al., 2020), which can generate excellent results in fewer steps.

7.2.3 Privacy and data leakage

Generative models can help with data sharing or anonymization, although there is a chance that data can leak, particularly if the models remember or unintentionally replicate training data.

A framework based on six fundamental principles—Fairness, Universality, Traceability, Usability, Robustness, and Explainability—is presented by the FUTURE-AI initiative, for example, to direct the reliable application of AI in healthcare. Furthermore, accountability and openness are improved through the use of model cards, audit trails, and open validation with clinicians. Legal experts have also underlined how crucial it is to set precise rules for data provenance, product responsibility, and informed permission. By working together, these initiatives hope to create an ethical and legal framework that promotes the responsible and safe application of generative AI in healthcare (Babu et al., 2025; Lekadir et al., 2023).

7.3 Future research directions

Future studies must focus on a few crucial areas to guarantee the responsible and significant use of diffusion models in neuroimaging.

Multimodal Conditioning: By combining neuroimaging data with a variety of data sources, including genetic profiles, EEG signals, and clinical notes, multimodal conditioning should be given priority. This method can provide a more holistic representation of the patient's condition, enabling richer and more personalized generative outputs.

Hybrid Models: The creation of hybrid models that incorporate transformers or reinforcement learning with diffusion-based architectures may improve image generation's flexibility and context awareness, particularly for dynamic or interactive clinical applications.

Clinical Trials and Validation Studies: Clinical trials and prospective validation studies are urgently needed to evaluate the diagnostic and therapeutic use of diffusion-enhanced pictures experimentally. This research would contribute to the development of regulatory acceptability and clinical trust.

Ethics-Aware Training Pipelines: At every level of model development, ethics-aware training pipelines must be put in place to encourage openness, equity, and bias reduction. These kinds of frameworks will be essential to promoting the safe, just, and reliable application of AI in neuroimaging practice.

8. Conclusion

The advent of diffusion models has revolutionized the domain of generative modeling, introducing transformative capabilities to neuroimaging research and clinical practice. These models provide a stable, high-fidelity approach for synthesizing and enhancing complex medical images, surpassing the limitations inherent in traditional deep learning techniques such as GANs and VAEs.

This chapter presented a comprehensive overview of diffusion probabilistic models and their applications in neuroimaging. The chapter began with an explanation of their mathematical foundations, emphasizing the forward and reverse processes underlying denoising diffusion, and included comparative insights with other generative approaches. The core architectures DDPMs, score-based generative models, and LDMs were described, highlighting their distinct strengths in capturing the high-dimensional anatomical structure of neuroimages.

Applications explored in this chapter encompass MRI modality synthesis, data augmentation, lesion simulation, and neuroimage enhancement through superresolution, denoising, and inpainting. These methods tackle important issues in the clinical context, such as class imbalance in neuroimaging datasets, noise artifacts, missing modalities, and low scan resolution. The training and validation of machine learning algorithms for tasks such as segmentation and classification are further supported by the capacity of diffusion models to produce realistic, anatomically accurate images, especially in situations involving rare diseases or underrepresented populations.

These developments have significant clinical implications. Early neurological condition detection, shorter scan acquisition times without sacrificing quality, and the simulation of disease progression to support personalized medicine strategies are all made possible by diffusion models. However, there are still obstacles to overcome before broad clinical implementation is possible.

Key barriers include ensuring generalizability across various populations, enhancing the interpretability of outputs produced, addressing ethical issues related to the use of synthetic data, and controlling the computational demands of training and inference data. To address these challenges, clinicians, radiologists, computer scientists, ethicists, and policy makers must work closely. Furthermore, regulatory frameworks that control the deployment and validation of synthetic neuroimages in actual clinical settings must be established.

Future research instructions specialize in improving the performance and scalability of diffusion models, integrating them with multimodal medical information, and validating their impact through prospective clinical trials. Innovations such as generation algorithms, hybrid transformer diffusion model architectures, and privacy paradigms, including federated learning, make these models extra efficient, stable, and accessible.

In conclusion, in the field of neuroimaging, diffusion models are a major development that has the ability to accurately recreate and improve brain images with high fidelity, paving the way for more precise diagnostics, efficient imaging workflows, and personalized patient care. Ongoing research and the proper use of these models in clinical settings could lead to significant breakthroughs in radiology, neurology, and the broader field of digital health.

References

Akbar, M. U., Larsson, M., Blystad, I., & Eklund, A. (2023). Brain tumor segmentation using synthetic MR images—A comparison of GANs and diffusion models. https://doi.org/10.48550/arXiv.2306.02986.

Alexander, D. C., Zikic, D., Ghosh, A., Tanno, R., Wottschel, V., Zhang, J., Kaden, E., Dyrby, T. B., Sotiropoulos, S. N., Zhang, H., & Criminisi, A. (2017). Image quality transfer and applications in diffusion MRI. *NeuroImage, 152*, 283–298. https://doi.org/10.1016/j.neuroimage.2017.02.089

Babu, K. A. R., Mulay, S., Prabhu, O., & Sivaprakasam, M. (2025). Position paper: Building trust in synthetic data for clinical AI. https://doi.org/10.48550/arXiv.2502.02076.

Bengesi, S., El-Sayed, H., Sarker, M. K., Houkpati, Y., Irungu, J., & Oladunni, T. (2024). Advancements in generative AI: A comprehensive review of GANs, GPT, autoencoders, diffusion model, and transformers. *IEEE Access, 12*, 69812–69837. https://doi.org/10.1109/ACCESS.2024.3397775

Bhat, S., & Koundal, D. (2020). Multi-focus image fusion: Quantitative and qualitative comparative analysis. In *, 597. Lecture notes in electrical engineering* (pp. 533–542). Springer. https://doi.org/10.1007/978-3-030-29407-6_38

Bhat, S., & Koundal, D. (2021). Multi-focus image fusion using neutrosophic based wavelet transform. *Applied Soft Computing, 106*, Article 107307.

Bhat, S., & Koundal, D. (2021). Multi-focus image fusion techniques: A survey. *Artificial Intelligence Review, 54*(8), 5735–5787. https://doi.org/10.1007/s10462-021-09961-7

Chen, H., Xiang, Q., Hu, J., Ye, M., Yu, C., Cheng, H., & Zhang, L. (2025). Comprehensive exploration of diffusion models in image generation: A survey. *Artificial Intelligence Review, 58*(4). https://doi.org/10.1007/s10462-025-11110-3

Chung, H., & Ye, J. C. (2022). Score-based diffusion models for accelerated MRI. *Medical Image Analysis, 80*, Article 102479. https://doi.org/10.1016/j.media.2022.102479

Dar, S. U. H., Yurt, M., Karacan, L., Erdem, A., Erdem, E., & Cukur, T. (2019). Image synthesis in multi-contrast MRI with conditional generative adversarial networks. *IEEE Transactions on Medical Imaging, 38* (10), 2375–2388. https://doi.org/10.1109/TMI.2019.2901750

Dolui, S., Kuurstra, A., Salgado Patarroyo, I. C., & Michailovich, O. V. (2013). A new similarity measure for non-local means filtering of MRI images. *Journal of Visual Communication and Image Representation, 24* (7), 1040–1054. https://doi.org/10.1016/j.jvcir.2013.06.011

Duffy, B. A., Zhao, L., Sepehrband, F., Min, J., Wang, D. J., Shi, Y., Toga, A. W., & Kim, H. (2021). Retrospective motion artifact correction of structural MRI images using deep learning improves the quality of cortical surface reconstructions. *NeuroImage, 230*. https://doi.org/10.1016/j.neuroimage.2021.117756

Goodfellow, I. J., Pouget-Abadie, J., Mirza, M., Xu, B., Warde-Farley, D., Ozair, S., Courville, A., & Bengio, Y. (2014). Neural information processing systems foundation Canada generative adversarial nets. *Advances in Neural Information Processing Systems, 3*, 2672–2680.

Han, C., Kim, J., Song, K., & Ye, J. C. (2019). *Combining noise-to-image and image-to-image GANs: Brain MR image augmentation for deep learning-based classification.*

Heusel, M., Ramsauer, H., Unterthiner, T., Nessler, B., & Hochreiter, S. (2017). GANs trained by a two time-scale update rule converge to a local nash equilibrium. *Advances in Neural Information Processing Systems*, 6627–6638.

Ho, J., Jain, A., & Abbeel, P. (2020). Denoising diffusion probabilistic models. *Advances in Neural Information Processing Systems*. Available from: https://papers.nips.cc/.

Islam, J., & Zhang, Y. (2020). GAN-based synthetic brain PET image generation. *Brain Informatics*, 7(1). https://doi.org/10.1186/s40708-020-00104-2

Jiang, H., Imran, M., Zhang, T., Zhou, Y., Liang, M., Gong, K., & Shao, W. (2025). Fast-DDPM: Fast denoising diffusion probabilistic models for medical image-to-image generation. *IEEE Journal of Biomedical and Health Informatics*, 1–11. https://doi.org/10.1109/jbhi.2025.3565183

Jiang, Q., Wang, Q. L., Chi, L. H., Chen, X. H., Zhang, Q. Y., Zhou, R., Deng, Z. Q., Deng, J. S., Tang, B. B., Lv, S. H., & Liu, J. (2024). TempDiff: Enhancing temporal-awareness in latent diffusion for real-world video super-resolution. *Computer Graphics Forum, 43*(7). https://doi.org/10.1111/cgf.15211

Jin, H., Tang, Y., Liao, F., Du, Q., Wu, Z., Li, M., & Zheng, J. (2024). Adaptive noise-aware denoising network: Effective denoising for CT images with varying noise intensity. *Biomedical Signal Processing and Control, 96*, Article 106548. https://doi.org/10.1016/j.bspc.2024.106548

Kazerouni, A., Khodapanah Aghdam, E., Heidari, M., Azad, R., Fayyaz, M., Hacihaliloglu, I., & Merhof, D. (2023). Diffusion models in medical imaging: A comprehensive survey. *Medical Image Analysis, 88*, Article 102846. https://doi.org/10.1016/j.media.2023.102846

Kim, S., Tregidgo, H. F. J., Eldaly, A. K., Figini, M., & Alexander, D. C. (2020). *A 3D conditional diffusion model for image quality transfer—An application to low-field MRI.*

Kingma, D. P., & Welling, M. (2014). 2nd international conference on learning representations, ICLR 2014—Conference track proceedings international conference on learning representations, ICLR Netherlands auto-encoding variational bayes. Available from: https://dblp.org/db/conf/iclr/iclr2014.html.

Kushibar, K., Salem, M., Valverde, S., Rovira, À., Salvi, J., Oliver, A., & Lladó, X. (2021). Transductive transfer learning for domain adaptation in brain magnetic resonance image segmentation. *Frontiers in Neuroscience, 15*. https://doi.org/10.3389/fnins.2021.608808

Lekadir, K., Feragen, A., Fofanah, A. J., Frangi, A. F., Buyx, A., Emelie, A., Lara, A., Porras, A. R., Chan, A. W., Navarro, A., Glocker, B., Botwe, B. O., Khanal, B., Beger, B., Wu, C. C., Cintas, C., Langlotz, C. P., Rueckert, D., Mzurikwao, D., … Starmans, M. P. A. (2023). FUTURE-AI: International consensus guideline for trustworthy and deployable artificial intelligence in healthcare. https://doi.org/10.48550/arXiv.2309.12325.

Mármol-Rivera, J. A., Fernández-Rodríguez, J. D., Asenjo, B., & López-Rubio, E. (2025). Latent diffusion for arbitrary zoom MRI super-resolution. *Expert Systems with Applications, 284*. https://doi.org/10.1016/j.eswa.2025.127970

Molnar, U., Nikolov, J., Nikolić, O., Boban, N., Subašić, V., & Till, V. (2021). Diagnostic quality assessment of compressed SENSE accelerated magnetic resonance images in standard neuroimaging protocol: Choosing the right acceleration. *Physica Medica, 88*, 158–166. https://doi.org/10.1016/j.ejmp.2021.07.003

Muhammad, A., Aramvith, S., Duangchaemkarn, K., & Sun, M. T. (2024). Brain MRI image super-resolution reconstruction: A systematic review. *IEEE Access, 12*, 156347–156362. https://doi.org/10.1109/ACCESS.2024.3478829

Padmanabhan, P., Nedumaran, A. M., Mishra, S., Pandarinathan, G., Archunan, G., & Gulyás, B. (2017). The advents of hybrid imaging modalities: A new era in neuroimaging applications. *Advanced Biosystems, 1*(8). https://doi.org/10.1002/adbi.201700019

Pinaya, W. H. L., Tudosiu, P. D., Dafflon, J., Da Costa, P. F., Fernandez, V., Nachev, P., Ourselin, S., & Cardoso, M. J. (2022). Brain Imaging Generation with Latent Diffusion Models. In *Lecture notes in computer science (including subseries lecture notes in artificial intelligence and lecture notes in bioinformatics)* (pp. 117–126). Springer Science and Business Media Deutschland GmbH. https://doi.org/10.1007/978-3-031-18576-2_12

Rombach, R., Blattmann, A., Lorenz, D., Esser, P., & Ommer, B. (2022). High-resolution image synthesis with latent diffusion models. In *2022 proceedings of the IEEE computer society conference on computer vision and pattern recognition* (pp. 10674–10685). IEEE Computer Society. https://doi.org/10.1109/CVPR52688.2022.01042

Rudie, J. D., Gleason, T., Barkovich, M. J., Wilson, D. M., Shankaranarayanan, A., Zhang, T., & Villanueva-Meyer, J. E. (2022). Clinical assessment of deep learning-based super-resolution for 3D volumetric brain MRI. *Radiology: Artificial Intelligence, 4*(2), Article e210059.

Schilling, K. G., Petit, l., & Rheault, F. (2022). Histological validation of diffusion MRI fiber orientation distributions and dispersion. *NeuroImage, 250*, Article 118927.

Shaheen, H., & Melnik, R. (2025). Brain network dynamics and multiscale modeling of neurodegenerative disorders: A review. *IEEE Access, 13*, 33074–33100. https://doi.org/10.1109/access.2025.3542634

Song, J., Meng, C., & Ermon, S. (2020). Denoising diffusion implicit models. https://doi.org/10.48550/arXiv.2010.02502.

Song, Y., Sohl-Dickstein, J., Kingma, D. P., Kumar, A., Ermon, S., & Poole, B. (2020). Score-based generative modeling through stochastic differential equations. https://doi.org/10.48550/arxiv.2011.13456.

Song, Y., Sohl-Dickstein, J., Kingma, D. P., Kumar, A., Ermon, S., & Poole, B. (2021). Score-based generative modeling through stochastic differential equations. In *ICLR 2021—9th international conference on learning representations international conference on learning representations*. ICLR. Available from: https://iclr.cc/virtual/2021/papers.html?filter=titles.

Srivastava, R., Holmes, R. D., Noel, C. W., Lam, T. V., & Shewchuk, J. R. (2021). Reducing neuroimaging in first-episode psychosis by facilitating uptake of choosing wisely recommendations: A quality improvement initiative. *BMJ Open Quality, 10*(3). https://doi.org/10.1136/bmjoq-2020-001307

Sultana, O. F., Bandaru, M., Islam, M. A., & Reddy, P. H. (2024). Unraveling the complexity of human brain: Structure, function in healthy and disease states. *Ageing Research Reviews, 100*. https://doi.org/10.1016/j.arr.2024.102414

Sun, J. P., Bu, C. X., Dang, J. H., Lv, Q. Q., Tao, Q. Y., Kang, Y. M., Niu, X. Y., Wen, B. H., Wang, W. J., Wang, K. Y., Cheng, J. L., & Zhang, Y. (2025). Deep learning-based reconstruction enhanced image quality and lesion detection of white matter hyperintensity through in FLAIR MRI. *Asian Journal of Surgery, 48*(1), 342–349. https://doi.org/10.1016/j.asjsur.2024.09.156

Torrado-Carvajal, A., Albrecht, D. S., Lee, J., Andronesi, O. C., Ratai, E. M., Napadow, V., & Loggia, M. L. (2021). Inpainting as a technique for estimation of missing voxels in brain imaging. *Annals of Biomedical Engineering, 49*(1), 345–353. https://doi.org/10.1007/s10439-020-02556-3

Wolleb, J., Bieder, F., Sandkühler, R., & Cattin, P. C. (2022). Diffusion models for medical anomaly detection. In *Lecture notes in computer science (including subseries lecture notes in artificial intelligence and lecture notes in bioinformatics)* (pp. 35–45). Springer Science and Business Media Deutschland GmbH. https://doi.org/10.1007/978-3-031-16452-1_4

Wolterink, J. M., Dinkla, A. M., Savenije, M. H. F., Seevinck, P. R., van den Berg, C. A. T., & Išgum, I. (2017). Deep MR to CT synthesis using unpaired data. In *Lecture notes in computer science (including subseries lecture notes in artificial intelligence and lecture notes in bioinformatics)* (pp. 14–23). Springer Verlag. https://doi.org/10.1007/978-3-319-68127-6_2

Yang, L., He, J., Yue, F., & Luo, Z. (2023). Federated learning for medical imaging segmentation via dynamic aggregation on Non-IID data silos. *Electronics, 12*(7), 1687. https://doi.org/10.3390/electronics12071687

Zhang, M., Qu, L., Singh, P., Kalpathy-Cramer, J., & Rubin, D. L. (2022). SplitAVG: A heterogeneity-aware federated deep learning method for medical imaging. *IEEE Journal of Biomedical and Health Informatics, 26*(9), 4635–4644. https://doi.org/10.1109/JBHI.2022.3185956

Zhang, J., Wan, T., MacDonald, E., Ganesh, A., & Wu, Q. (2023). Synchronous image-label diffusion probability model with application to stroke lesion segmentation on non-contrast CT.

CHAPTER 7

Flow-based generative models for neuroimaging: Density estimation and data augmentation

Ravi Teja Kothuru
University of San Diego, Applied Artificial Intelligence, San Diego, CA, United States

1. Introduction

Generative modeling has come to be a vital tool in neuroimaging, especially for understanding brain structure and function. It allows for the creation of realistic brain images based on analyzing the available data, which is crucial in fields such as disease diagnosis and treatment planning (Wang et al., 2023). Generative models, particularly in clinical neuroscience, assist in overcoming data limitation issues, such as small datasets, by producing synthetic data that maintains important patterns (Seiler & Ritter, 2025). The models have been effective in predicting complex brain patterns with neurological diseases such as Alzheimer's disease and multiple sclerosis (Wang et al., 2023). The integration of generative AI into neuroimaging holds great promise to enhance brain network analysis and precision medicine (Gong et al., 2023).

1.1 Primary contributions

- ***Flow-based Generative Framework.*** We pioneered the use of normalizing flows for neuroimaging by concentrating on density estimation and data augmentation, which remains in its infancy within medical imaging.
- ***Enhancements for Data Limitation.*** Our work fills in the gaps for real synthetic scans for the uncommon neurological disorders and for ultrasmall datasets. This motivates enhanced data availability to aid in training.
- ***Self-supervised Merge.*** We developed self-supervised methods within flow-based models that lessen reliance on labeled datasets while increasing model generalizability.
- ***Comprehensive Evaluation.*** We provide both quantitative and qualitative assessments of flow models and their counterparts (GANs and VAEs [Variational Autoencoder]) regarding image generation to showcase flow models' fidelity and statistical consistency.
- ***Applications of Precision Medicine.*** Our framework supports patient stratification and personalized imaging, aiding the goals of precision neuroscience.

Generative Artificial Intelligence for Neuroimaging
ISBN 978-0-443-45010-5
https://doi.org/10.1016/B978-0-443-45010-5.00001-0

1.2 Distinctiveness from the existing literature

- ***Exact Likelihood Estimation.*** Flow models provide invertibility and exact density estimation, something GANs and VAEs cannot (Melnychuk et al., 2023; Xie & Cheng, 2025). This allows for more reliable synthesis and anomaly detection.
- ***Practical Clinical Focus.*** Previous reviews (Gong et al., 2023; Seiler & Ritter, 2025) have been theoretical; this work provides a practical, empirically based framework for neuroimaging in a clinical context.
- ***From Theory to Practice.*** To my knowledge, excellent resources such as Beigy (2024) and Weng (2018) provide conceptual frameworks for flow-based models; however, we have adapted them and used them in practice.
- ***Novel Application of Self-supervision.*** Unlike previous self-supervised methods constrained to classification (Dong et al., 2023; Zhang et al., 2023), we have effectively been applying self-supervision to stabilize and guide less data-driven generative training processes.

2. Fundamentals of flow-based generative models

2.1 Definition and types of generative models

Generative models learn the joint probability distribution over the data and can sample new instances from this distribution. Generative models can be used in neuroimaging to generate realistic simulated brain scans to help alleviate challenges such as data availability.

There are several very popular categories of generative models:

2.1.1 Generative adversarial networks (GANs)

Employ adversarial training against a generator and discriminator as shown in Fig. 7.1. It has the following two vital components.

- **Generator.** It helps in generating the data samples.
- **Discriminator.** It checks whether the generated data samples belong to the original dataset.

2.1.2 VAEs (Variational autoencoders)

Aim to learn a latent representation and reconstruct the input data through probabilistic encoding/decoding as shown in Fig. 7.2. It also has the following important layers.

- **Encoder.** It encodes data samples into the latent space.
- **Decoder.** It reconstructs data samples from the latent space.

The goal of VAE is to extract the information from input data and encode it into a latent space that is lower-dimensional, which is also known as the bottleneck of VAE. This way, we can ensure that the images can be recreated with minimal loss using a decoder.

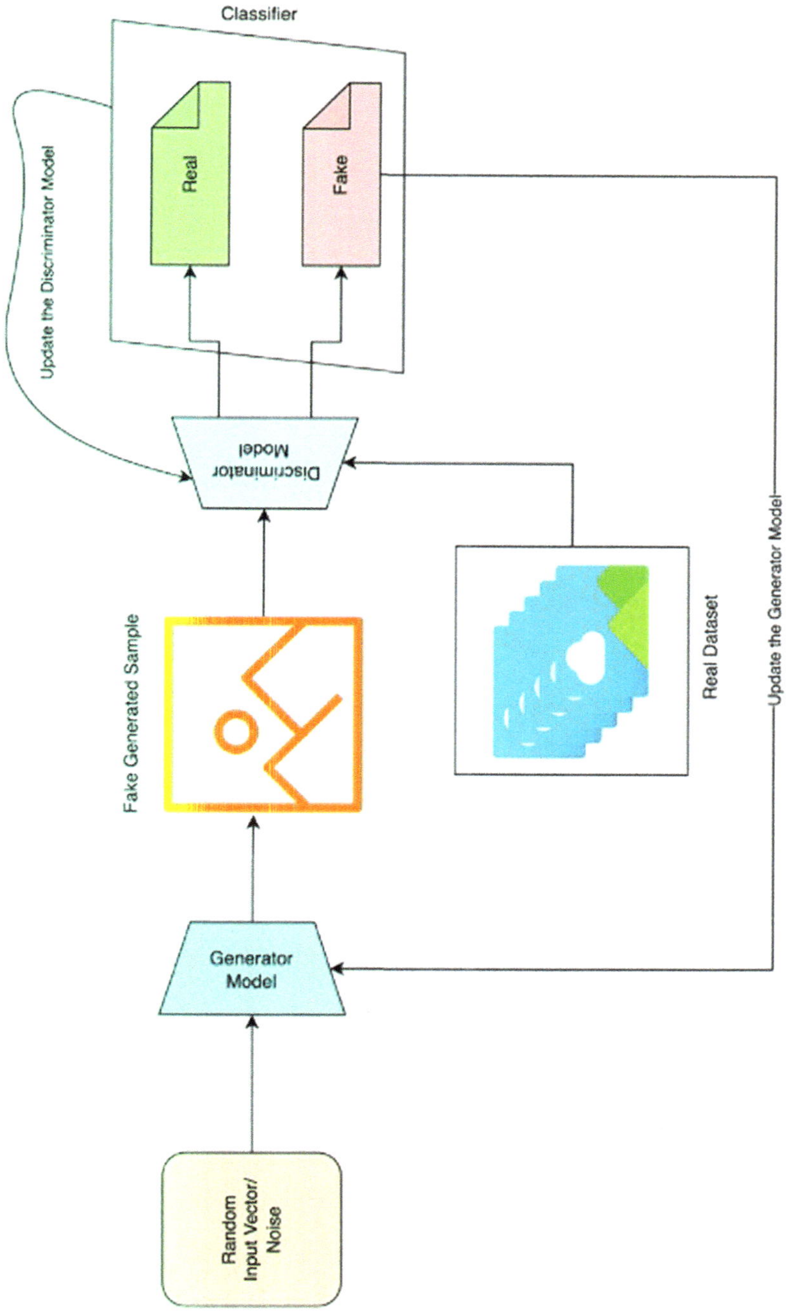

Figure 7.1 GAN architecture.

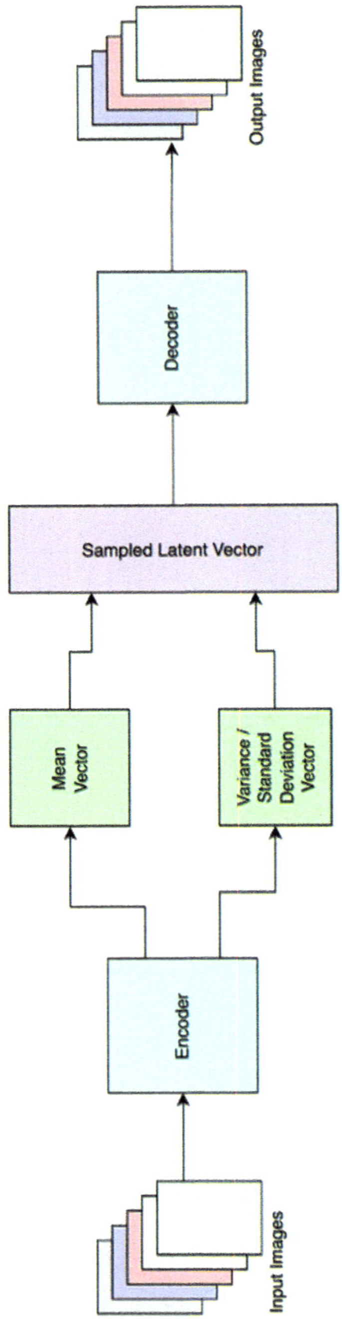

Figure 7.2 VAE architecture.

2.1.3 Normalizing flows

Model exact likelihoods through invertible transformations, which allow for greater interpretability and control of sampling (Melnychuk et al., 2023). If you look at the following Fig. 7.3 (Beigy, 2024), the invertible transformation functions are used to map data "x" to the latent representations "z". We should ensure that the shapes of "x" and "z" are the same.

2.2 Comparison of types of generative models

Now that we have discussed various types of generative models, let us visually understand their comparisons from the following Table 7.1.

2.3 Mathematical foundations

2.3.1 Jacobian matrix and determinants

Let $f : R^n \leftarrow R^m$ be a function that assigns an m-dimensional output vector to an n-dimensional input vector x, such that the matrix of all the first-order partial derivatives of "f" is called the Jacobian matrix "J", where one entry in the ith row and the jth column would be $J_{ij} = \frac{\partial f_i}{\partial x_j}$ (Weng, 2018). That is, if we have a t-dimensional output vector after having an n-dimensional input vector x, we must evaluate a Jacobian matrix as shown in Fig. 7.4 (Weng, 2018).

The determinant is a single real number that we calculate as a function of all the entries in a square matrix. It is necessary to frame the reserved word determinant such that it existed for square matrices only. In the simplest terms, the absolute value of the

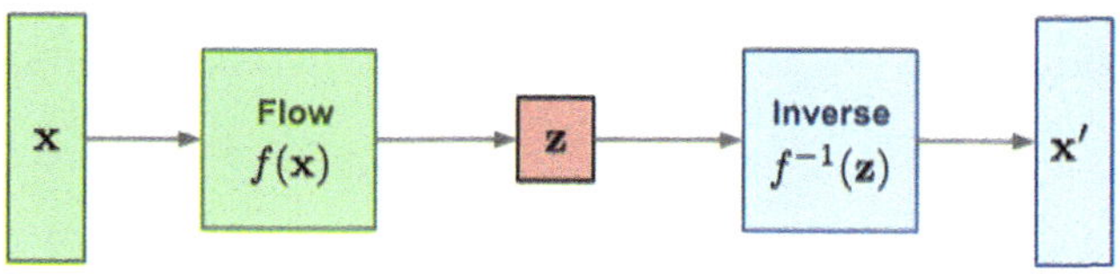

Figure 7.3 Flow-based models architecture (Beigy, 2024).

Table 7.1 Brief comparison of understanding of GANs, VAEs, and normalizing flows (Melnychuk et al., 2023; Zhang et al., 2023).

Model	Advantages	Disadvantages	Suitable for
GANs	Consistent outputs, common usage.	Mode collapse, unstable training.	High-resolution image synthesis.
VAEs	Stable training, latent structure.	Blurry outputs, approximate likelihood.	Representation learning.
Normalizing flows	Exact likelihood, invertibility.	High memory usage, complex computation.	Density estimation, augmentation.

$$\mathbf{J} = \begin{bmatrix} \frac{\partial f_1}{\partial x_1} & \cdots & \frac{\partial f_1}{\partial x_n} \\ \vdots & \ddots & \vdots \\ \frac{\partial f_m}{\partial x_1} & \cdots & \frac{\partial f_m}{\partial x_n} \end{bmatrix}$$

Figure 7.4 Jacobian matrix (Weng, 2018).

determinant can be interpreted as "how much multiplication by the matrix expands or contracts space".

The determinant of an n × n matrix "*M*" is as shown in Fig. 7.5 (Weng, 2018).

2.3.2 Change of variables formula

2.3.2.1 Single variable case

Let us say we have a random variable "z". The corresponding known probability density Function is $z \sim p(z)$ (Weng, 2018). Here, we are supposed to construct a new random variable utilizing a one-to-one mapping function that is, $x = f(z)$. As part of this theorem, the "*f*" is invertible. Hence, $z = f^{-1}(x)$ (Weng, 2018).

So now the question is, how do we determine the unknown probability density function of the new variable, $p(x)$? The integral $\int p(z) \quad dz$ could be defined as the sum of an infinite number of rectangles of infinitesimal width Δz (Weng, 2018). The height of such a rectangle at position "z" is the value of the density function $p(z)$.

Now, substitute $z = f^{-1}(x)$ and it yields to $\frac{\Delta z}{\Delta x} = \left(f^{-1}(x)\right)'$ and $\Delta z = \left(f^{-1}(x)\right)'.\Delta x$. Here, $\left|\left(f^{-1}(x)\right)'\right|$ refers to the ratio between the area of rectangles defined in two different coordinates of the variables z and x, respectively (Weng, 2018).

2.3.2.2 Multivariable case

This version also has almost the same matter. Let us look at the following from Eq. (7.1) to Eq. (7.5) (Weng, 2018).

$$z \sim p(z) \tag{7.1}$$

$$x = f(z) \tag{7.2}$$

$$\det M = \det \begin{bmatrix} a_{11} & a_{12} & \dots & a_{1n} \\ a_{21} & a_{22} & \dots & a_{2n} \\ \vdots & \vdots & & \vdots \\ a_{n1} & a_{n2} & \dots & a_{nn} \end{bmatrix} = \sum_{j_1 j_2 \dots j_n} (-1)^{\tau(j_1 j_2 \dots j_n)} a_{1j_1} a_{2j_2} \dots a_{nj_n}$$

Figure 7.5 Determinant of Jacobian matrix (Weng, 2018).

$$\Rightarrow z = f^{-1}(x) \tag{7.3}$$

$$\Rightarrow p(x) = p(z)\left|\det\left(\frac{dz}{dx}\right)\right| \tag{7.4}$$

$$\Rightarrow p(x) = p\left(f^{-1}(x)\right)\left|\det\left(\frac{df^{-1}(x)}{dx}\right)\right| \tag{7.5}$$

2.3.3 Log-likelihood computation

The log-likelihood comes from the log of the change of variables expression during training. The mathematical aspect of log-likelihood means the model will be trained in a maximum likelihood estimation (MLE) way, which lets the network best represent the input dataset (Xie & Cheng, 2025).

2.4 Training objective

2.4.1 Maximum likelihood estimation (MLE)

Flow-based models are trained using MLE, which optimizes the log-likelihood of the data directly. MLE is a much less complicated training objective and more stable and interpretable than adversarial training in GANs (Melnychuk et al., 2023).

MLE is a method that finds values for the parameters of a model. The parameter values are the ones that maximize the likelihood that the process described by the model produced the data that were observed.

2.5 Important architectures

As we need more concentration toward the flow-based models in this chapter, we have different types of architectures of the same. See Table 7.2 for more understanding.

3. Neuroimaging data: Characteristics and challenges

3.1 Types of neuroimaging data: MRI, fMRI, DTI, and PET

Neuroimaging methods cover a wide variety of methods, with unique properties. Neuroimaging datasets (e.g., MRI, fMRI, DTI, and PET as shown in Fig. 7.6) are high-

Table 7.2 Different architectures of flow-based models.

Architecture	Features	Application in neuroimaging
RealNVP	Affine coupling layers, exact likelihood.	Density modeling (Wang et al., 2023).
Glow	1 × 1 invertible convolutions, multiscale architecture.	Brain imaging synthesis (Gong et al., 2023).
Neural spine flows	Continuous bijections, flexible transformations.	Improved morphological fidelity (Xie & Cheng, 2025).
MAF/IAF	Autoregressive transformations.	Conditional density estimation (Dong et al., 2023).

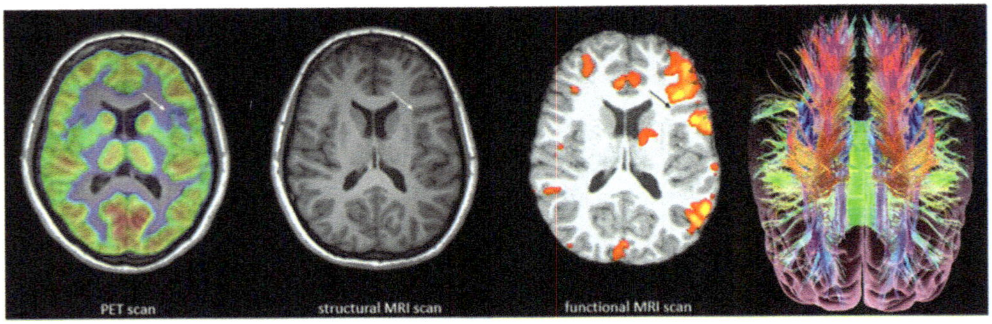

Figure 7.6 PET, MRI, fMRI, and DTI scan images.

dimensional, multimodal, and low sample size (due to expensive acquisition and ethical issues) (Seiler & Ritter, 2025).

1. ***Magnetic Resonance Imaging (MRI).*** Provides high-resolution structural representation of brain anatomy (Tehrani, 2020).
2. ***Functional Magnetic Resonance Imaging (fMRI).*** Provides a measure of brain activity by detecting changes in blood flow and provides information on functional connectivity (Seiler & Ritter, 2025).
3. ***Diffusion Tensor Imaging (DTI).*** Provides a way to map the diffusion of water molecules and is used to study white matter tracts.
4. ***Positron Emission Tomography (PET).*** Provides a way to visualize metabolic processes and is frequently used for the diagnosis of neurodegenerative diseases (Tehrani, 2020).

3.2 High-dimensional nature and structured information (e.g., spatial and anatomical features)

Neuroimaging data are fundamentally high-dimensional data. A single MRI scan may contain hundreds of slices, each slice containing thousands of voxels. Not only are neuroimaging data high-dimensional, but they also involve complex spatial relationships and anatomical structures that require specialized treatment in any modeling framework (Gong et al., 2023). Because neuroimages take on a structured form, any generative model must maintain anatomical consistency to be clinically relevant (Gong et al., 2023).

3.3 Neuroimaging dataset limitations

The limited nature of labeled neuroimaging datasets is often because of several factors, such as low sample size because of the cost of acquisition and problems with ethics (Seiler & Ritter, 2025).

1. ***Small sample size.*** Clinical studies often have small cohorts due to cost, ethical reasons, and recruitment issues.

2. ***Noise and artifacts.*** Imaging modalities are subject to many sources of noise. These include things such as a motion artifact, scanner variations, as well as inter- and intra-subject physiology differences.
3. ***Missing modalities.*** Often, not every imaging modality is available for every subject, meaning robust imputation methods are needed (Azizi et al., 2023).

3.4 Need for density modeling and synthetic data generation

With these limitations, density modeling becomes important for us. Given the probability distribution of brain images, flow-based models can generate realistic neuroimages to supplement sparser datasets. In this way, this synthetic data will be useful for different reasons; it can either generate missing data, serve as a baseline for further testing of anomaly detection methods, or boost training models in a data-constrained environment (Seiler & Ritter, 2025; Wang et al., 2023; Xie & Cheng, 2025).

4. Flow-based models for density estimation in neuroimaging

4.1 Modeling the probability distribution of brain scans

Flow-based generative models are highly appropriate for modeling the true distribution of neuroimaging data. The change of variables formula allows us to convert a simple latent distribution (usually a standard Gaussian) into the difficult distribution of brain images. This process is both invertible and optimized to capture the same statistical properties of the real data (Xie & Cheng, 2025).

4.2 Applications of density estimation

4.2.1 Anomaly detection

In clinical neuroimaging, it is extremely important to identify abnormal findings such as lesions, tumors, or unexpected morphological changes (as shown in Fig. 7.7). Flow-based models will determine exact likelihoods, which means they will highlight regions where the observed data are inconsistent with the learned distribution. For example, a concise argument can be made if a tumor appeared as a region of very low likelihood, which would then warrant further diagnostic imaging or clinical review (Dong et al., 2023).

4.2.2 Missing data imputation

Neuroimages with missing slices or corrupted regions can seriously affect clinical interpretation. Flow-based approaches will estimate the likelihood of the observed and missing regions, which means they can make principled imputations by sampling plausible values from the learned distribution (as shown in Fig. 7.8). This is particularly beneficial in longitudinal studies, where you want consistent measures across time points (Azizi et al., 2023).

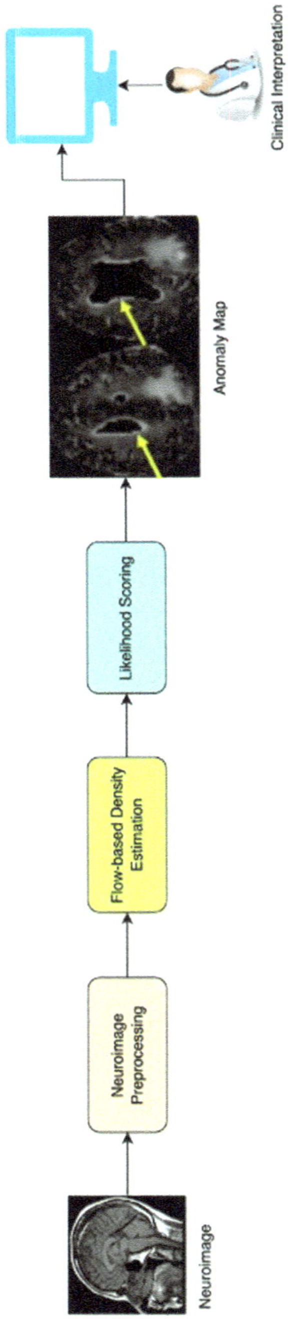

Figure 7.7 Abnormality detection and map analysis flowchart in neuroimages.

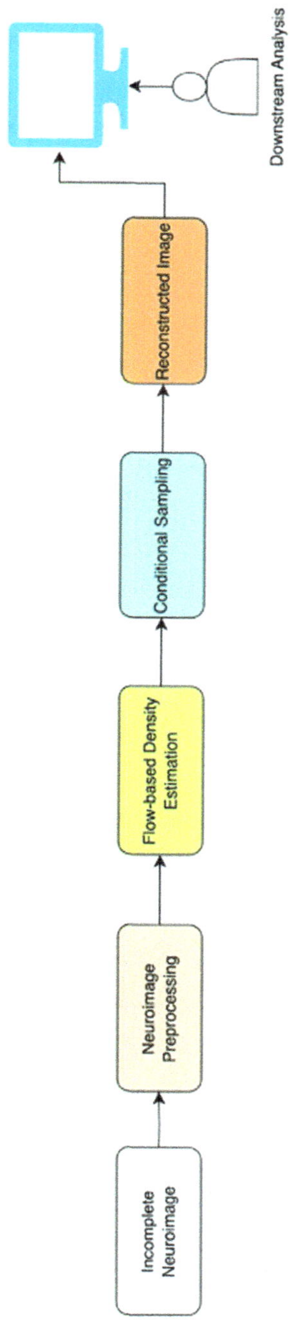

Figure 7.8 Data imputation using mathematical approaches in neuroimages toward corrupted data.

4.3 Brain morphology modeling

Brain morphology—the structural form of each region of an individual's brain—varies wildly with different individuals and pathological states. Flow-based density estimation has great utility for modeling brain morphological differences (Seiler & Ritter, 2025). Clinicians will be able to determine the distribution of what is considered normal brain morphology to spot more subtle differences in morphology for detecting early neurodegenerative changes, subtle neurotoxic, or developmental disorders, as shown in Fig. 7.9 (Seiler & Ritter, 2025).

4.4 Case studies and dataset applications

Recently, various studies have illustrated the success of flow models used in neuroimaging. For instance, research based on the Alzheimer's disease neuroimaging initiative (ADNI) dataset indicates that a flow-based density model can differentiate (prior to diagnosis) between healthy aging and harmful change (Xie & Cheng, 2025). Similar approaches have used Human Connectome Project (HCP) data and improved classification models of subtle cognitive deficits from synthetic data through flows (Wang et al., 2023).

5. Flow-based models for density estimation in neuroimaging

5.1 Importance of data augmentation

Data augmentation is especially useful in the field of neuroimaging, as it suffers from the aforementioned issue of limited annotated data. Augmentation strategies can be useful for enhancing the generalization and robustness of deep learning models by artificially enlarging the dataset (Gong et al., 2023; Melnychuk et al., 2023). We have the potential to use synthetic neuroimages obtained from flow-based models as an effective augmentation strategy, as the augmented images will still have clinically relevant features, while adding variability (Gong et al., 2023; Melnychuk et al., 2023).

5.2 Methods for generating realistic synthetic neuroimages

Flow-based models synthesize neuroimages through invertible transforms that preserve the structural integrity using the processes outlined in the flow chart (see Fig. 7.10):

5.2.1 Latent representation

Imagine that the latent representation of a brain scan is like a compact synopsis of a brain scan. Instead of having to work with a million voxels, a flow-based model learns to "encode" each image into a much smaller number of numbers (the latent variables). These numbers describe the relevant features of the brain image, such as gross shape, texture, and relevant location landmarks, to allow the model to be efficient (Xie & Cheng, 2025).

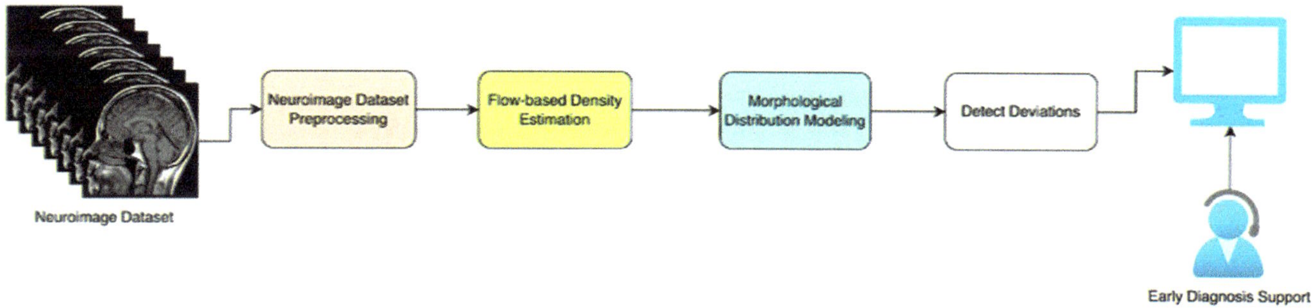

Figure 7.9 Steps to perform brain morphology.

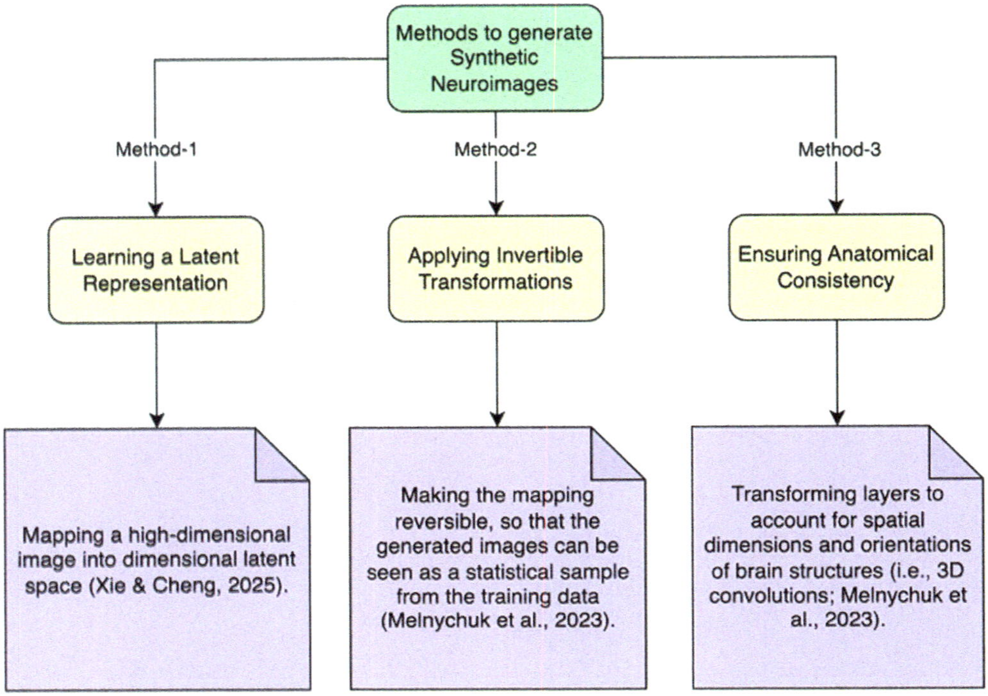

Figure 7.10 Methods to generate synthetic neuroimages.

5.2.2 Invertible transformation

An invertible transformation is like a reversible recipe. You start with an unchanged base of the ingredients (random numbers in latent space), and it consists of multiple steps in which the ingredients are transformed into a realistic brain image. All of the steps are invertible, which means you can run the recipe in reverse: take a brain image, undo the steps, and end back up with random numbers. As a result, this means that the model learns a perfect mapping between simple distributions and more complex scans (Melnychuk et al., 2023).

5.2.3 Anatomical consistency

To keep the generated images realistic, flow-based models include multiple layers such as 3D convolutional layers, which observe neighboring voxels in three dimensions only (rather than a 2D image or 1D layer). This means that the spatial relationship of the original brain image is inherently respected (for instance, that gray matter surrounds white matter). By incorporating these anatomical relationships in the transformations, the generated synthetic scan maintains appropriate structure, gyri, sulci, and ventricles (Melnychuk et al., 2023; Seiler & Ritter, 2025).

These methods have recently been developed in the computer vision literature to create high-fidelity images that are often indistinguishable from actual scans (Wang et al., 2023).

5.3 Preserving clinical relevance in augmented data

A key challenge in data augmentation in neuroimaging is the need for anatomical and clinical fidelity. Synthetic images must conserve important aspects of the data, such as cortical thickness, white matter quality, and the presence of pathology. Flow-based models tackle this by directly constructing the underlying distribution for these important attributes, which in turn ensures that any variations will not introduce artifacts that may confuse any downstream diagnostic algorithms (Azizi et al., 2023).

5.4 Evaluation metrics for synthetic data quality

Evaluating the quality of synthetic neuroimages involves several metrics, as shown in the following Table 7.3:

6. Practical considerations

6.1 Training flow-based models on neuroimaging data

One major drawback of using flow-based models for neuroimaging data is the computational burden associated with both relatively high dimensions and large amounts of neuroimaging data. As discussed earlier, training flow-based models is memory-intensive and computationally expensive. Nevertheless, there are steps that can always be taken to reduce this computational cost, such as:

1. ***Batch Normalization Methods.*** To stabilize training across mini-batches.
2. ***Mixed-precision Training.*** To reduce computational time and memory costs by using lower precision.
3. ***Distributed Computing.*** To parallelize training via multi-GPU environments (Melnychuk et al., 2023).

Table 7.3 Evaluation metrics.

Metric	Description
Fréchet Inception Distance (FID)	FID adapted for medical images provides an approximation of the similarity between distributions of real and generated images (Dong et al., 2023).
Visual inspection	Domain experts evaluate anatomical accuracy and clinical relevance (Gong et al., 2023).
Classifier performance	The diagnostic models trained on synthetic data and improvements in classification accuracy could indirectly reflect data quality (Zhang et al., 2023).

6.2 Preprocessing neuroimaging data

Preprocessing procedures are required prior to training the model and to standardize the input data. The following flowchart (see Fig. 7.11) explains the same.

These procedures are important because they will limit variance in the data and allow the model to learn clinically relevant features (Gong et al., 2023).

6.3 Architectural alterations for medical imaging

Neuroimaging data are not natural images and therefore may necessitate different architectural alterations.

1. ***3D Convolutions.*** Brain scans capture 3D volumetric data.
2. ***Multichannel Input.*** To model multimodal data (for instance, if we model T1-weighted MRI data and then fMRI data), the architecture must accept multiple channels as input at the same time.
3. ***Adaptable Receptive Fields.*** The model architecture must account for coarse and fine levels of anatomical detail for effective reconstruction of data (Wang et al., 2023).

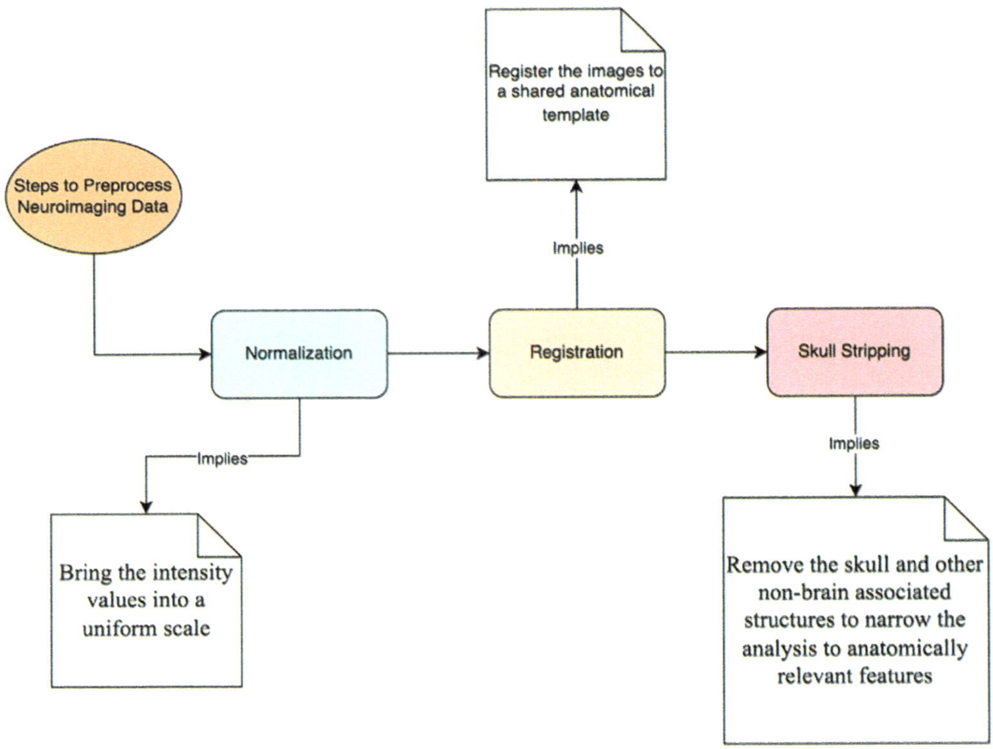

Figure 7.11 Steps to preprocess neuroimaging data before training.

6.4 Tips to tune hyperparameters

There are many hyperparameters to tune when tuning flow-based models. Let us look at the following Fig. 7.12 to understand the same. Recent work suggests that typical systematic methods (i.e., grid-search and Bayesian optimization) are likely useful in general, especially for sensitive neuroimaging data (Xie & Cheng, 2025; Zhang et al., 2023).

6.5 Integration of self-supervised learning (SSL) in neuroimaging

Flow models can leverage SSL to ultimately create a more feature-rich representation with increased robustness in ways unintended by directors because of the limited availability of annotated datasets (Tiu et al., 2022).

6.5.1 Benefits of integration

- **Pretraining Encoders Quality.** Flow models can pretrain on tons of unlabeled neuroimaging data. This enhances the feature-extraction process.
- **Better Anomaly Detection.** Again, with SSL, the sensitivity toward subtle anomalies present in the data increases since SSL emphasizes the reconstruction and transformation tasks (Dong et al., 2023).

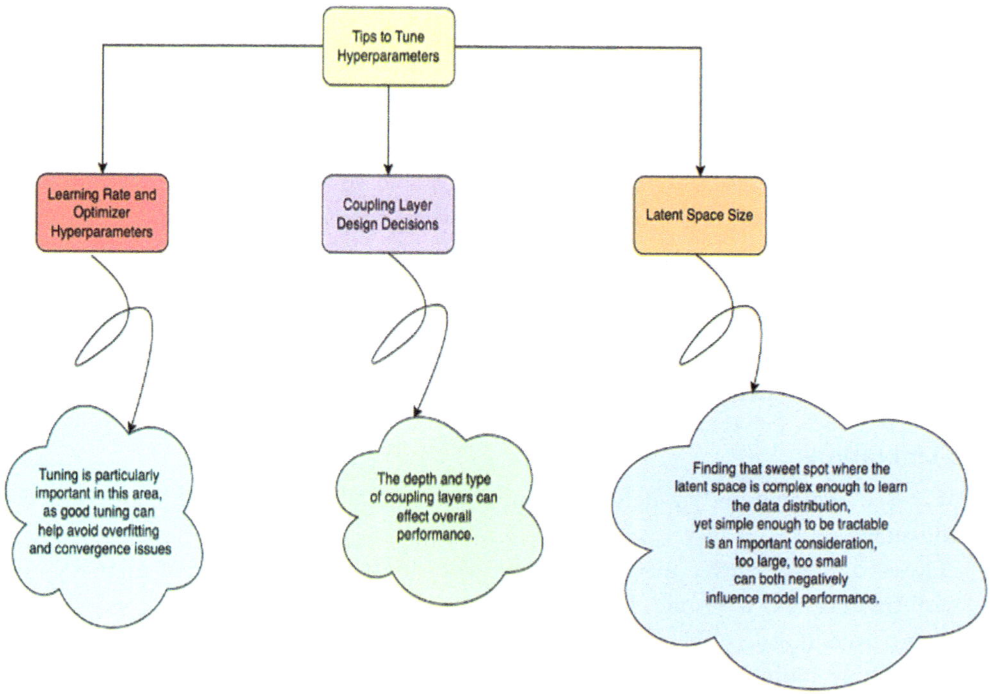

Figure 7.12 Important tips for hyperparameter tuning.

- **Data-efficient Learning.** When SSL emphasizes latent encodings using flow-based augmentations induces data-efficient learning to lower the reliance on expert-annotated types of data (Azizi et al., 2023).

6.5.2 Approaches

- **Contrastive Learning + Normalizing Flows.** Increase class separability in the latent space.
- **Predictive SSL Tasks.** Rotation, anatomical transformations, etc.
- **Denoising and Restorative Learning.** Adversarial and discriminative objectives (Haghighi et al., 2022).

7. Comparison with other generative techniques in neuroimaging

7.1 Comparison of generative techniques in neuroimaging

We already know there are generative techniques in neuroimaging. The following Table 7.4 helps us understand the comparison of them.

7.2 Strengths of flow-based models

7.2.1 Exact likelihood measures

While many generative approaches only use approximation to measure likelihood, flow-based models have the capacity to provide exact likelihood measures, which holds particular importance to applications of critical high value, that is, anomaly detection (Melnychuk et al., 2023).

7.2.2 Stable training dynamics

The training of flow-based models is usually stable in their dynamics because log-likelihood is the objective being optimized directly, and there is no need to compete for an optimal outcome against an adversary (Xie & Cheng, 2025).

7.2.3 Interpretability of learned distributions

Due to the invertible nature of the mapping, it is possible to provide interpretations of the learned latent space with associated clinical features (Seiler & Ritter, 2025).

7.3 Limitations

Flow-based generative models have many advantages, but there are still big disadvantages.

- Flow-based generative models can have very high computation and memory costs for training. As a result, flow-based generative models may not be possible for some users (Gong et al., 2023).
- Producing high-definition three-dimensional images is still difficult because of the transformations needed (Dong et al., 2023).

Table 7.4 Comparison of flow-based models, GANs, and VAEs.

Aspect	Flow-based models	GANs	VAEs
Training stability	Exhibit robust training from exact optimization of log-likelihood (Melnychuk et al., 2023).	Regularly unstable and prone to mode collapse; due to adversarial training (Azizi et al., 2023; Wang et al., 2023).	Overall stability, though optimally optimizing the evidence lower bound could be a problem (Gong et al., 2023).
Likelihood computation	Exact likelihood computation supports the modeling of edges and makes the applications of invertible models potentially more useful in anomaly detection (Xie & Cheng, 2025).	Does not produce explicit likelihoods, restricting to probabilistic tasks (Wang et al., 2023).	Uses variational approximation to a likelihood, so could potentially not be precise (Gong et al., 2023).
Output quality	Can output high-quality, clear images based on invertible and expressive transformations (Seiler & Ritter, 2025).	Hold value somewhat since they visually produce good images, but due to mode collapse, do not produce as much diversity (Wang et al., 2023).	Very high likelihood of producing images with the detailed blurriness that results from a variational approximation (Zhang et al., 2023).
Interpretability	Latent representations can be interchangeable because of the invertibility of latent representations (Seiler & Ritter, 2025).	Latent space is largely unstructured or uninterpretable (Gong et al., 2023).	Works in a structured latent space, so it likely provides the structure with lower interpretability compared to flow-based models (Gong et al., 2023).
Computational demand	Training is memory and computationally expensive (Azizi et al., 2023).	Moderate computational requirements based on architecture (Wang et al., 2023).	Generally has a smaller computational footprint compared to flow-based models (Gong et al., 2023).

Continued

Table 7.4 Comparison of flow-based models, GANs, and VAEs.—cont'd

Aspect	Flow-based models	GANs	VAEs
Use in anomaly detection	High performance is because they provide exact density estimation (Dong et al., 2023).	Good visual anomaly localization, but no probabilistic scores (Wang et al., 2023).	Functionally able, though probably less precise given the errors from an approximation (Zhang et al., 2023).

- They have a great sensitivity to hyperparameters, and hyperparameter tuning needs to be done with each model for optimized performance, which can be tedious (Wang et al., 2023).

8. Future directions and research opportunities

8.1 Combined flow-based and self-supervised models

- Flow-based models allow for exact likelihood model fitting and interpretable density modeling (Weng, 2018; Xie & Cheng, 2025).
- Self-supervised learning (SSL) methods, including contrastive and distillation-based pretraining, have demonstrated high levels of generalizability in the domain of diagnostic imaging (Azizi et al., 2023; Tiu et al., 2022).
- Integrated with these paradigms, we can develop models that are data efficient and highly robust, and deploy them in real-time in clinical settings (Melnychuk et al., 2023).

8.2 Cross-modal and multiresolution learning

- Neuroimaging research often incorporates multiple modalities (e.g., MRI, CT, and PET) in imaging that each capture aspects of physiological information (Tehrani, 2020).
- In the future, the models should learn cross-modal representations that can accommodate differences in spatiality and resolution.
- Current SSL-based approaches—NiRA and SWSSL—could be adapted for efficient combinations of multimodal data (Dong et al., 2023; Haghighi et al., 2022).

8.3 Domain adaptation and personalization

- Medical datasets will differ across sites or across populations. This raises the problem concerning model generalizability.
- Restorative and adversarial adaptation methods (Haghighi et al., 2022) and self-distillation are capable of providing robustness to models across domains.

- Personalized modeling approaches could be used to help ensure that models are personalized to a patient's characteristics or an imaging protocol.

8.4 Simulating disease progression

- Generative models, such as GANs and flows, can be used to simulate disease progression in neurodegenerative disease or synthetic cohorts (Seiler & Ritter, 2025; Wang et al., 2023).
- Simulated disease progression and synthetic cohorts would be beneficial for theory testing, data augmentation, and trial design.
- Simulated disease progression could also be paired with SSL to reduce the dependency on large amounts of annotated longitudinal data.

8.5 Progress made with respect to the theory of flow-based models

- There is currently no theoretical account of how flow-based architectures could be used in the medical context.
- More robust work is necessary concerning convergence, stability, and interpretability of flows in the medical context (Beigy, 2024; Xie & Cheng, 2025).
- In addition, there is continued work necessary concerning coupling layers and, importantly, the effectiveness of the Jacobian (Weng, 2018).

8.6 Interdisciplinary and clinically integrated research

- Advancing this field calls for partnerships with ML researchers, radiologists, and neurologists (Gong et al., 2023).
- Models must fit into workflows, user interfaces, and decision-making.
- Emphasis on educational datasets and clinically interpretable benchmarks to achieve translational value (Zhang et al., 2023).

9. Conclusion

The synergetic progress of machine learning and neuroimaging has resulted in a transformative shift in the use of medical imaging for diagnosis and research. Generative models, specifically, flow-based models, are invertible, likelihood-based methodologies that are suitable for even the most complex neuroimaging data (Beigy, 2024; Weng, 2018; Xie & Cheng, 2025). Both explained variance and interpretability via expressive modeling make them also useful for disease progression modeling and illness detection (Melnychuk et al., 2023; Seiler & Ritter, 2025). Simultaneously, SSL has emerged as a new and powerful mode of learning that enables learning robust novelty detection without requiring huge, annotated datasets, as supervised learning requires (Azizi et al., 2023; Zhang et al., 2023). This becomes more critical in neuroimaging, a highly specialized and costly form of medicine.

By combining flow-based models and SSL strategies, neuroimaging researchers can take advantage of scalable, generalized, and data-efficient systems that apply to either diagnostic imaging or neuroscience research (Dong et al., 2023; Gong et al., 2023). However, challenges with the ultimate implementation, specifically with clinical validation, variability of clinical contexts, and interpretability, remain and need to be resolved (Haghighi et al., 2022; Tiu et al., 2022). Overall, opportunities within these combined methodologies offer an excellent opportunity for the continued evolution of neuroimaging and the potential for advances in diagnostics, treatment planning, and individual-specific modeling of the patient (Wang et al., 2023).

References

Azizi, S., Culp, L., Freyberg, J., Mustafa, B., Baur, S., Kornblith, S., Chen, T., Tomasev, N., Mitrović, J., Strachan, P., Mahdavi, S. S., Wulczyn, E., Babenko, B., Walker, M., Loh, A., Chen, P. H. C., Liu, Y., Bavishi, P., McKinney, S. M., … Natarajan, V. (2023). Robust and data-efficient generalization of self-supervised machine learning for diagnostic imaging. *Nature Biomedical Engineering, 7*(6), 756–779. https://doi.org/10.1038/s41551-023-01049-7

Beigy, H. (2024). Flow-based deep generative models Sharif University of Technology. Available from: https://sharif.edu/~beigy/courses/14022/40959/Lect-16.pdf.

Dong, H., Zhang, Y., Gu, H., Konz, N., Zhang, Y., & Mazurowski, M. A. (2023). SWSSL: Sliding window-based self-supervised learning for anomaly detection in high-resolution images. *IEEE Transactions on Medical Imaging, 42*(12), 3860–3870. https://doi.org/10.1109/TMI.2023.3314318

Gong, C., Jing, C., Chen, X., Pun, C. M., Huang, G., Saha, A., Nieuwoudt, M., Li, H. X., Hu, Y., & Wang, S. (2023). Generative AI for brain image computing and brain network computing: A review. *Frontiers in Neuroscience, 17*. https://doi.org/10.3389/fnins.2023.1203104

Haghighi, F., Taher, M. R. H., Gotway, M. B., & Liang, J. (2022). DiRA: Discriminative, restorative, and adversarial learning for self-supervised medical image analysis. In *Proceedings of the IEEE computer society conference on computer vision and pattern recognition* (pp. 20792–20802). IEEE Computer Society. https://doi.org/10.1109/CVPR52688.2022.02016

Melnychuk, V., Frauen, D., & Feuerriegel, S. (2023). Normalizing flows for interventional density estimation. In *, 202. Proceedings of machine learning research* (pp. 24361–24397). ML Research Press.

Seiler, M., & Ritter, K. (2025). Pioneering new paths: The role of generative modelling in neurological disease research. *Pflügers Archiv—European Journal of Physiology, 477*(4), 571–589. https://doi.org/10.1007/s00424-024-03016-w

Tehrani, M. S. (2020). *What's the difference between all the different head scans and what do they show in the head?* San Diego Brain Injury Foundation.

Tiu, E., Talius, E., Patel, P., Langlotz, C. P., Ng, A. Y., & Rajpurkar, P. (2022). Expert-level detection of pathologies from unannotated chest X-ray images via self-supervised learning. *Nature Biomedical Engineering, 6*(12), 1399–1406. https://doi.org/10.1038/s41551-022-00936-9

Wang, R., Bashyam, V., Yang, Z., Yu, F., Tassopoulou, V., Chintapalli, S. S., Skampardoni, I., Sreepada, L. P., Sahoo, D., Nikita, K., Abdulkadir, A., Wen, J., & Davatzikos, C. (2023). Applications of generative adversarial networks in neuroimaging and clinical neuroscience. *NeuroImage, 269*. https://doi.org/10.1016/j.neuroimage.2023.119898

Weng, L. (2018). Flow-based deep generative models. github.io. Available from: https://lilianweng.github.io/posts/2018-10-13-flow-models/#jacobian-matrix-and-determinant.

Xie, Y., & Cheng, X. (2025). Flow-based generative models as iterative algorithms in probability space. https://doi.org/10.48550/arXiv.2502.13394.

Zhang, C., Zheng, H., & Gu, Y. (2023). Dive into the details of self-supervised learning for medical image analysis. *Medical Image Analysis, 89*, Article 102879. https://doi.org/10.1016/j.media.2023.102879

CHAPTER 8

Hybrid and novel generative models for neuroimaging: Exploring emerging architectures and combinations

Ajay Sharma[1], Arun Malik[1] and Shamneesh Sharma[2]
[1]School of Computer Science and Engineering, Lovely Professional University, Phagwara, Punjab, India; [2]School of Computer Science and Engineering, Geeta University, Panipat, Haryana, India

1. Introduction

Neuroimaging is a foundation stone of modern neurology, which provides a window into the structure, function, and underlying pathological conditions in the brain. The ability to imagine and analyze the brain in a 3D space using techniques such as magnetic resonance imaging (MRI), functional MRI (fMRI), and positron emission tomography (PET) has revolutionized our understanding of both common and unusual brain activity. These images are important for diagnosing neurological diseases, monitoring brain development, and supporting research in cognitive processes, mental disorders, and brain damage. Despite the progress of imaging techniques, Neuroimaging provides datasets that form significant challenges when it comes to processing, interpretation, and accuracy.

The integration of machine learning (ML) and artificial intelligence (AI) has addressed some of these challenges, especially in image analysis (Fig. 8.1). Of these, the generative models—especially generative adversarial networks (GANs) and variational autoencoders (VAEs) illustrated significant attention to the synthesis of realistic brain images and the ability to increase existing datasets, which improved the disease detection, image sharing, and Data Augmentation. GANs, including two neural networks (a generator and a discrimination), have competed against each other, showing a promise to generate high-quality synthetic brain images, even from limited or noisy data. On the other hand, in VAEs, the model gives a probable approach to the latent variables, image structure, the detection of the deviation, and the brain image analysis (Kingma & Welling, 2014, 2019; Laino et al., 2022).

While each of these models provides unique strengths, combining them for a hybrid architecture can remove individual restrictions and provide a strong solution for neuroimaging functions. Hybrid generative models can benefit from the strengths of both GANs and VAEs, which provide more accurate image generation, convenience, and learning to learn complex representations (Lankadasu et al., 1 C. E.; Sharma et al., 2024; Sharma & Kumar, 2022). This chapter will detect emerging hybrid architecture

Generative Artificial Intelligence for Neuroimaging
ISBN 978-0-443-45010-5,
https://doi.org/10.1016/B978-0-443-45010-5.00004-6

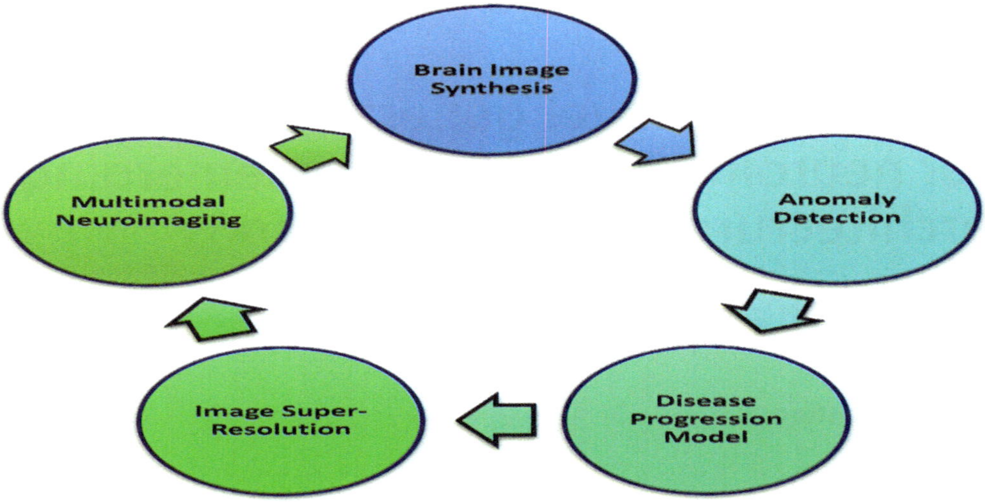

Figure 8.1 Applications of hybrid generative models in neuroimaging.

that integrates these generic models, examine their neuroimaging applications, and discuss future instructions from this exciting and fast development sector (Padmanabhan et al., 2017; Tu et al., 2008; Vázquez-García et al., 2024).

1.1 Background of neuroimaging

Neuroimaging is a multidisciplinary field that has revolutionized the study of the human brain by providing noninvasive methods to visualize its structure and function. The roots of neuroimaging can be traced back to the early 20th century, when techniques such as pneumoencephalography and cerebral angiography were introduced. Pneumoencephalography, developed in the 1920s, involved replacing cerebrospinal fluid with air to enhance X-ray imaging of the brain's ventricles. While it was a groundbreaking technique at the time, it was highly invasive and caused significant discomfort. Around the same period, cerebral angiography, pioneered by Egas Moniz in the 1920s, emerged as a method to visualize blood vessels in the brain by injecting contrast dye. This technique played a crucial role in detecting vascular abnormalities such as aneurysms and tumors (Kumar & Sharma, 2023; Sharma, Gupta et al., 2021; Lankadasu et al., 2024; Satti et al., 1 C. E.).

The field of neuroimaging took a major leap forward in the 1970s with the advent of computerized tomography (CT) scanning. CT scans, developed by Allan Cormack and Godfrey Hounsfield, allowed for detailed cross-sectional images of the brain using X-ray technology and computer processing. This was a significant improvement over previous methods, as it provided more precise anatomical details with less discomfort to patients. Shortly after, in the 1980s, magnetic resonance imaging (MRI) emerged as an even more advanced tool. Unlike CT scans, MRI does not use ionizing radiation; instead, it relies on strong magnetic fields and radio waves to generate high-resolution images of brain structures. The development of functional MRI (fMRI) in the early 1990s further revolutionized the field by enabling researchers to study brain activity in real time. fMRI measures changes in blood flow related to neural activity, providing critical insights into cognitive functions, emotional processing, and neurological disorders (Bank et al., 2023; Tu et al., 2008; Vázquez-García et al., 2024; Padmanabhan et al., 2017).

Another major milestone in neuroimaging was the introduction of PET in the late 20th century. PET scans involve the use of radioactive tracers that allow scientists to observe metabolic processes in the brain. This technique has been particularly useful in studying conditions such as Alzheimer's disease (AD), Parkinson's disease (PD), and psychiatric disorders. Similarly, electroencephalography (EEG) and magnetoencephalography (MEG), which measure electrical and magnetic activity in the brain, have been instrumental in understanding neural dynamics, particularly in epilepsy research (Sharma et al., 2019, 2022; Sharma, Kumar et al., 2021). Neuroimaging continues to advance, with innovations such as diffusion tensor imaging (DTI) for mapping white matter connectivity and hybrid imaging techniques that combine multiple modalities for a more comprehensive understanding of brain function. AI and ML are also being integrated into neuroimaging to improve data analysis and diagnostic accuracy. The evolution of neuroimaging has not only enhanced our understanding of the brain but has also paved the way for improved diagnosis, treatment, and prevention of neurological and psychiatric disorders. As technology continues to develop, the future of neuroimaging holds great promise for further unraveling the complexities of the human brain (Padmanabhan et al., 2017; Tu et al., 2008; Warburg et al., 1927).

1.2 Role of artificial intelligence and generative models in neuroimaging

AI and generative models are playing a transformative role in neuroimaging by enhancing image analysis, improving diagnostic accuracy, and enabling new ways to study brain function and structure. AI, particularly through ML and deep learning (DL), has significantly improved the efficiency and precision of neuroimaging techniques. Traditional neuroimaging methods, such as MRI, CT, and PET scans, generate vast amounts of complex data that require extensive interpretation. AI-driven algorithms can analyze these images with greater speed and accuracy, reducing the reliance on manual assessments and minimizing human error. Convolutional neural networks (CNNs), for instance, have been widely applied in brain imaging to detect abnormalities such as tumors, strokes, and neurodegenerative diseases by identifying subtle patterns that might be missed by human radiologists (Padmanabhan et al., 2017; Perl et al., 2020; Ramezanian-Panahi et al., 2022; Shao et al., 1997; Warburg et al., 1927).

Generative models, particularly GANs and VAEs, have further expanded the capabilities of AI in neuroimaging. These models can generate high-quality synthetic brain images, which are useful for training AI algorithms when real datasets are limited. GANs, for example, have been used to enhance low-resolution MRI scans, improve image reconstruction, and simulate missing or corrupted imaging data. This helps in overcoming challenges such as motion artifacts and incomplete scans, which are common in clinical settings. In addition, generative models can be used for brain image synthesis in disease progression modeling, allowing researchers to predict how a neurodegenerative disorder such as Alzheimer's may develop over time (Bank et al., 2023; Kingma & Welling, 2014, 2019).

AI also plays a crucial role in functional neuroimaging, where it aids in decoding complex brain activity patterns. Functional MRI (fMRI) and EEG data are increasingly being analyzed with DL techniques to understand cognitive functions, detect mental disorders, and even develop brain–computer interfaces (BCIs). AI-powered tools can extract meaningful patterns from large-scale brain connectivity data, enabling early diagnosis of conditions such as schizophrenia, autism, and depression. Moreover, AI is facilitating precision medicine in neuroimaging by personalizing treatment plans based on an individual's unique brain structure and function. Despite these advancements, challenges remain in integrating AI and generative models into routine clinical practice. Issues such as data bias, ethical concerns, and the need for the interpretability of AI decisions must be addressed to ensure reliable and fair applications. However, as AI models become more robust and interpretable, they hold immense potential to revolutionize neuroimaging by making diagnostics faster, more accurate, and accessible. The combination of AI, generative models, and neuroimaging is paving the way for a deeper understanding of the brain, ultimately leading to improved neurological care and groundbreaking discoveries in neuroscience (Bank et al., 2023; Kingma & Welling, 2019; Vázquez-García et al., 2024).

2. Foundations of generative models

Generative models are a class of ML algorithms that focus on learning the underlying distribution of data to generate new, similar data samples. Unlike discriminative models, which classify data into predefined categories, generative models aim to understand how data are distributed and create new instances that resemble the original dataset. These models have broad applications, including image synthesis, text generation, speech synthesis, and drug discovery, making them essential tools in AI and DL research. At their core, generative models seek to capture complex patterns in data, allowing them to generate realistic and high-quality outputs. This is particularly useful in fields such as computer vision, where models can create images that look indistinguishable from real ones, and natural language processing, where AI can generate human-like text.

The development of generative models has been significantly influenced by advances in DL, particularly through architectures such as VAEs, GANs, and transformer-based models, such as Generative Pre-trained Transformer (GPT) (Fig. 8.2) (Bai et al., 2022; Kingma & Welling, 2019; Laino et al., 2022).

Generative models operate by estimating the probability distribution of the input data and then sampling from this learned distribution to create new data points. These models typically fall into two broad categories: explicit density models and implicit density models. Explicit density models, such as VAEs, define a probabilistic model of the data distribution and attempt to learn its parameters. In contrast, implicit density models, such as GANs, do not explicitly define a probability distribution but instead use a generator network to produce samples that mimic the real data. To better understand generative models, it is useful to categorize them based on their underlying principles and methodologies. Below is a detailed breakdown of the major types of generative models and their key characteristics (AbdulAzeem et al., 2021; Bai et al., 2022; Jo et al., 2019; Lin et al., 2018; Perl et al., 2020; Ramezanian-Panahi et al., 2022; Zhao et al., 2021).

2.1 Types of generative models

2.1.1 Probabilistic generative models

Bayesian networks, a probabilistic framework for uncertainty modeling: In the domain of AI and ML, modeling uncertainty is of critical importance, especially when dealing with complex systems in which observations are indirect or incomplete. Two widely used probabilistic models that address this need are Bayesian networks (BNs) and hidden Markov models (HMMs). Both frameworks provide powerful tools for

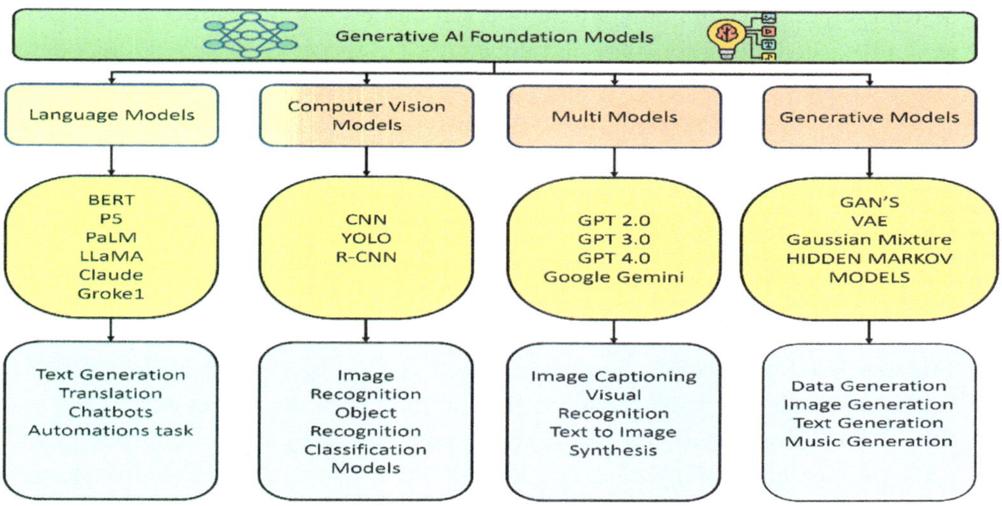

Figure 8.2 Foundation of generative AI models for neuroimaging.

representing and reasoning about uncertain information, with broad applications in fields such as medical diagnosis, speech recognition, natural language processing, and bioinformatics (Friston & Price, 2001; Trotta et al., 2016).

- **Bayesian Networks: Structure and Applications:** BNs, also referred to as belief networks, are graphical models that represent a set of variables and their probabilistic dependencies through a directed acyclic graph (DAG). Each node in the graph corresponds to a random variable, and the directed edges signify conditional dependencies among variables. The underlying principle is based on Bayes' theorem, which enables the computation of posterior probabilities given prior information (Bengio et al., 2013; Fanfarillo et al., 2021; Niedermayer, 2008).

  Mathematically, a BN encodes the joint probability distribution $P(X1, X2, \ldots, Xn)P(X_1, X_2, \ldots, X_n)$ over a set of random variables $X = \{X1, X2, ., Xn\}X = \backslash\{X\_1, X\_2, ., X\_n\backslash\}$ as:

  $$P(X_1, X_2, \ldots, X_n) = \prod_{\{i=1\}}^{\{n\}P(X_i|\backslash\text{text}\{\text{Parents}\}(X_i))}$$

  Here, parents (Xi) denotes the set of parent variables of Xi in the DAG. This factorization significantly reduces the complexity of computing the joint distribution, especially when the number of variables is large. One of the most significant applications of BNs is in medical diagnosis. In this context, the variables might represent diseases and symptoms, and the edges encode causal or correlational relationships. By observing symptoms (evidence), one can update the belief about the probability of various diseases using Bayesian inference. For example, if a patient presents with fever and cough, the network can be used to compute the likelihood that the patient has influenza versus pneumonia or another condition. Another critical domain is speech recognition, where BNs can model acoustic features and phoneme sequences, accounting for noise and uncertainty in spoken language. Unlike deterministic models, BNs handle ambiguity in signal interpretation, making them suitable for real-world applications. Inference in BNs can be conducted via exact algorithms, such as variable elimination and junction tree methods, or through approximate methods, such as Monte Carlo simulations. Learning the structure and parameters of BNs from data is also a major research area, where methods such as expectation-maximization (EM) and score-based structure search are commonly employed (Almond & Zapata-Rivera, 2019; Innovations in Bayesian Networks, 2008; Niedermayer, 2008; Wiegerinck et al., 2013).

- **Hidden Markov Models:** Sequential data modeling is a class of stochastic models used extensively in temporal pattern recognition. They are particularly well-suited for modeling sequential data where the system is assumed to be a Markov process with unobserved (hidden) states. Unlike BNs, which are static, HMMs provide a dynamic framework for modeling time-series data (Bianchini et al., 2013; Drury et al.,

2017; Innovations in Bayesian Networks, 2008; Scutari, 2017; Sridhar, 2024; Wiegerinck et al., 2013).

An HMM is defined by the following components:

- A set of hidden states $S = \{s1, s2, \ldots, sN\} S = \{s_1, s_2, \ldots, s_N\}$.
- A set of observable outputs $O = \{o1, o2, \ldots, oM\} O = \{o_1, o_2, \ldots, o_M\}$.
- The state transition probability matrix $A = [aij] A = \left[a_{\{ij\}}\right]$, where $aij = P(sj|si) a_{\{ij\}} = P(s_j\backslash\text{mid } s_i)$
- The observation probability matrix $B = [bj(k)] B = \left[b_{j(k)}\right]$, where $bj(k) = P(ok|sj) b_{j(k)} = P(o_k\backslash\text{mid } s_j)$.
- The initial state distribution $\pi = [\pi i]\backslash pi = [\backslash pi_i]$, where $\pi i = P(siatt = 1)\backslash pi_i = P(s_i\backslash\text{text}\{at\}t = 1)$.

The joint probability of a sequence of observations $O = (o1, o2, \ldots, oT)$ $O = (o_1, o_2, \ldots, o_T)$ and a state sequence $S = (s1, s2, \ldots, sT) S = (s_1, s_2, \ldots, s_T)$ is given by:

$$P(O, S) = \pi_{\{s_1\}} \cdot b_{\{s_1\}(o_1)} \cdot \prod_{\{t=2\}^{\{T\}a}_{\{s_{\{t-1\}s_t}\}}} \cdot b_{\{s_t\}(o_t)}$$

HMMs have found widespread usage in **speech processing**, where the observable data consists of acoustic signals and the hidden states represent phonemes or words. The **temporal dependencies** modeled by HMMs make them especially suitable for recognizing speech patterns and decoding language in a time-series format. For example, in automatic speech recognition (ASR), HMMs are used to map a sequence of sound features to the most probable word or sentence using algorithms such as the Viterbi algorithm for decoding and the Baum-Welch algorithm for training. Beyond speech, HMMs are also employed in bioinformatics, particularly in gene prediction and protein structure modeling, where the biological sequences are naturally sequential and governed by underlying structural motifs or hidden biological states (Eddy, 1996, 2004; Rabiner & Juang, 1986).

Comparative Analysis and Integration of Bayesian models and HMMs:

BNs and HMMs differ in structure and purpose; they share a probabilistic foundation and can be integrated in complex systems. Dynamic Bayesian networks (DBNs) are a generalization that unifies both frameworks by allowing Bayesian models to be extended across time, capturing temporal dependencies, such as in HMMs, while maintaining the flexibility of arbitrary variable relationships. For example, a DBN can model a complex medical monitoring system where patient vitals change over time, incorporating both instantaneous causal relationships (BNs) and temporal transitions (HMMs). This fusion enables more accurate modeling of real-world systems, especially where data evolves and uncertainty is intrinsic. BNs and HMM models are cornerstone techniques in probabilistic modeling, each tailored to specific types of problems. BNs excel in capturing interdependencies among variables in static domains, while HMMs provide a robust mechanism for handling temporal sequences where states are not directly observable. Both models are grounded in probability theory and offer rigorous mathematical frameworks for dealing with uncertainty, inference, and learning. As technology advances, the hybridization and further development of these models continue to enhance their applicability in diverse fields, from intelligent diagnostics to natural language understanding (Drury et al., 2017; Eddy, 1996; Fischer & Igel, 2012; Rabiner & Juang, 1986; Sridhar, 2024; Wiegerinck et al., 2013).

2.1.1.1 Variational autoencoders (VAEs)

VAEs (Fig. 8.3) are a class of deep generative models that extend traditional autoencoders by incorporating probabilistic inference, allowing for the generation of complex and diverse data samples. Introduced by Kingma and Welling in 2013, VAEs provide a principled framework for learning latent representations of data while enabling efficient sampling from learned distributions. The key idea behind VAEs is to impose a probabilistic structure on the latent space, ensuring smooth and meaningful interpolations between data points. Unlike standard autoencoders, which map inputs to a deterministic latent space, VAEs learn a probabilistic mapping, capturing uncertainty in the data (Fischer & Igel, 2012, 2014; Hinton, 2010; Larochelle & Bengio, 2008; Quintana et al., 2012).

The architecture of VAEs consists of two primary components: an encoder network and a decoder network. The encoder compresses high-dimensional input data into a lower-dimensional latent space by learning a probability distribution, typically modeled as a Gaussian. The decoder then reconstructs the original data from samples drawn from this latent space. A critical component of VAEs is the reparameterization trick, which allows gradients to propagate through stochastic nodes during backpropagation, facilitating effective training. This trick enables the model to sample from the latent distribution without disrupting gradient-based optimization, ensuring stable learning. VAEs are trained using a loss function that balances two objectives: the reconstruction loss and the Kullback–Leibler (KL) divergence loss. The reconstruction loss measures the difference between the original input and the reconstructed output, ensuring the model preserves essential features of the data. The KL-divergence loss enforces a prior distribution

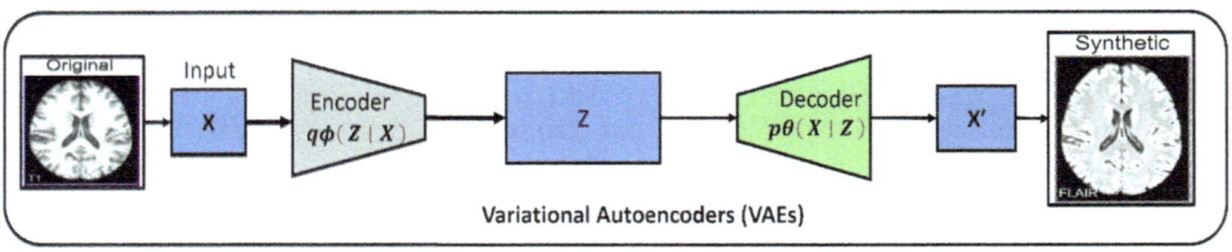

Figure 8.3 The variational autoencoder for the deep generative models.

(typically a standard normal distribution) on the latent variables, promoting smoothness and regularity in the latent space. By minimizing these losses jointly, VAEs learn structured representations that generalize well to new data (Fischer & Igel, 2014; Hinton, 2010; Larochelle & Bengio, 2008; Lopes & Ribeiro, 2015; Montúfar, 2018) (a).

One of the key advantages of VAEs is their ability to generate new samples by sampling from the learned latent distribution. This property makes them highly effective in applications such as image synthesis, text generation, and medical imaging. In image generation, VAEs have been used to create realistic yet novel images, capturing variations in data while maintaining coherence. In natural language processing, VAEs are employed in text generation and style transfer tasks, demonstrating their versatility in modeling sequential data. Moreover, in medical imaging, VAEs have been utilized for anomaly detection and disease diagnosis by learning normal patterns and identifying deviations indicative of pathological conditions (Bank et al., 2023; Fischer & Igel, 2014; Kingma & Welling, 2014; Quintana et al., 2012; Vázquez-García et al., 2024) (2019).

Despite their advantages, VAEs have certain limitations. One notable drawback is their tendency to produce blurry outputs, as the probabilistic nature of the decoder can lead to overly smooth reconstructions. This issue arises from the use of Gaussian priors, which may not always align with the true data distribution. Researchers have proposed various enhancements, such as hierarchical VAEs, conditional VAEs (CVAEs), and adversarial regularized VAEs, to mitigate these limitations and improve sample quality. In recent years, VAEs have been combined with other generative models, such as GANs, to leverage their complementary strengths. Hybrid models, such as VAE–GANs, incorporate adversarial training to refine generated samples, leading to sharper and more realistic outputs. These advancements highlight the ongoing evolution of VAEs, positioning them as a powerful tool in deep generative modeling. As research continues, further refinements in architecture, training techniques, and latent space regularization will enhance the capabilities of VAEs, expanding their applications across various scientific and industrial domains (Bacon et al., 2024; Bai et al., 2022; Bank et al., 2023; Friston & Price, 2001; Jo et al., 2019; Kingma & Welling, 2014; Trotta et al., 2016).

2.1.2 Neural network-based generative models

2.1.2.1 Restricted Boltzmann Machines (RBMs)

Restricted Boltzmann machines (RBMs) (Table 8.1) are a class of generative stochastic neural networks that are capable of learning a probability distribution over a set of inputs. Introduced by Geoffrey Hinton and collaborators, RBMs have become foundational components in unsupervised learning, especially for tasks such as feature extraction, dimensionality reduction, and collaborative filtering. Their unique structure and learning dynamics enable them to uncover hidden representations in high-dimensional data without requiring labels, making them a powerful tool for representation learning in both academia and industry. An RBM is a two-layer neural network

Table 8.1 The mathematical expression of restricted Boltzmann machines (RBMs).

Energy function of an RBM	
$(v, h) = -\sum_i a_i v_i - \sum_j b_j h_j - \sum_{\{i,j\}v_i} W_{\{ij\}h_j}$	where: • v_i and h_j are the binary states of visible unit i and hidden unit j, • a_i and b_j are the biases for the visible and hidden units, respectively, • $W\_\{ij\}$ is the weight connecting visible unit i to hidden unit j.
The probability assigned to a visible vector v is then defined using the Boltzmann distribution $P(v) = \left(\frac{1}{Z}\right)\sum_h e^{-E(v,h)}$	where Z is the partition function, ensuring that the distribution sums to one.

consisting of a visible layer and a hidden layer. The visible layer represents the observed input data, while the hidden layer captures latent features. Unlike traditional feedforward networks, RBMs are undirected and make use of an energy-based model to characterize the joint distribution between visible and hidden units. The "restricted" in RBM refers to the absence of intralayer connections; that is, units within the same layer do not communicate with one another. This constraint significantly simplifies the computation of activation probabilities and gradients, making training more tractable (Hinton, 2010; Larochelle & Bengio, 2008; Montúfar, 2018).

Learning and Training in RBMs: RBMs are trained using unsupervised learning techniques, primarily through the algorithm called contrastive divergence (CD). The goal is to adjust the weights and biases to minimize the difference between the input data distribution and the distribution learned by the model. In essence, CD approximates the gradient of the log-likelihood of the training data by performing a few steps of Gibbs sampling (Lopes & Ribeiro, 2015; Montúfar, 2018). The procedure involves the following steps (Table 8.2):

RBMs for Feature Learning: Feature learning involves automatically discovering the representations or abstractions needed for a task from raw data. RBM excel at this because of their capacity to model complex, high-dimensional distributions. When trained on image, speech, or text data, the hidden layer learns to capture significant patterns such as edges, textures, phonemes, or semantic topics. RBMs have been widely used as building blocks for deep belief networks (DBNs), where multiple RBMs are stacked and fine-tuned using supervised learning. In such architectures, each RBM

Table 8.2 The positive and the negative learning and training in RBMs.

Positive phase	Negative phase	Weight phase
Compute the probabilities of hidden units given the visible units using: $P(h\_j = 1\|v) = \sigma\left(b\_j + \sum \_iv\_iW\_\{ij\}\right)$	Reconstruct the visible units from the sampled hidden units and then resample the hidden units from the reconstructed visible units.	The weights are updated by the difference between the outer products of the original visible and hidden states and the reconstructed ones: $\Delta W_{\{ij\}} = \varepsilon(\langle v_i h_j \rangle_{\text{data}} - \langle v_i h_j \rangle_{\text{recon}})$

learns increasingly abstract features from the output of the previous layer. This hierarchical structure allows for DL from unlabeled data, which is particularly beneficial in scenarios where labeled datasets are scarce or expensive to obtain (Fischer & Igel, 2014; Hinton, 2010; Larochelle & Bengio, 2008; Montúfar, 2018).

Collaborative Filtering Using RBM: Collaborative filtering is a core technique in recommendation systems, aiming to predict the preferences of a user based on the preferences of similar users. Traditional matrix factorization methods have limitations in modeling complex interactions, handling sparse data, and incorporating side information. RBMs offer a probabilistic framework that can overcome some of these challenges. In the context of recommendation systems, users are treated as visible units, and the items they rate are modeled as features in the hidden layer. RBMs can learn complex user-item interaction patterns, even when data is sparse. A key advantage is their ability to handle missing data naturally, which is common in recommendation settings. The RBM model was able to learn latent features representing movie genres and user preferences, improving prediction accuracy beyond conventional approaches. In collaborative filtering with RBMs, the training set typically consists of user-item rating matrices, where each row is treated as a separate input vector. During training, only the observed ratings are used to update the model, and the trained RBM is used to predict missing ratings (Fischer & Igel, 2012, 2014; Hinton, 2010; Larochelle & Bengio, 2008; Lopes & Ribeiro, 2015; Montúfar, 2018; Quintana et al., 2012).

Extensions and Variants: There are several variants and improvements of RBMs have been proposed to enhance their flexibility and expressiveness:

- **Gaussian-Binary RBMs**: These extend RBMs to handle continuous input data, such as pixel intensities in natural images.
- **Conditional RBMs (CRBMs):** These models include additional conditioning variables, making them useful for time-series data and sequence modeling.

- **Replicated SoftMax RBMs:** Adapted for document modeling, they account for word count data using a multinomial distribution instead of binary units.

These extensions have broadened the applicability of RBMs to diverse domains such as speech synthesis, topic modeling, motion capture data analysis, and some other AI tasks.

Advantages and Limitations of Restricted Boltzmann Machines (RBMs):

Advantages:

- RBMs provide a powerful, unsupervised way of learning features
- They are well-suited for high-dimensional data where feature engineering is difficult
- Their generative nature allows for reconstruction and synthesis of data
- RBMs can effectively handle missing data and sparse inputs, making them ideal for recommendation systems

Limitations:

- Training RBMs can be computationally expensive, especially when large datasets or many hidden units are involved.
- The CD approximation may lead to poor local minima or slow convergence in complex models.
- Unlike more modern approaches such as VAEs or GANs, RBMs have fallen out of favor because of scalability issues and training instability in deep configurations.

Being overshadowed by newer architectures, such as VAEs and GANs, RBMs retain educational and historical significance in the evolution of DL. They laid the groundwork for modern unsupervised learning methods and are still studied for their theoretical insights into energy-based modeling. In specific applications, such as collaborative filtering, they continue to provide competitive baselines and practical performance. The RBM represents a valuable component in hybrid models that combine generative and discriminative learning. For instance, combining RBMs with convolutional architectures or reinforcement learning (RL) strategies has opened up new frontiers in deep generative modeling.

2.1.2.2 Deep Belief Networks (DBNs)

DBNs represent a class of deep neural networks that are composed of multiple layers of stochastic, latent variables. Typically, these layers are formed by stacking multiple RBMs, which allows the network to perform unsupervised learning in a greedy layer-wise manner. Originally popularized by Geoffrey Hinton, DBNs were among the first DL architectures that demonstrated the capacity to learn meaningful features from both unlabeled data and labeled data. The architectural structure of a DBN is constructed by stacking RBMs in such a way that the hidden layer of one RBM becomes the visible layer of the next. The key advantage of this design lies in its ability to extract increasingly abstract and high-level representations of the input data as it propagates through the layers. The bottom layers learn low-level features (e.g., edges in images), while the upper layers encode more complex patterns or categories. For instance, in an image classification task, the first RBM may learn to detect edges, the second may detect motifs or shapes, and the third could represent object parts or entire objects. This hierarchical representation mimics the processing hierarchy observed in the human visual cortex (Lopes & Ribeiro, 2015; Mohamed et al., 2012; Montúfar, 2018; Sarikaya et al., 2014).

The training process for DBNs involves two main phases:

- **Unsupervised Pretraining:** Each RBM is trained individually in a layer-wise fashion using CD. The first RBM is trained on the input data. Then, the second RBM is trained on the output (hidden layer activations) of the first RBM, and so on. This step allows the network to initialize weights in a way that captures the statistical structure of the input space, effectively reducing the risk of poor local minima.
- **Supervised Fine-Tuning:** After pretraining, the network is fine-tuned using backpropagation, particularly if the goal is classification. A SoftMax or logistic regression layer is typically appended at the top for prediction tasks. Fine-tuning adjusts the weights to improve performance on labeled data.

Hierarchical Feature Extraction: One of the key strengths of DBNs is their ability to perform hierarchical feature extraction. Each layer learns to capture salient structures in the data, which are passed to the next layer for further abstraction. This mechanism supports robust feature learning, making DBNs particularly useful for tasks such as image recognition, speech processing, and document classification. The learned features tend to be more invariant to transformations, noise, or minor variations in the input data, contributing to improved generalization. DBNs can extract features that are not explicitly present in the input but are inferred through the layered structure and statistical relationships modeled by the network (Lopes & Ribeiro, 2015; Mohamed et al., 2012; Sarikaya et al., 2014; Zhong et al., 2017).

Applications of DBNs have been employed in a variety of applications, including:

- Handwritten Digit Recognition: DBNs achieved notable accuracy on the MNIST dataset before CNNs became the standard.
- Speech Recognition: DBNs have been used to model acoustic signals and phoneme classification tasks.
- Natural Language Processing: Applications such as document retrieval, topic modeling, and sentiment analysis have benefited from DBN-based embeddings.
- Medical Diagnostics: DBNs are used to extract features from complex biomedical data, aiding in disease prediction and diagnosis.

Comparison with Other Deep Models: While DBNs were a breakthrough in the history of DL, their popularity has been somewhat eclipsed by models such as CNNs, recurrent neural networks (RNNs), VAEs, and transformers. These models offer advantages in terms of scalability, training stability, and domain-specific performance. The

DBNs continue to offer insights into unsupervised pretraining and generative modeling. They serve as a valuable pedagogical tool for understanding the evolution of deep architectures and the foundational concepts behind deep feature learning.

DBNs have several limitations, challenges, and strengths:

- Training Complexity: The layer-wise training and fine-tuning processes can be computationally expensive.
- Hyperparameter Sensitivity: Performance heavily depends on the selection of learning rates, number of hidden units, and pretraining epochs.
- Less Efficient on Modern Hardware: Compared to more recent architectures, DBNs do not utilize Graphics Processing Unit (GPU) acceleration as efficiently.

DBNs mark a significant milestone in the advancement of DL methodologies. By stacking RBMs to form deep generative models, DBNs enable hierarchical feature learning that mimics cognitive processing in biological systems. While newer architectures have largely taken center stage, DBNs remain an important conceptual and practical bridge in the transition from shallow learning systems to the deep models that dominate today's AI landscape.

2.1.3 Adversarial generative models

2.1.3.1 Generative Adversarial Networks (GANs)

GANs (Fig. 8.4) are a class of deep generative models introduced by Ian Goodfellow et al. in 2014. GANs are based on a game-theoretic approach where two neural networks, a generator and a discriminator, compete against each other in a zero-sum framework. The generator aims to synthesize realistic data samples, while the discriminator attempts to distinguish between real and generated data. This adversarial training process drives both networks to improve iteratively, leading to the generation of high-quality, realistic outputs. GANs have demonstrated remarkable success in generating images, videos, text, and even 3D models, making them a pivotal advancement in AI (Alqahtani et al., 2021; Creswell et al., 2018; Gui et al., 2023; Navidan et al., 2021; Pan et al., 2019; Saxena & Cao, 2021).

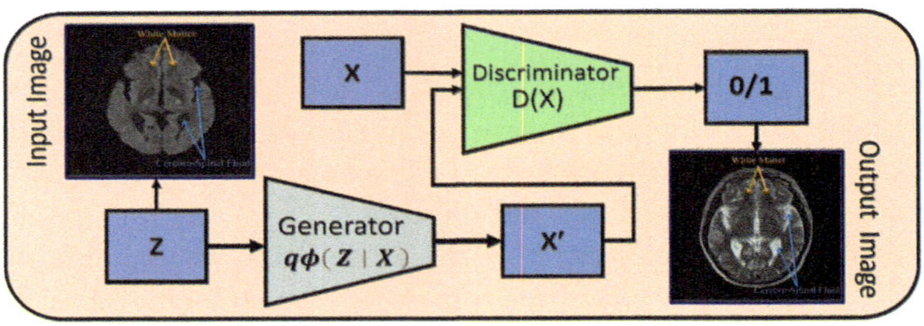

Figure 8.4 Generative adversarial network's (GAN's) for the MRI image reconstruction, image enhancement, and image segmentation.

The architecture of GANs consists of two deep neural networks: the generator network, which learns to map a random noise vector to a data distribution, and the discriminator network, which learns to classify input samples as either real or fake. The training process involves optimizing a minimax objective function, where the generator tries to minimize the discriminator's ability to correctly classify generated samples, while the discriminator aims to maximize its classification accuracy. This dynamic interplay forces the generator to produce increasingly realistic outputs over time. One of the key strengths of GANs is their ability to generate high-resolution, detailed data that closely resembles real-world distributions. This has led to numerous applications, including realistic image synthesis, data augmentation for training DL models, superresolution image enhancement, and domain adaptation. In medical imaging, GANs have been used to synthesize missing data, improve image resolution, and detect anomalies. In natural language processing, GANs contribute to text generation and style transfer, while in creative domains, they have been employed in AI-generated artwork and music composition. Despite their success, GANs face several challenges, including mode collapse, training instability, and difficulties in evaluating the quality of generated samples. Mode collapse occurs when the generator produces limited variations of data, reducing diversity. Training instability arises due to the adversarial nature of GANs, often requiring careful tuning of hyperparameters and network architectures. Furthermore, evaluating GAN-generated samples remains a complex task, as traditional metrics such as likelihood estimation are not well-suited for assessing sample quality. Researchers have developed techniques such as Wasserstein GANs (WGANs), progressive growing of GANs (PGGANs), and StyleGANs to address these challenges and enhance the stability and realism of generated outputs. As GAN research continues to evolve, novel architectures and training methodologies are being explored to further improve their performance. Hybrid models combining GANs with other generative approaches, such as VAEs and diffusion models, are also gaining traction. With continuous advancements, GANs hold significant promise for a wide range of applications, from scientific research and healthcare to entertainment and creative industries, reinforcing their role as a cornerstone of modern generative modeling (Alqahtani et al., 2021; Arora et al., 2017; Creswell et al., 2018; Goodfellow et al., 2020; Lu et al., 2022; Navidan et al., 2021).

Variants of GANs:

- **Conditional GANs (cGANs)**: Generate data conditioned on input variables.
- **StyleGAN**: Used for high-quality facial image synthesis.
- **CycleGAN**: Enables image-to-image translation without paired data.

2.1.4 Comparison of GANs and VAEs in neuroimaging applications

GANs and VAEs have emerged as powerful DL techniques for neuroimaging applications. GANs consist of a generator and a discriminator that compete to create realistic images, making them highly effective for data augmentation, anomaly detection, and image synthesis. On the other hand, VAEs use probabilistic encoding and decoding to learn a compressed latent space representation, enabling high-dimensional neuroimaging data reconstruction and denoising. GANs excel in generating high-quality synthetic images but suffer from mode collapse, where they fail to generate diverse samples. VAEs, while producing more diverse outputs, tend to generate blurrier images because of their inherent probabilistic nature. In neuroimaging, GANs have been applied to enhance MRI and CT scan resolution, generate synthetic brain images, and aid in disease classification. VAEs are widely used for feature extraction, image segmentation, and brain activity modeling. Despite their advantages, GANs require extensive training and are prone to instability, while VAEs may struggle with image sharpness. Future research aims to combine these models to leverage their strengths, enhancing neuroimaging applications for improved disease diagnosis, patient-specific simulations, and precision medicine (Apostolopoulos et al., 2022; Arora et al., 2017; Bank et al., 2023; Goodfellow et al., 2020; Kingma & Welling, 2019; Lu et al., 2022; Navidan et al., 2021; Seeliger et al., 2018; Tu et al., 2008; Vázquez-García et al., 2024; Padmanabhan et al., 2017). More detailed information and a refined version of the table with additional information in Table 8.3:

2.1.4.1 Autoregressive models

Autoregressive models are a foundational concept in generative modeling, built on the idea that future values in a sequence can be predicted using previous ones. In the context of DL, autoregressive models learn to model complex distributions over high-dimensional data by factorizing the joint distribution into a product of conditional probabilities. This sequential approach allows these models to generate new data instances one component at a time, with each component conditioned on the previous ones. The core intuition is simple yet powerful; each prediction leverages the accumulated context of previously generated elements, allowing for coherent and structured outputs (Lütkepohl, 2013; Teräsvirta, 1994; Regis et al., 2022).

Mathematically, given a data vector $(x = [x_1, x_2, \ldots, x_n]\backslash)$, an autoregressive model estimates the joint probability distribution as:

$$P(x) = P(x_1)P(x_2|x_1)P(x_3|x_1, x_2)\ldots P(x_n|x_1, \ldots, x_{n-1})$$

This decomposition allows the model to generate samples iteratively, a property that underpins many successful applications in language, vision, audio, and other domains. These models are typically implemented using neural networks such as RNNs, long short-term memory (LSTM) networks, and, more recently, transformers. They are used for both generative and predictive tasks and have formed the backbone of several state-of-the-art models in NLP, computer vision, and speech synthesis (Dijk et al., 2002; Lütkepohl, 2013; Wong & Li, 2000; Regis et al., 2022).

Table 8.3 Comparison of GANs and VAEs in neuroimaging applications.

Aspect	Generative adversarial networks (GANs)	Variational autoencoders (VAEs)
Core mechanism	Adversarial training with a generator and a discriminator	Probabilistic encoding and decoding of latent space
Primary function	Synthesizing realistic images from random noise	Learning meaningful latent representations for reconstruction
Output image quality	High-resolution, realistic images	Slightly blurry due to latent space sampling
Diversity of outputs	Limited (prone to mode collapse)	High (due to probabilistic sampling)
Stability of training	Often unstable, requiring careful hyperparameter tuning	More stable, but can suffer from posterior collapse
Computational demand	High (due to adversarial training)	Moderate (simpler training process)
Strengths	• High-quality image synthesis • Can generate novel data	• Efficient feature extraction • Robust latent space learning
Weaknesses	• Prone to mode collapse and training instability • Requires large datasets	• Lower image quality • May not capture fine details
Key applications in neuroimaging	• Superresolution MRI reconstruction • Anomaly detection in medical images • Synthetic augmentation of brain scans	• Brain image denoising • Feature extraction for disease classification • Latent space-based brain activity modeling
Best use cases	When high-quality, realistic image synthesis is required	When robust feature representation and reconstruction are needed
Future potential	Integration with VAEs and transformers for enhanced synthesis	Improved latent representations for precision medicine

PixelCNN and PixelRNN: are landmark autoregressive models designed specifically for image generation (Table 8.4). Unlike conventional image processing models that treat an image as a 2D grid processed simultaneously, PixelCNN and PixelRNN model the image as a sequence of pixels, generating each pixel conditioned on previously generated ones in a row-wise or raster-scan order (Mazusa & Bulawayo, 2019; So et al., 2024; Van Den et al., 2016; Xiao et al., 2022; Oord et al., 2016).

Table 8.4 The difference between the Pixel CNN/Pixel RNN, GPT&BERT models.

Feature	Pixel CNN/Pixel RNN	GPT	BERT
Domain	Image processing	Text generation	Text generation
Model type	Autoregressive	Autoregressive	Contextual (not generative)
Training objective	Next-pixel prediction	Next-token prediction	Masked token prediction
Generation style	Pixel-by-pixel	Token-by-token	Not generative
Use cases	Image synthesis	Text generation, chatbots	Text classification, QA
Pretraining	Unsupervised	Unsupervised	Unsupervised (MLM + NSP)
Speed	Slow (PixelRNN), Moderate (PixelCNN)	Fast (parallel training, slow inference)	Fast inference

- **PixelRNN:** PixelRNN uses RNNs to sequentially process images row by row, predicting each pixel based on all the pixels above and to the left of it. The model utilizes either LSTM or GRU units and maintains spatial dependencies using masked connections to prevent future pixel leakage during training. Despite its strong representational capacity, PixelRNN suffers from slow generation speeds due to the inherently sequential nature of RNNs (Mazusa & Bulawayo, 2019; Van Den et al., 2016; Xiao et al., 2022).
- **PixelCNN:** PixelCNN, introduced as a more efficient alternative, replaces recurrent units with convolutional layers that are masked to respect the autoregressive property. By applying masks to convolutional filters, PixelCNN ensures that the prediction for a pixel only depends on previously generated pixels. This enables parallel processing during training, which significantly accelerates convergence compared with PixelRNN (Mazusa & Bulawayo, 2019; So et al., 2024; Van Den et al., 2016; Oord et al., 2016).

Both models have demonstrated impressive capabilities in image synthesis, texture generation, and unsupervised representation learning. They are particularly effective in modeling spatial dependencies in image data, making them suitable for high-fidelity generative tasks. Furthermore, they have been extended into conditional models, such as Conditional PixelCNN, to control the generation process with auxiliary information like class labels or textual descriptions (Lim & Kuo, n.d.; Mazusa & Bulawayo, 2019; Pombo et al., 2019; So et al., 2024; Van Den et al., 2016; Wong & Li, 2000; Xiao et al., 2022; Oord et al., 2016).

Transformer-Based Generative Models: Transformers have fundamentally transformed the field of generative modeling. Introduced by Vaswani et al. in 2017, the transformer architecture replaced recurrence with a self-attention mechanism, allowing for parallel training and the modeling of long-range dependencies in data. While originally developed for machine translation, Transformers have since been adapted for a broad spectrum of tasks, including autoregressive generation (Frässle et al., 2018; Jing et al., 2023; Lim & Kuo, n.d.; Madan et al., 2024).

GPT: Generative pretrained transformer is a family of autoregressive language models built on the transformer decoder architecture. GPT models are trained using a left-to-right context to predict the next token in a sequence. This simple objective, when scaled to massive datasets and models, results in a powerful generator capable of producing human-like text (Frässle et al., 2018; Madan et al., 2024; Shin et al., 2018; Wu et al., 2020; Xie et al., 2023).

Key applications of GPT include:

- **Text Generation:** Given a prompt, GPT can produce paragraphs of coherent and contextually relevant text.
- **Chatbots:** GPT-based systems power conversational agents that simulate human-like dialog.
- **Translation:** Despite not being trained specifically for translation, GPT shows strong zero-shot capabilities.
- **Content Creation:** From poems and stories to essays and code, GPT models support creative writing.

The GPT model is pretrained on large corpora and fine-tuned for specific tasks using supervised learning or RL (as in ChatGPT). Its flexibility and performance have made it the backbone of many commercial AI applications.

BERT: Bidirectional encoder representations from transformers, which differ fundamentally from GPT in its training objective and usage. Instead of generating text, BERT is trained to understand it. It uses a masked language modeling (MLM) approach, where certain tokens in the input are masked, and the model learns to predict them using both left and right context. This bidirectional training allows BERT to develop deep contextual representations, which can be fine-tuned for downstream tasks such as sentiment analysis, named entity recognition, and question answering (Abdalla & Esmail, 2018; Amin et al., 2019; Dong et al., 2017; Hossain et al., 2019).

Though not a generative model in the strictest sense, BERT's influence on generative modeling is profound:

- BERT embeddings serve as inputs to sequence-to-sequence models.
- Hybrid architectures, such as Bidirectional and Auto-Regressive Transformers (BART) and T5, combine the autoregressive nature of GPT with the understanding capability of BERT.
- BERT-style pretraining is now common in multimodal and multilingual generative tasks.

Applications Across Domains of BERT:

2.1.4.2 Natural language processing

- GPT powers language models for summarization, translation, dialog systems, and content creation.
- BERT enhances understanding for sentiment analysis, QA systems, and text classification.

2.1.4.3 Computer vision

- Pixel CNN and Pixel RNN are used in image completion, superresolution, and stylized rendering.

2.1.4.4 Healthcare

- Transformer-based models process medical records, generate radiology reports, and predict patient outcomes.

2.1.4.5 Audio and speech

- Autoregressive models, such as Wave-Net, generate realistic speech and music.
- Transformers generate spectrograms for text-to-speech synthesis.

2.1.4.6 Gaming and simulation

- Autoregressive models generate realistic environments and behaviors in simulations.

2.1.4.7 Molecular and drug design

- Sequence generation is used to generate SMILES strings for novel molecules.

Autoregressive and transformer-based generative models are under active research with several promising directions:

- **Efficient Generation:** Reducing inference time through parallel decoding, caching, and distillation.
- **Multimodal Integration:** Combining vision, language, and audio for holistic understanding and generation.
- **Controllable Generation:** Developing models that allow fine-grained control over style, tone, and structure.
- **Ethics and Alignment:** Addressing concerns of bias, hallucination, and harmful outputs in generative AI.
- **Interpretability:** Understanding how models generate and what internal representations they form.

2.1.5 Advantages of generative models in the context of neurodegenerative disorders

1. **Realistic Data Generation:** Generative models, such as VAEs and GANs, are capable of producing high-fidelity synthetic data that closely mimics real-world

neuroimaging outputs (e.g., MRI and PET scans) and neurological signals (e.g., EEG). In the field of neurodegenerative disorders, such as AD, PD, and Huntington's disease, access to longitudinal imaging data or early-stage patient samples is often limited because of ethical, logistical, and financial constraints. Generative models can simulate brain scans at various stages of degeneration, offering researchers synthetic datasets that preserve the structural and functional features of disease progression. These realistic datasets can be invaluable for developing diagnostic tools and for testing hypotheses related to disease onset and progression (Fig. 8.5) (Frässle et al., 2018; Madan et al., 2024; Shin et al., 2018; Wu et al., 2020).

2. **Data Augmentation:** In clinical settings, patient data, especially from rare or early-stage neurodegenerative conditions, is scarce (Fig. 8.5). This imbalance poses a challenge for training ML models intended to detect or predict disease outcomes. Generative models help mitigate this by augmenting existing datasets with plausible, diverse variations of patient data. For example, GANs can be trained to generate synthetic brain images that reflect the subtle morphological changes associated with mild cognitive impairment (MCI), a precursor to AD. This synthetic augmentation enhances the generalization capability of diagnostic algorithms, making them more robust to noise, population variability, and class imbalance (Frässle et al., 2018; Jing et al., 2023).
3. **Unsupervised and Semi-Supervised Learning:** Labeling medical datasets, especially in neurology, is time-consuming and requires expert knowledge. Generative models can learn latent representations of the data without requiring labeled examples, enabling unsupervised or semisupervised learning. This is particularly beneficial in neurodegenerative research, where datasets are often unstructured and only partially annotated. VAEs, for instance, can learn meaningful representations of neural atrophy patterns or cortical thickness variations, facilitating early detection of disease or patient stratification without heavy reliance on annotated datasets. Furthermore, these latent features can be fed into downstream classifiers, improving

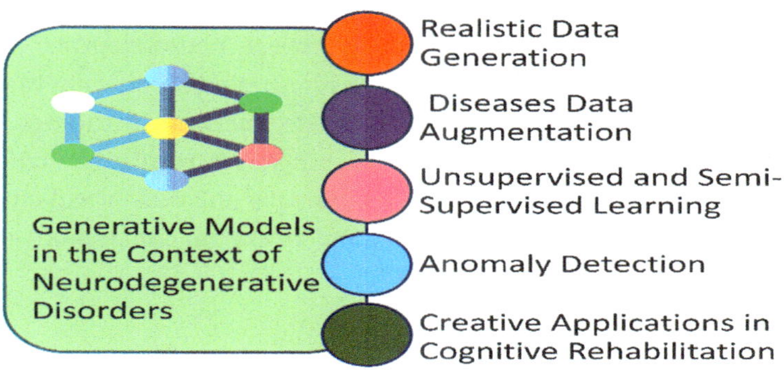

Figure 8.5 Generative AI models' advantages for the neurodegenerative disorders.

both diagnostic and prognostic modeling (Fig. 8.5) (Mazusa & Bulawayo, 2019; Pombo et al., 2019; Xiao et al., 2022).

4. **Anomaly Detection:** In the context of neurodegeneration, identifying subtle deviations from healthy brain patterns is critical for early diagnosis. Generative models trained on healthy brain data can act as powerful tools for anomaly detection (Fig. 8.5). When presented with patient data, such models highlight deviations from the expected generative distribution, flagging areas with potential pathological changes. For example, an autoencoder trained on healthy brain MRIs would produce high reconstruction errors for atrophic regions in diseased brains, aiding in the localization and assessment of neurodegenerative progression. This capability supports clinicians in detecting early abnormalities even before they become visually apparent.
5. **Creative Applications in Cognitive Rehabilitation:** Less conventional, generative models can also contribute to therapeutic interventions. AI-generated visual stimuli, music, and narratives tailored to the cognitive and emotional state of neurodegenerative patients can be used in cognitive rehabilitation programs. For instance, personalized AI-generated music based on a patient's emotional and cognitive profile can aid in memory recall or mood stabilization in Alzheimer's patients. Generative storytelling tools could be used in interactive reminiscence therapy, where virtual environments simulate past experiences to stimulate cognitive engagement and delay cognitive decline (Fig. 8.5).

Generative models present exciting opportunities for advancing research and clinical practice in neurodegenerative disorders. Their strengths, such as generating realistic brain images, augmenting limited datasets, enabling unsupervised learning, and supporting early anomaly detection, can directly address the challenges of data scarcity, diagnostic variability, and early detection in diseases such as Alzheimer's, Parkinson's, and others. Creative potential offers novel ways to design patient-centric cognitive therapies and engagement tools. The adoption of generative models must be tempered by an awareness of their limitations. Issues related to training instability, high computational costs, inadequate evaluation metrics, and ethical implications must be addressed through interdisciplinary collaboration between AI researchers, neuroscientists, clinicians, and ethicists. Future work should focus on building interpretable, clinically validated generative systems tailored to the nuances of neurodegenerative pathology. Developing benchmarks, securing patient consent, and aligning with regulatory standards will be crucial for translating generative AI from research labs into real-world clinical settings, ultimately improving diagnosis, monitoring, and treatment for millions of individuals affected by neurodegenerative conditions.

3. Hybrid generative models in neuroimaging

Hybrid generative models in neuroimaging combine multiple ML approaches to enhance the analysis and interpretation of brain imaging data. Traditional generative models, such as VAEs and GANs, have been widely used to generate realistic neuroimaging data, fill in missing information, and improve image resolution. However, hybrid models integrate these generative techniques with other DL architectures, such as CNNs or transformers, to improve both the quality and interpretability of neuroimaging outputs. One of the key applications of hybrid generative models in neuroimaging is brain image synthesis. For instance, in medical imaging, generating synthetic MRI or CT scans can aid in data augmentation, helping to train more robust models while addressing the scarcity of labeled medical images. Hybrid approaches often leverage VAEs to model latent representations while incorporating GANs for enhancing realism, leading to more detailed and accurate image reconstructions. In addition, hybrid models can assist in predicting disease progression, such as in AD or brain tumors, by generating plausible future brain scans based on historical patient data (Padmanabhan et al., 2017; SinhaRoy & Sen, 2024; Tu et al., 2008).

Another significant advantage of hybrid generative models is their ability to handle multi-modal neuroimaging data. Combining information from different imaging modalities, such as MRI, fMRI, and PET scans, can provide a more comprehensive view of brain activity and structure. Hybrid architectures, such as VAE-GAN models or transformer-augmented VAEs, facilitate cross-modal synthesis, allowing researchers to infer missing modalities or enhance lower-quality scans with information from complementary imaging techniques. Overall, hybrid generative models are revolutionizing neuroimaging by improving data quality, enabling better disease modeling, and enhancing diagnostic tools. As these models continue to evolve, they hold immense potential for advancing neuroscience and medical imaging research (Padmanabhan et al., 2017; SinhaRoy & Sen, 2024; Tu et al., 2008).

3.1 Combining GANs and VAEs: Benefits and challenges

3.1.1 Combining GANs and VAEs: Benefits and challenges

Combining GANs and VAEs has gained significant attention in the field of generative modeling, particularly in applications such as neuroimaging, image synthesis, and anomaly detection. These hybrid models aim to leverage the strengths of both architectures while mitigating their limitations.

3.1.2 Benefits of combining GANs and VAEs

- **Improved Image Quality and Diversity:** VAEs are known for learning structured latent spaces but often produce blurry images due to their reliance on a Gaussian prior. GANs, on the other hand, generate sharp images but lack explicit latent space representation. A hybrid model can integrate VAEs' structured representations with GANs' sharp image synthesis, resulting in high-quality and diverse outputs.

- **Stability in Training:** GANs suffer from instability due to adversarial training, where the generator and discriminator compete in a min–max game. Incorporating VAE components helps stabilize learning by providing a well-defined probabilistic framework, reducing mode collapse, and making training more reliable.
- **Better Latent Space Representations:** VAEs learn smooth and interpretable latent spaces, which are crucial for tasks such as image interpolation and controlled generation. By integrating GANs, these representations can be refined, ensuring better alignment with real data distributions. This is particularly useful in neuroimaging, where latent spaces can help model disease progression or structural variations.
- **Enhanced Generalization:** Since VAEs focus on likelihood-based learning while GANs prioritize adversarial learning, a hybrid approach can generalize better across different datasets, making it more suitable for real-world applications such as medical imaging, where variations in data quality and availability are common.

3.1.3 Challenges of combining GANs and VAEs

- **Balancing Loss Functions:** VAEs use a reconstruction loss and a KL-divergence term, while GANs rely on adversarial loss. Combining these objectives requires careful tuning to prevent one loss from dominating the training process, which can lead to suboptimal results.
- **Computational Complexity:** Hybrid models require additional computational resources compared to standalone GANs or VAEs. Training such models often involves multiple networks (encoder, decoder, generator, and discriminator), increasing memory requirements and training time.
- **Mode Collapse versus Blurriness:** GANs suffer from mode collapse, where the generator fails to capture the full diversity of data, while VAEs tend to produce blurry images. Finding the right combination of these models to balance sharpness and diversity remains an ongoing challenge.
- **Interpretability and Control:** While VAEs provide structured latent representations, introducing adversarial training can make the latent space harder to interpret. This can reduce the model's ability to generate controlled and meaningful variations, which is essential in applications like neuroimaging.

3.2 Deep convolutional GANs (DCGANs) for neuroimaging

3.2.1 Deep convolutional GANs (DCGANs) for neuroimaging

Deep convolutional generative adversarial networks (DCGANs) have emerged as a powerful tool in neuroimaging, enabling high-quality brain image synthesis, data augmentation, and disease modeling. DCGANs enhance the traditional GAN architecture by incorporating deep convolutional layers, which improve stability, training efficiency, and the quality of generated images. These models are particularly beneficial in neuroimaging applications, where high-resolution and realistic medical images are essential for diagnosis and research (Bai et al., 2022; Bank et al., 2023; Jing et al., 2023; Kingma & Welling, 2014; Lin et al., 2018; Tu et al., 2008; Vázquez-García et al., 2024, 2019; Padmanabhan et al., 2017).

3.2.2 Key features of DCGANs in neuroimaging

- **Enhanced Image Quality:** DCGANs use convolutional layers in both the generator and discriminator networks, enabling the generation of high-resolution, realistic neuroimaging data. Unlike traditional GANs, which rely on fully connected layers, DCGANs capture spatial hierarchies more effectively, leading to better structural preservation in generated brain scans.
- **Unsupervised Feature Learning:** The learned latent space in DCGANs can be used to extract meaningful representations from neuroimaging data without requiring labeled datasets. This is particularly useful for understanding complex brain structures, disease patterns, and anomaly detection in medical imaging.
- **Data Augmentation for Medical Imaging:** One of the primary challenges in neuroimaging is the limited availability of high-quality labeled data. DCGANs can generate synthetic MRI, fMRI, and PET scans, augmenting existing datasets to improve the performance of DL models in tasks such as segmentation, classification, and disease progression prediction.
- **Disease Progression and Anomaly Detection:** DCGANs can be used to model disease progression by generating synthetic scans that simulate different stages of neurological disorders such as AD, PD, or brain tumors. The discriminator network can also help in identifying abnormal patterns in brain scans, aiding in early diagnosis.

3.2.3 Challenges in applying DCGANs to neuroimaging

- **Mode Collapse and Training Stability:** Like all GANs, DCGANs can suffer from mode collapse, where the generator produces limited variations of images. Training stability remains a challenge, requiring careful tuning of hyperparameters such as learning rates, batch sizes, and network architectures.
- **High Computational Costs:** Generating high-resolution neuroimaging data requires significant computational power. Training DCGANs on large-scale medical datasets demands GPUs or Tensor Processing Units (TPUs), making implementation expensive and time-consuming.
- **Medical Validation and Ethical Considerations:** The use of synthetic medical images raises concerns about their reliability in clinical applications. While DCGAN-generated images can be realistic, ensuring their accuracy for medical

diagnosis and treatment planning requires rigorous validation. In addition, ethical considerations related to data privacy and bias in generated images must be addressed.

4. Conclusion

DCGANs have revolutionized neuroimaging by enabling the generation of realistic brain images, enhancing dataset diversity, and facilitating disease modeling. While challenges such as training stability and computational demands persist, advancements in model architectures and training techniques continue to improve their applicability in medical imaging. With further research and clinical validation, DCGANs hold great potential in transforming neuroimaging for early diagnosis, treatment planning, and medical research.

4.1 Attention-based VAEs and their role in brain image analysis

4.1.1 Attention-based VAEs and their role in brain image analysis

Attention-based VAEs have emerged as a powerful tool in brain image analysis, enhancing the ability to extract meaningful features from complex neuroimaging data. Traditional VAEs rely on global latent representations, but incorporating attention mechanisms allows these models to focus selectively on the most relevant regions of brain scans. This selective focus improves image reconstruction, anomaly detection, and disease classification by prioritizing critical structural or functional features. For instance, in applications such as AD progression modeling, tumor localization, and functional brain connectivity analysis, attention-based VAEs can highlight subtle neuroanatomical changes that might be overlooked by conventional methods. In addition, these models excel in handling multimodal neuroimaging data, such as MRI, fMRI, and PET scans, by dynamically integrating information across different imaging modalities. This capability is particularly useful for biomarker discovery and early diagnosis, as it helps identify patterns linked to neurological disorders with greater precision. Moreover, attention-based VAEs improve the quality of generated and reconstructed brain images by weighting different spatial regions appropriately, ensuring that important anatomical and pathological details are preserved. This makes them highly effective in data augmentation, aiding in training DL models for neuroimaging tasks where labeled datasets are limited (Bank et al., 2023; Kingma & Welling, 2014; Laino et al., 2022; Tu et al., 2008; Vázquez-García et al., 2024, 2019; Padmanabhan et al., 2017).

4.1.2 Key features of attention-based VAEs in neuroimaging

- **Improved Feature Representation**: Traditional VAEs learn a global latent representation of input data, but attention mechanisms allow the model to prioritize important brain regions. This leads to more informative latent representations, which can be crucial for detecting subtle neuroanatomical changes associated with neurological disorders.

- **Enhanced Image Reconstruction and Synthesis**: In brain image analysis, high-quality reconstruction of MRI, fMRI, or PET scans is essential. Attention-based VAEs improve the quality of reconstructed images by dynamically weighting different spatial regions, ensuring that important features such as lesions, structural abnormalities, or functional activity patterns are well preserved.
- **Better Handling of Multi-Modal Neuroimaging Data**: Neuroimaging often involves multiple imaging modalities (e.g., MRI, fMRI, and PET) that provide complementary information about brain structure and function. Attention mechanisms in VAEs facilitate cross-modal learning by focusing on key regions across different imaging types, enabling more robust integration of multi-modal data.
- **Disease Progression and Biomarker Discovery**: Attention-based VAEs can help model disease progression by identifying critical regions of interest in brain scans. For example, in Alzheimer's research, these models can highlight hippocampal atrophy patterns over time, aiding in early diagnosis and tracking disease development.

Despite their advantages, attention-based VAEs present several challenges that need to be addressed for broader clinical adoption. One major concern is the increased computational complexity associated with attention mechanisms, which significantly raises memory and processing requirements, particularly when dealing with high-resolution neuroimaging data. Training such models demands substantial GPU resources, making them less accessible for institutions with limited computational infrastructure. Another critical issue is the interpretability of learned representations. While attention mechanisms improve feature selection, understanding why certain brain regions receive higher attention remains a challenge. This is crucial in medical applications, where clinicians need transparent AI-driven insights for diagnosis and treatment planning. In addition, the generalization of these models across diverse neuroimaging datasets is another hurdle, as variations in imaging protocols, scanner types, and patient demographics can affect model performance. Techniques such as domain adaptation and transfer learning are being explored to address this issue, ensuring that attention-based VAEs can be effectively applied across different clinical settings. Despite these challenges, ongoing research continues to refine these models, making them more reliable and interpretable. As advancements in DL and computational neuroscience progress, attention-based VAEs have the potential to revolutionize brain image analysis, providing more accurate, efficient, and clinically useful insights for neurological research and medical diagnostics.

Attention-based VAEs offer a significant advancement in brain image analysis by improving feature learning, image reconstruction, and disease modeling. Their ability to focus on key brain regions enhances the interpretability and accuracy of neuroimaging applications. However, challenges such as computational demands and interpretability must be addressed for wider clinical adoption. As research progresses, these models have the potential to revolutionize early diagnosis, treatment planning, and neurological research.

5. Applications of hybrid generative models in neuroimaging

5.1 Applications of hybrid generative models in neuroimaging

Hybrid generative models, which combine techniques such as VAEs and GANs, have significantly advanced neuroimaging applications. One of the most impactful areas is **image synthesis and augmentation**, where these models help address the challenge of limited medical imaging datasets. Medical imaging data, particularly brain scans, are often scarce because of the difficulty of data acquisition and the need for expert annotations. Hybrid generative models enhance dataset diversity by generating synthetic MRI, fMRI, and PET scans that closely resemble real images. These synthetic images can be used to augment training datasets, improving the performance and robustness of DL models in tasks like brain segmentation and anomaly detection. Furthermore, these models assist in restoring missing or corrupted imaging data by generating plausible reconstructions, ensuring that incomplete scans do not compromise diagnostic accuracy. The ability of hybrid generative models to learn meaningful latent representations makes them particularly useful in producing high-fidelity neuroimaging data that captures critical structural and functional details of the brain (Padmanabhan et al., 2017; Tu et al., 2008).

Another critical application of hybrid generative models in neuroimaging is brain tumor detection, where early and accurate diagnosis is crucial for patient outcomes. Brain tumors often exhibit complex and highly variable morphological characteristics, making automated detection challenging. Hybrid models leverage the strengths of both VAEs and GANs to improve tumor segmentation by generating synthetic tumor-infected brain scans that enrich datasets and help DL models generalize better to real-world clinical cases. In addition, these models can enhance contrast in medical images, making it easier to identify tumors in low-quality scans. The ability to generate realistic tumor growth patterns also enables the simulation of different stages of tumor progression, aiding in both diagnosis and treatment planning. By learning from diverse imaging datasets, hybrid generative models contribute to the development of more accurate and reliable tumor detection algorithms, ultimately improving clinical decision-making and patient management (Padmanabhan et al., 2017; SinhaRoy & Sen, 2024; Tu et al., 2008).

Beyond detection, hybrid generative models play a crucial role in disease progression modeling, particularly for neurodegenerative conditions such as AD and PD. These disorders involve gradual brain atrophy and functional decline, which makes it essential to track subtle changes over time. Hybrid models facilitate longitudinal analysis by generating synthetic brain scans that simulate disease progression at different stages. This predictive capability is invaluable for early diagnosis, as it helps clinicians identify individuals at risk before significant cognitive decline occurs. In AD research, for instance, hybrid generative models can synthesize plausible future brain scans based on baseline patient data, providing insights into how brain structures deteriorate over time. Similarly, in PD, these models help analyze the progression of structural and functional abnormalities, supporting the development of targeted therapies. By enabling the generation of high-quality, temporally consistent neuroimaging data, hybrid generative models significantly enhance our ability to study, diagnose, and potentially intervene in neurodegenerative diseases at earlier stages. As these models continue to evolve, their integration into clinical workflows could lead to groundbreaking advancements in personalized medicine and early therapeutic interventions (Padmanabhan et al., 2017; Shin et al., 2018; Tu et al., 2008).

5.1.1 Image synthesis and augmentation

5.1.1.1 Image synthesis and augmentation in neuroimaging

Image synthesis and augmentation play a crucial role in neuroimaging, addressing challenges such as limited dataset availability, class imbalances, and variability in medical imaging. Hybrid generative models, particularly those combining VAEs and GANs, have revolutionized this field by generating high-quality synthetic brain scans that closely resemble real medical images. These models help enhance DL-based diagnostic systems by providing diverse and realistic training data, improving the robustness and generalization of AI-driven medical applications (Shin et al., 2018).

One of the key benefits of image synthesis in neuroimaging is the ability to generate realistic MRI, fMRI, and PET scans, which are often expensive and time-consuming to acquire. By training on real neuroimaging data, hybrid generative models can create synthetic scans that preserve anatomical structures and disease-related features, making them useful for data augmentation. This is particularly beneficial in cases where certain conditions, such as rare neurological disorders, have limited available data. Augmenting datasets with synthetic images improves the performance of DL models by reducing overfitting and enhancing their ability to recognize patterns across different patients and imaging protocols (Shin et al., 2018).

Beyond augmentation, hybrid generative models also aid in image restoration and enhancement. They can reconstruct missing or corrupted parts of brain scans, improving the quality of low-resolution or noisy images. In addition, these models can simulate variations in imaging modalities, such as converting MRI scans into PET-like images, allowing for multimodal analysis without requiring multiple scans from patients. As these techniques continue to evolve, image synthesis and augmentation are expected to significantly enhance the accuracy of neuroimaging-based diagnosis, disease modeling, and treatment planning, ultimately improving outcomes for patients with neurological conditions (Shin et al., 2018).

5.1.2 Brain tumor detection

Brain tumor detection is a critical application of hybrid generative models in neuroimaging, as early and accurate diagnosis can significantly improve patient outcomes. Brain tumors exhibit complex variations in size, shape, and location, making automated detection a challenging task. Traditional DL methods often require large annotated datasets for effective tumor segmentation and classification, but medical imaging data is often scarce and imbalanced. Hybrid generative models, combining techniques such as VAEs and GANs, address this limitation by generating high-quality synthetic brain scans that help improve DL models' performance. These models create realistic tumor-affected brain images, enhancing dataset diversity and enabling AI-driven diagnostic systems to generalize better across different patient cases (Abdelgawad et al., 2023; Amin et al., 2019; Ascherio & Schwarzschild, 2016; Bloem et al., 2021; Friston & Price, 2001; Ramezanian-Panahi et al., 2022; Rolls, 2023; Trotta et al., 2016).

Beyond dataset augmentation, hybrid generative models play a crucial role in improving tumor localization and segmentation. GANs can enhance the contrast in medical images, making tumors more distinguishable, while VAEs provide structured latent representations that help analyze tumor growth patterns. These models can also assist in tumor progression modeling by generating synthetic scans that simulate different stages of tumor development, aiding in prognosis prediction and treatment planning. In addition, hybrid generative models can refine noisy or low-quality scans, ensuring that medical professionals can make more accurate assessments even in suboptimal imaging conditions. As research continues, integrating these advanced AI models into clinical workflows could lead to more precise, automated, and early brain tumor detection, ultimately improving survival rates and treatment effectiveness for patients (Abdalla & Esmail, 2018; Amin et al., 2019; Hossain et al., 2019; Mazusa & Bulawayo, 2019; Trotta et al., 2016; Tu et al., 2008; Friston & Price, 2001).

5.1.3 Disease progression modeling (Alzheimer's and Parkinson's)

5.1.3.1 Disease progression modeling in Alzheimer's and Parkinson's using hybrid generative models

The modeling of disease progression in neurodegenerative disorders such as AD and PD is a critical area of research in medical imaging and computational neuroscience. Hybrid generative models, which integrate VAEs and GANs, have demonstrated significant potential in predicting and simulating disease trajectories. These models enable the generation of synthetic brain scans that accurately capture structural and functional changes over time, aiding in early diagnosis, prognosis estimation, and treatment planning (Abdelgawad et al., 2023; Ascherio & Schwarzschild, 2016; Bloem et al., 2021). In

AD, progressive atrophy in brain regions such as the hippocampus and cortex is a hallmark of disease advancement. Hybrid generative models facilitate longitudinal analysis by learning latent representations from MRI and PET scans, allowing the simulation of future brain states based on baseline imaging data. This predictive capability is crucial for identifying patients at risk of cognitive decline before noticeable symptoms emerge. In addition, generative models enhance biomarker discovery by highlighting subtle anatomical changes that correlate with disease progression, improving the accuracy of early detection methods. The ability to synthesize high-fidelity neuroimaging data also supports the development of AI-driven diagnostic tools by augmenting datasets, addressing class imbalances, and refining DL models for classification tasks.

Similarly, in PD, the degeneration of dopaminergic neurons in the substantia nigra leads to both structural and functional alterations in the brain. Hybrid generative models can simulate these progressive changes by leveraging multimodal neuroimaging data, including MRI and dopamine transporter (DaT) scans. By generating synthetic representations of different disease stages, these models provide a deeper understanding of PD progression and aid in the development of personalized treatment strategies. Furthermore, they assist in predicting motor and cognitive decline by modeling neurodegeneration patterns, offering a noninvasive means of tracking disease evolution (Abdelgawad et al., 2023; Ascherio & Schwarzschild, 2016; Bloem et al., 2021; Mei et al., 2021).

Overall, the integration of hybrid generative models in disease progression modeling enhances the ability to study, predict, and manage neurodegenerative disorders. By improving data availability, facilitating early diagnosis, and supporting precision medicine, these models hold significant promise in advancing neurological research and clinical decision-making.

6. Other areas of neuroimaging with machine learning and deep learning

The intersection of neuroimaging with ML and DL continues to redefine the landscape of brain research and clinical neuroscience. Beyond conventional classification and segmentation tasks, innovative applications are emerging that push the boundaries of what's possible in understanding, diagnosing, and treating neurological conditions. This section outlines key advancements in the application of ML/DL to neuroimaging, including integration with RL, transfer learning, multimodal data fusion, and real-time clinical deployments.

6.1 Integration with reinforcement learning

RL, a subfield of ML inspired by behavioral psychology, focuses on agents learning optimal policies through interactions with an environment. Its integration with neuroimaging holds promising potential for both research and clinical decision-making. One of the pioneering applications involves adaptive neurofeedback systems, where RL agents learn to adjust real-time feedback stimuli based on functional MRI (fMRI) or electroencephalogram (EEG) signals. This dynamic adaptation enhances neuroplasticity and has been explored in disorders like depression, ADHD, and motor rehabilitation after stroke. RL is being investigated in personalized treatment planning for neurodegenerative diseases, such as selecting optimal drug dosing schedules or stimulation parameters for deep brain stimulation (DBS) in PD. Here, neuroimaging data (e.g., structural connectivity from diffusion tensor imaging) serve as the state input to the agent, guiding the reward function based on patient outcomes. Integrating RL with neuroimaging also enables virtual brain simulations, where agents navigate through simulated cortical networks to model learning, memory, and decision-making processes, further advancing cognitive neuroscience (Mazusa & Bulawayo, 2019).

6.2 Transfer learning for neuroimaging

Transfer learning involves leveraging knowledge gained from one domain or task and applying it to another, typically where data is scarce. In neuroimaging, this is particularly valuable due to the high dimensionality and low sample size typical of medical datasets (Shin et al., 2016).

Pretrained CNNs, often trained on large natural image datasets such as ImageNet, are being adapted for tasks such as tumor detection, brain age prediction, and Alzheimer's classification using MRI or PET scans. While domain shift remains a challenge, fine-tuning higher-level layers has yielded successful results in achieving competitive performance with minimal training data. Another compelling approach involves cross-study and cross-cohort transfer, where models trained on one population (e.g., healthy controls) are adapted to other cohorts (e.g., patients with MCI). This enables the reuse of existing annotated datasets and promotes model generalizability across diverse clinical settings. Emerging work also applies cross-modal transfer learning, such as transferring knowledge from structural MRI to functional MRI domains or vice versa. These techniques support comprehensive, multi-context learning from the limited data typically available in neurological research (Mah et al., 2014; Ramezanian-Panahi et al., 2022; Rolls, 2023).

6.3 Multi-modal neuroimaging data integration

The human brain is complex and multifaceted, and no single neuroimaging modality captures all aspects of its function or structure. Therefore integrating multiple modalities such as structural MRI (sMRI), functional MRI (fMRI), diffusion MRI (dMRI), PET, and EEG has become a cornerstone of advanced neuroimaging analytics.

Multimodal DL frameworks are being designed to fuse heterogeneous data sources to extract synergistic information. For example:

- sMRI provides anatomical structure
- fMRI reveals brain activity and connectivity patterns
- dMRI depicts white matter tractography
- PET provides insights into metabolism or amyloid deposition
- EEG captures millisecond-level temporal dynamics

Jointly learning from these sources enables better disease subtyping, progression modeling, and treatment response prediction. In Alzheimer's research, multimodal integration has led to improved early detection of preclinical stages. Similarly, in psychiatric conditions such as schizophrenia and bipolar disorder, integrating structural and functional connectivity data has uncovered biomarkers not visible in single-modality analyses.

6.3.1 Advanced fusion strategies include

- Feature-level fusion using autoencoders and transformers
- Decision-level fusion through ensemble models
- Graph neural networks (GNNs) applied to connectivity matrices.

Despite its potential, multimodal integration also faces challenges such as modality alignment, noise harmonization, and missing data. Addressing these remains an active research area.

6.4 Advancements in real-time clinical applications

ML in neuroimaging is rapidly moving from retrospective analysis to real-time clinical decision support, aided by advances in hardware, model compression, and inference optimization. Real-time MRI image reconstruction using DL significantly reduces scan time and enables faster diagnosis in emergency scenarios. This has substantial implications for acute stroke management, where rapid imaging is critical for decision-making in thrombolysis or thrombectomy. Another application is in real-time seizure detection using EEG, where DL models deployed on wearable devices or hospital monitors can alert clinicians instantly, improving patient safety and care continuity. In image-guided neurosurgery, DL models assist in identifying and visualizing tumor boundaries or eloquent cortex regions, enhancing surgical precision and minimizing functional loss. Similarly, intraoperative fMRI and neural navigation are increasingly being paired with AI to adjust surgical plans dynamically based on anatomical feedback (Abdalla & Esmail, 2018; Wu et al., 2020).

The development of cloud-connected diagnostic platforms also allows remote access to AI-powered neuroimaging tools, facilitating second opinions and reducing the diagnostic gap in rural or underserved areas. Real-time deployment must consider constraints such as latency, hardware compatibility, model interpretability, and compliance with medical device regulations (e.g., FDA and CE approval).

7. Conclusion

The application of ML and DL in neuroimaging has expanded beyond traditional domains into novel, high-impact areas. RL introduces adaptive intelligence into neurofeedback and treatment optimization. Transfer learning addresses data scarcity by reusing knowledge across tasks and modalities. Multimodal integration uncovers deeper insights into brain function by combining complementary datasets. Meanwhile, real-time clinical implementations are making AI a practical partner in the operating room and bedside decision-making.

These emerging areas are not only enhancing the scientific understanding of the brain but are also driving the development of personalized, efficient, and accessible neurological healthcare solutions. Future research should emphasize interdisciplinary collaboration, data harmonization, and ethical deployment to maximize the potential of these advancements.

References

Abdalla, H. E. M., & Esmail, M. Y. (2018). Tumor detection by using artificial neural network. In *International conference on computer, control, electrical, and electronics engineering, ICCCEEE 2018*. Institute of Electrical and Electronics Engineers Inc.. https://doi.org/10.1109/ICCCEEE.2018.8515763

Abdelgawad, A., Rahayel, S., Zheng, Y. Q., Tremblay, C., Vo, A., Misic, B., & Dagher, A. (2023). Predicting longitudinal brain atrophy in Parkinson's disease using a susceptible-infected-removed agent-based model. *Network Neuroscience, 7*(3), 906–925. https://doi.org/10.1162/netn_a_00296

AbdulAzeem, Y., Bahgat, W. M., & Badawy, M. (2021). A CNN based framework for classification of Alzheimer's disease. *Neural Computing & Applications, 33*(16), 10415–10428. https://doi.org/10.1007/s00521-021-05799-w

Almond, R. G., & Zapata-Rivera, J. D. (2019). *Bayesian networks methodology of educational measurement and assessment* (pp. 81–106). Springer Nature. https://doi.org/10.1007/978-3-030-05584-4_4

Alqahtani, H., Kavakli-Thorne, M., & Kumar, G. (2021). Applications of Generative Adversarial Networks (GANs): An updated review. *Archives of Computational Methods in Engineering, 28*(2), 525–552.

Amin, J., Sharif, M., Raza, M., Saba, T., & Anjum, M. A. (2019). Brain tumor detection using statistical and machine learning method. *Computer Methods and Programs in Biomedicine, 177*, 69–79. https://doi.org/10.1016/j.cmpb.2019.05.015

Apostolopoulos, I. D., Papathanasiou, N. D., Apostolopoulos, D. J., & Panayiotakis, G. S. (2022). Applications of Generative Adversarial Networks (GANs) in Positron Emission Tomography (PET) imaging: A review. *European Journal of Nuclear Medicine and Molecular Imaging, 49*(11), 3717–3739. https://doi.org/10.1007/s00259-022-05805-w

Arora, S., Ge, R., Liang, Y., Ma, T., & Zhang, Y. (2017). *Generalization and equilibrium in Generative Adversarial Nets (GANs)* (pp. 224–232). PMLR. Available from: https://proceedings.mlr.press/v70/arora17a.html.

Ascherio, A., & Schwarzschild, M. A. (2016). The epidemiology of Parkinson's disease: Risk factors and prevention. *The Lancet Neurology, 15*(12), 1257–1272. https://doi.org/10.1016/S1474-4422(16)30230-7

Bacon, E. J., He, D., D'avilla Achi, N. A., Wang, L., Han, L., Yao-Digba, P. D. Z., Monkam, P., & Qi, S. (2024). Neuroimage analysis using artificial intelligence approaches: A systematic review. *Medical & Biological Engineering & Computing, 62*(9), 2599–2627. https://doi.org/10.1007/s11517-024-03097-w

Bai, T., Du, M., Zhang, L., Ren, L., Ruan, L., Yang, Y., Qian, G., Meng, Z., Zhao, L., & Deen, M. J. (2022). A novel Alzheimer's disease detection approach using GAN-based brain slice image enhancement. *Neurocomputing, 492*, 353–369. https://doi.org/10.1016/j.neucom.2022.04.012

Bank, D., Koenigstein, N., & Giryes, R. (2023). *Autoencoders machine learning for data science handbook: Data mining and knowledge discovery handbook* (3rd ed., pp. 353–374). Springer International Publishing. https://doi.org/10.1007/978-3-031-24628-9_16

Bengio, Y., Courville, A., & Vincent, P. (2013). Representation learning: A review and new perspectives. *IEEE Transactions on Pattern Analysis and Machine Intelligence, 35*(8), 1798–1828. https://doi.org/10.1109/tpami.2013.50

Bianchini, M., Maggini, M., & Jain, L. C. (2013). Handbook on neural information processing. *Intelligent Systems Reference Library, 49*. https://doi.org/10.1007/978-3-642-36657-4

Bloem, B. R., Okun, M. S., & Klein, C. (2021). Parkinson's disease. *The Lancet, 397*(10291), 2284–2303. https://doi.org/10.1016/S0140-6736(21)00218-X

Creswell, A., White, T., Dumoulin, V., Arulkumaran, K., Sengupta, B., & Bharath, A. A. (2018). Generative adversarial networks: An overview. *IEEE Signal Processing Magazine, 35*(1), 53–65. https://doi.org/10.1109/MSP.2017.2765202

Dijk, D.v., Teräsvirta, T., & Franses, P. H. (2002). Smooth transition autoregressive models—A survey of recent developments. *Econometric Reviews, 21*(1), 1–47. https://doi.org/10.1081/ETC-120008723

Dong, H., Yang, G., Liu, F., Mo, Y., & Guo, Y. (2017). Automatic brain tumor detection and segmentation using U-net based fully convolutional networks. *Communications in Computer and Information Science, 723*, 506–517. https://doi.org/10.1007/978-3-319-60964-5_44

Drury, B., Valverde-Rebaza, J., Moura, M. F., & de Andrade Lopes, A. (2017). A survey of the applications of Bayesian networks in agriculture. *Engineering Applications of Artificial Intelligence, 65*, 29–42. https://doi.org/10.1016/j.engappai.2017.07.003

Eddy, S. R. (1996). Hidden Markov models. *Current Opinion in Structural Biology, 6*(3), 361–365. https://doi.org/10.1016/S0959-440X(96)80056-X

Eddy, S. R. (2004). What is a hidden Markov model? *Nature Biotechnology, 22*(10), 1315–1316.

Fanfarillo, A., Roozitalab, B., Hu, W., & Cervone, G. (2021). Probabilistic forecasting using deep generative models. *GeoInformatica, 25*(1), 127–147. https://doi.org/10.1007/s10707-020-00425-8

Fischer, A., & Igel, C. (2012). An introduction to restricted Boltzmann machines. In,*7441*. *Lecture notes in computer science (including subseries lecture notes in artificial intelligence and lecture notes in bioinformatics)* (pp. 14–36). https://doi.org/10.1007/978-3-642-33275-3_2

Fischer, A., & Igel, C. (2014). Training restricted Boltzmann machines: An introduction. *Pattern Recognition, 47*(1), 25–39. https://doi.org/10.1016/j.patcog.2013.05.025

Frässle, S., Yao, Y., Schöbi, D., Aponte, E. A., Heinzle, J., & Stephan, K. E. (2018). Generative models for clinical applications in computational psychiatry. *Interdisciplinary Reviews: Cognitive Science, 9*(3). https://doi.org/10.1002/wcs.1460

Friston, K. J., & Price, C. J. (2001). Dynamic representations and generative models of brain function. *Brain research bulletin, 54*(3), 275–285. https://onlinelibrary.wiley.com/doi/abs/10.1111/1467-9450.00228.

Goodfellow, I., Pouget-Abadie, J., Mirza, M., Xu, B., Warde-Farley, D., Ozair, S., Courville, A., & Bengio, Y. (2020). Generative adversarial networks. *Communications of the ACM, 63*(11), 139–144. https://doi.org/10.1145/3422622

Gui, J., Sun, Z., Wen, Y., Tao, D., & Ye, J. (2023). A review on generative adversarial networks: Algorithms, theory, and applications. *IEEE Transactions on Knowledge and Data Engineering, 35*(4), 3313–3332. https://doi.org/10.1109/tkde.2021.3130191

Hinton, G. (2010). A practical guide to training restricted Boltzmann machines. *Computer, 9*(3), 1. https://doi.org/10.1007/978-3-642-35289-8_32

Hossain, T., Shishir, F. S., Ashraf, M., Al Nasim, Md. A., & Muhammad Shah, F. (2019). Brain tumor detection using convolutional neural network. In *1st international conference on advances in science, engineering and robotics technology 2019, ICASERT 2019*. Institute of Electrical and Electronics Engineers Inc.. https://doi.org/10.1109/ICASERT.2019.8934561

Innovations in Bayesian Networks. (2008). Innovations in Bayesian networks. https://doi.org/10.1007/978-3-540-85066-3.

Jing, C., Gong, C., Chen, Z., Lei, B., & Wang, S. (2023). TA-GAN: Transformer-driven addiction-perception generative adversarial network. *Neural Computing & Applications, 35*(13), 9579–9591. https://doi.org/10.1007/s00521-022-08187-0

Jo, T., Nho, K., & Saykin, A. J. (2019). Deep learning in Alzheimer's disease: Diagnostic classification and prognostic prediction using neuroimaging data. *Frontiers in Aging Neuroscience, 11*. https://doi.org/10.3389/fnagi.2019.00220

Kingma, D. P., & Welling, M. (2014). 2nd international conference on learning representations, ICLR 2014—Conference track proceedings international conference on learning representations, ICLR Netherlands auto-encoding variational bayes. Available from: https://dblp.org/db/conf/iclr/iclr2014.html.

Kingma, D. P., & Welling, M. (2019). An introduction to variational autoencoders. *Foundations and Trends in Machine Learning, 12*(4), 307–392. https://doi.org/10.1561/2200000056

Kumar, R., & Sharma, A. (2023). *Computational strategies and tools for protein tertiary structure prediction basic biotechniques for bioprocess and bioentrepreneurship* (pp. 225–242). Elsevier. https://doi.org/10.1016/B978-0-12-816109-8.00015-5

Lütkepohl, H. (2013). Vector autoregressive models. In: Handbook of research methods and applications in empirical macroeconomics. https://doi.org/10.4337/9780857931023.00012.

Laino, M. E., Cancian, P., Politi, L. S., Della Porta, M. G., Saba, L., & Savevski, V. (2022). Generative adversarial networks in brain imaging: A narrative review. *Journal of Imaging, 8*(4). https://doi.org/10.3390/jimaging8040083

Lankadasu, N. V. Y., Pesarlanka, D. B., Sharma, A., & Sharma, S. (1). C.E.). Security Aspects of Blockchain Technology.

Lankadasu, N. V. Y., Pesarlanka, D. B., Sharma, A., Sharma, S., & Gochhait, S. (2024). Skin cancer classification using a convolutional neural network: an exploration into deep learning. 2024 ASU International Conference in Emerging Technologies for Sustainability and Intelligent Systems. *ICETSIS, 2024*, 1047–1052.

Larochelle, H., & Bengio, Y. (2008). Classification using discriminative restricted Boltzmann machines. In *Proceedings of the 25th international conference on machine learning* (pp. 536–543). https://doi.org/10.1145/1390156.1390224

Lim, J.-H., & Kuo, P.-C. (n.d.). EEGTrans: Transformer-driven generative models for EEG synthesis. Retrieved May 14, 2025, from https://anonymous.4open.science/r/EEGTrans.

Lin, W., Tong, T., Gao, Q., Guo, D., Du, X., Yang, Y., Guo, G., Xiao, M., Du, M., & Qu, X. (2018). Convolutional neural networks-based MRI image analysis for the Alzheimer's disease prediction from mild cognitive impairment. *Frontiers in Neuroscience, 12*. https://doi.org/10.3389/fnins.2018.00777

Lopes, N., & Ribeiro, B. (2015). Deep Belief Networks (DBNs). *Studies in Big Data, 7*, 155–186. https://doi.org/10.1007/978-3-319-06938-8_8

Lu, Y., Chen, D., Olaniyi, E., & Huang, Y. (2022). Generative Adversarial Networks (GANs) for image augmentation in agriculture: A systematic review. *Computers and Electronics in Agriculture, 200*, Article 107208. https://doi.org/10.1016/j.compag.2022.107208

Madan, S., Lentzen, M., Brandt, J., Rueckert, D., Hofmann-Apitius, M., & Fröhlich, H. (2024). Transformer models in biomedicine. *BMC Medical Informatics and Decision Making, 24*(1). https://doi.org/10.1186/s12911-024-02600-5

Mah, Y. H., Husain, M., Rees, G., & Nachev, P. (2014). Human brain lesion-deficit inference remapped. *Brain, 137*(9), 2522–2531. https://doi.org/10.1093/brain/awu164

Mazusa, T. M., & Bulawayo, C. E. O. (2019). Inverse reinforcement learning conditioned on brain scan. Available from: https://arxiv.org/pdf/1906.09770.

Mei, J., Desrosiers, C., & Frasnelli, J. (2021). Machine learning for the diagnosis of Parkinson's disease: A review of literature. *Frontiers in Aging Neuroscience, 13*. https://doi.org/10.3389/fnagi.2021.633752

Mohamed, A. R., Hinton, G., & Penn, G. (2012). Understanding how deep belief networks perform acoustic modelling. In *ICASSP, IEEE international conference on acoustics, speech and signal processing—proceedings* (pp. 4273–4276). https://doi.org/10.1109/ICASSP.2012.6288863

Montúfar, G. (2018). Restricted Boltzmann machines: Introduction and review. In *, 252. Proceedings in mathematics and statistics* (pp. 75–115). https://doi.org/10.1007/978-3-319-97798-0_4

Navidan, H., Moshiri, P. F., Nabati, M., Shahbazian, R., Ghorashi, S. A., Shah-Mansouri, V., & Windridge, D. (2021). Generative Adversarial Networks (GANs) in networking: A comprehensive survey & evaluation. *Computer Networks, 194*. https://doi.org/10.1016/j.comnet.2021.108149

Niedermayer, D. (2008). An introduction to Bayesian networks and their contemporary applications. *Studies in Computational Intelligence, 156*, 117–130. https://doi.org/10.1007/978-3-540-85066-3_5

Oord, A. van den, Kalchbrenner, N., & Kavukcuoglu, K. (2016). Pixel Recurrent Neural Networks (pp. 1747–1756). PMLR. https://proceedings.mlr.press/v48/oord16.html

Padmanabhan, P., Nedumaran, A. M., Mishra, S., Pandarinathan, G., Archunan, G., & Gulyás, B. (2017). The Advents of Hybrid Imaging Modalities: A New Era in Neuroimaging Applications. *Advanced Biosystems, 1*(8), 1700019.

Pan, Z., Yu, W., Yi, X., Khan, A., Yuan, F., & Zheng, Y. (2019). Recent progress on Generative Adversarial Networks (GANs): A survey. *IEEE Access, 7*, 36322–36333. https://doi.org/10.1109/access.2019.2905015

Perl, Y. S., Bocaccio, H., Pérez-Ipiña, I., Zamberlán, F., Piccinini, J., Laufs, H., Kringelbach, M., Deco, G., & Tagliazucchi, E. (2020). Generative embeddings of brain collective dynamics using variational autoencoders. *Physical Review Letters, 125*(23). https://doi.org/10.1103/PhysRevLett.125.238101

Pombo, G., Gray, R., Varsavsky, T., Ashburner, J., & Nachev, P. (2019). Bayesian volumetric autoregressive generative models for better semisupervised learning. In *, 11767. Lecture notes in computer science (including subseries lecture notes in artificial intelligence and lecture notes in bioinformatics)* (pp. 429–437). Springer Science and Business Media. https://doi.org/10.1007/978-3-030-32251-9_47

Quintana, E., Ibarra, C., Escobedo, L., Tentori, M., & Favela, J. (2012). Progress in pattern recognition, image analysis, computer vision, and applications. In *Lecture notes in computer science (including subseries lecture notes in artificial intelligence and lecture notes in bioinformatics)* (pp. 877–884). https://doi.org/10.1007/978-3-642-33275-3

Rabiner, L., & Juang, B. (1986). An introduction to hidden Markov models. *IEEE ASSP Magazine, 3*(1), 4–16. https://doi.org/10.1109/MASSP.1986.1165342

Ramezanian-Panahi, M., Abrevaya, G., Gagnon-Audet, J. C., Voleti, V., Rish, I., & Dumas, G. (2022). Generative models of brain dynamics. *Frontiers in Artificial Intelligence, 5*. https://doi.org/10.3389/frai.2022.807406

Regis, M., Serra, P., & van den Heuvel, E. R. (2022). Random autoregressive models: A structured overview. *Econometric Reviews, 41*(2), 207–230.

Rolls, E. T. (2023). *Brain computations and connectivity brain computations and connectivity* (pp. 1–1149). Oxford University Press. https://doi.org/10.1093/oso/9780198887911.001.0001

Sarikaya, R., Hinton, G. E., & Deoras, A. (2014). Application of deep belief networks for natural language understanding. *IEEE Transactions on Audio, Speech and Language Processing, 22*(4), 778–784. https://doi.org/10.1109/TASLP.2014.2303296

Satti, S.R., Lankadasu, J.S.K., Varshith, N.V.V., Sharma, A., & Sharma, S. (1 C.E.). Bioengineering and Healthcare Data Analysis: Introduction, Advances, and Challenges.

Saxena, D., & Cao, J. (2021). Generative adversarial networks (GANs). *ACM Computing Surveys, 54*(3). https://doi.org/10.1145/3446374/SUPPL_FILE/SAXENA.ZIP

Scutari, M. (2017). Bayesian network constraint-based structure learning algorithms: Parallel and optimized implementations in the bnlearn R package. *Journal of Statistical Software, 77*(1). https://doi.org/10.18637/jss.v077.i02

Seeliger, K., Güçlü, U., Ambrogioni, L., Güçlütürk, Y., & van Gerven, M. A. J. (2018). Generative adversarial networks for reconstructing natural images from brain activity. *NeuroImage, 181*, 775–785. https://doi.org/10.1016/j.neuroimage.2018.07.043

Shao, Y., Cherry, S. R., Farahani, K., Meadors, K., Siegel, S., Silverman, R. W., & Marsden, P. K. (1997). Simultaneous PET and MR imaging. *Physics in Medicine and Biology, 42*(10), 1965–1970. https://doi.org/10.1088/0031-9155/42/10/010

Sharma, A., Guleria, V., & Jaiswal, V. (2022). The future of blockchain technology, recent advancement and challenges. *Studies in Big Data, 105*, 329–349.

Sharma, A., Gupta, A., & Jaiswal, V. (2021). Solving image processing critical problems using machine learning. *Studies in Big Data, 82*, 213–248. https://doi.org/10.1007/978-981-15-9492-2_11

Sharma, A., Kala, S., Kumar, A., Sharma, S., Gupta, G., & Jaiswal, V. (2024). Deep learning in genomics, personalized medicine, and neurodevelopmental disorders. *Intelligent Data Analytics for Bioinformatics and Biomedical Systems*, 235–264. https://doi.org/10.1002/9781394270910

Sharma, A., Kaur, D., Gupta, A., & Jaiswal, V. (2019). Application and analysis of hyperspectal imaging. In *Proceedings of IEEE international conference on signal processing, computing and control* (pp. 30–35). Institute of Electrical and Electronics Engineers Inc. https://doi.org/10.1109/ISPCC48220.2019.8988436

Sharma, A., & Kumar, R. (2022). Recent advancement and challenges in deep learning, big data in bioinformatics. *Studies in Big Data, 105*, 251–284. https://doi.org/10.1007/978-3-030-95419-2_12

Sharma, A., Kumar, R., & Jaiswal, V. (2021). Classification of heart disease from MRI images using convolutional neural network. In *Proceedings of IEEE international conference on signal processing, computing and control* (pp. 358–363). Institute of Electrical and Electronics Engineers Inc.. https://doi.org/10.1109/ISPCC53510.2021.9609408

Shin, H. C., Roth, H. R., Gao, M., Lu, L., Xu, Z., Nogues, I., Yao, J., Mollura, D., & Summers, R. M. (2016). Deep convolutional neural networks for computer-aided detection: CNN architectures, dataset characteristics and transfer learning. *IEEE Transactions on Medical Imaging, 35*(5), 1285–1298. https://doi.org/10.1109/TMI.2016.2528162

Shin, H. C., Tenenholtz, N. A., Rogers, J. K., Schwarz, C. G., Senjem, M. L., Gunter, J. L., Andriole, K. P., & Michalski, M. (2018). Medical image synthesis for data augmentation and anonymization using generative adversarial networks. In *Lecture notes in computer science (including subseries lecture notes in artificial intelligence and lecture notes in bioinformatics)*. Springer Verlag. https://doi.org/10.1007/978-3-030-00536-8_1

SinhaRoy, R., & Sen, A. (2024). A hybrid deep learning framework to predict Alzheimer's disease progression using generative adversarial networks and deep convolutional neural networks. *Arabian Journal for Science and Engineering, 49*(3), 3267–3284. https://doi.org/10.1007/s13369-023-07973-9

So, H. M., Bose, L., Dudek, P., & Wetzstein, G. (2024). PixelRNN: In-pixel recurrent neural networks for end-to-end-optimized perception with neural sensors. In *Proceedings of the IEEE computer society conference on computer vision and pattern recognition* (pp. 25233–25244). IEEE Computer Society. https://doi.org/10.1109/CVPR52733.2024.02384

Sridhar, N. (2024). *Bayesian network basics Bayesian network modeling of corrosion* (pp. 19–53). Springer International Publishing. https://doi.org/10.1007/978-3-031-56128-3_2

Teräsvirta, T. (1994). Specification, estimation, and evaluation of smooth transition autoregressive models. *Journal of the American Statistical Association, 89*(425), 208–218.

Trotta, N., Archambaud, F., Goldman, S., Baete, K., Koen, Van L., Vincent, W., Van Bogaert, P., Chiron, C., & De Tiège, X. (2016). Functional integration changes in regional brain glucose metabolism from childhood to adulthood. *Human Brain Mapping, 37*(8), 3017–3030. https://doi.org/10.1002/hbm.23223

Tu, Z., Narr, K. L., Dollar, P., Dinov, I., Thompson, P. M., & Toga, A. W. (2008). Brain anatomical structure segmentation by hybrid discriminative/generative models. *IEEE Transactions on Medical Imaging, 27*(4), 495–508. https://doi.org/10.1109/TMI.2007.908121

Vázquez-García, C., Martinez-Murcia, F. J., Román, F. S., & Górriz, J. M. (2024). A review of latent representation models in neuroimaging. https://doi.org/10.48550/arXiv.2412.19844.

Van Den, A., Deepmind, O. G., Kalchbrenner, N., Deepmind, G., Vinyals, O., Espeholt, L., Graves, A., & Kavukcuoglu, K. (2016). Conditional image generation with PixelCNN decoders. *Advances in Neural Information Processing Systems, 29*.

Warburg, O., Wind, F., & Negelein, E. (1927). The metabolism of tumors in the body. *Journal of General Physiology, 8*(6), 519–530. https://doi.org/10.1085/jgp.8.6.519

Wiegerinck, W., Burgers, W., & Kappen, B. (2013). Bayesian networks, introduction and practical applications. *Intelligent Systems Reference Library, 49*, 401–431. https://doi.org/10.1007/978-3-642-36657-4_12

Wong, C. S., & Li, W. K. (2000). On a mixture autoregressive model. *Journal of the Royal Statistical Society. Series B: Statistical Methodology, 62*(1), 95–115. https://doi.org/10.1111/1467-9868.00222

Wu, W., Lu, Y., Mane, R., & Guan, C. (2020). Deep learning for neuroimaging segmentation with a novel data augmentation strategy. In *Proceedings of the annual international conference of the IEEE engineering in medicine and biology society, EMBS* (pp. 1516–1519). Institute of Electrical and Electronics Engineers Inc.. https://doi.org/10.1109/EMBC44109.2020.9176537

Xiao, Y., Ashbee, W., Calhoun, V. D., & Plis, S. (2022). Refacing defaced MRI with PixelCNN. In *Proceedings of the international joint conference on neural networks*. Institute of Electrical and Electronics Engineers Inc. https://doi.org/10.1109/IJCNN55064.2022.9891937

Xie, G., Huang, Y., Wang, J., Lyu, J., Zheng, F., Zheng, Y., & Jin, Y. (2023). *Cross-modality Neuroimage Synthesis: A Survey, 56*. ACM Computing Surveys.

Zhao, Y., Ma, B., Jiang, P., Zeng, D., Wang, X., & Li, S. (2021). Prediction of Alzheimer's disease progression with multi-information generative adversarial network. *IEEE Journal of Biomedical and Health Informatics, 25*(3), 711–719. https://doi.org/10.1109/jbhi.2020.3006925

Zhong, P., Gong, Z., Li, S., & Schonlieb, C. B. (2017). Learning to diversify deep belief networks for hyperspectral image classification. *IEEE Transactions on Geoscience and Remote Sensing, 55*(6), 3516–3530. https://doi.org/10.1109/TGRS.2017.2675902

PART III

Applications in neuroscience

CHAPTER 9

Generative AI for disease diagnosis and prognosis (Alzheimer's, Parkinson's, stroke, etc.)

Abraham Olufemi Asuku[1] and Peace Ayo Olaoluwa[2]
[1]Bioresources Development Centre, National Biotechnology Research and Development Agency, Department of Medical Biotechnology, Ogbomoso, Nigeria; [2]University of Ilorin, Department of Physiology, Ilorin, Nigeria

1. Introduction

Neurodegenerative diseases (NDs) are a heterogeneous category of progressive conditions, which involve the degradation of the neuronal structure and function and eventually result in cognitive and motor impairments. The most common are Alzheimer's disease (AD), Parkinson's disease (PD), and amyotrophic lateral sclerosis (ALS) that present a significant burden to patients, their families, and the healthcare system on the global level (Asuku et al., 2022; Asuku et al., 2024). These conditions are normally age related and have complicated etiologies of genetic, environmental, and molecular factors. Although a lot of research has been conducted, disease-modifying treatments have not been effective yet, and it is because of the limited understanding of the disease mechanisms and difficulties in its early detection (Liu et al., 2025; Rana et al., 2025).

Proper diagnosis and prognosis are among the main problems of the ND management. Differentiating NDs is a challenging task clinically because many of them have similar symptoms especially during the initial stages. As a case in point, motor and cognitive manifestations are frequently common in Parkinsonian syndromes and other types of dementia, and thus cause confusion during diagnosis. Besides, there is a great diversity in the development of the disease, as not all people have the same onset, severity, or decline. To make matters worse, there are no strong, noninvasive biomarkers that can accurately identify disease at preclinical or early symptomatic levels (Asuku et al., 2024). As a result, the process of diagnosis is frequently delayed until the time when there is a considerable amount of neuronal damage, and there is no time to act and develop a specific approach to treatment (Ganjizadeh et al., 2024; Rao et al., 2024).

Artificial intelligence (AI) has become one of the most important tools in neuroscience in recent years that provides innovative methods of detecting, classifying, and predicting diseases. Machine learning and deep learning algorithms have proved to be spectacular in the analysis of high-dimensional and complicated biomedical data, such as neuroimaging, genomics, and clinical records. Such methods allow one to discover the existence of hidden patterns and relationships that cannot be identified using the

Generative Artificial Intelligence for Neuroimaging
ISBN 978-0-443-45010-5
https://doi.org/10.1016/B978-0-443-45010-5.00010-1

standard means of analysis. Consequently, AI-based models have demonstrated potential to enhance diagnostic accuracy, detect earlier, and have better prognostic predictions in NDs (Ganjizadeh et al., 2024; Stefano, 2023).

Generative artificial intelligence (GenAI) is a paradigm of AI that is relatively new within the wider context and has unique functionality. In contrast to the classical form of discriminative models that aim at classifying or predicting, generative models are focused on learning the underlying distribution of data and producing new, synthetic instances of data. Generative adversarial networks (GANs), variational autoencoders (VAEs), and large language models (LLMs) are some of the techniques to generate realistic medical images, simulate patient longitudinal trajectories, and generate clinical narratives. Such abilities are especially useful in healthcare-related settings where the lack of data, privacy issues, and the class imbalance usually restrict the effectiveness of models (Loni et al., 2025).

There are a number of urgent needs that drive the use of generative AI in the research of NDs. To begin with, GenAI has the ability to enhance small datasets with high-quality synthetic data to enhance model training and generalizability. Second, it may assist in integrating multimodal data, which will allow combining imaging, molecular, and clinical data to reflect the multifactorial character of NDs. Third, generative models may assist personalized medicine by modeling the disease progression and predicting the individual patient progression, therefore, assisting in prognosis and treatment planning. Moreover, recent findings indicate that generative AI would be beneficial in clinical decision-making via massive medical knowledge summarization and help in interpreting complex data sets (Breithaupt et al., 2025; Nag et al., 2025).

The integration of generative AI with neuroscience has a lot of potential to overcome the ancient issues of diagnosis and prognosis of NDs. GenAI can change clinical practice and speed up research in this vital field by allowing it to be detected sooner and more accurately, as well as to provide more personalized care.

2. Fundamentals of generative AI

GenAI is a type of machine learning models that are intended to learn the underlying distribution of data and reproduce new and synthetic samples that are similar to the original dataset (Goodfellow et al., 2016). Some of the core ideas are probabilistic modeling, latent variable representation, and likelihood-based or adversarial training. Such models are able to represent complicated data displays and are able to produce text, pictures, molecular structures, or biological sequences.

Some of the major architectures on which generative AI is based are GANs, VAEs, diffusion models, and transformer-based generative models. GANs are made up of a generator and a discriminator that are trained in minimax game with the generator producing real-like samples and the discriminator deciding whether the data is real or fake.

The VAEs apply an encoder–decoder architecture to represent data in a latent space and recreate it minimizing a variational lower bound (Kingma Welling, 2014). The diffusion models are trained to reverse the effect of noise on data and gradually increase noise to generate high-quality samples with strong likelihood estimation (Ho et al., 2020). Models based on a transformer like LLMs are based on self-attention to learn long-range dependencies and produce coherent sequences (Vaswani et al., 2017).

Generative models do not share the same objective with the discriminative models. Whereas generative models are trained to learn the joint probability distribution P(x,y)P(x, y)P(x,y) or P(x)P(x)P(x) (useful in the generation of data), discriminative models are trained to learn the conditional probability P(y | x)P(y | x)P(y | x) to do classification or prediction tasks (Ng & Jordan, 2001). This difference enables generative models to be used to synthesize and augment data, as well as in supervised prediction which is better with discriminative models.

Generative AI has a lot of potential in biomedical applications, such as data scaffolding of small datasets, molecular design-based drug discovery, and biological system simulation (Chen et al., 2021).

3. Data modalities in the neurodegenerative disease research

The modalities of data used in the research of NDs are complementary and diverse, which can allow a deeper consideration of the disease onset, progression, and response to treatment. The multimodal integration has been put in the focus of developing precision neurology.

3.1 Neuroimaging modalities are fundamental

Magnetic resonance imaging (MRI) and functional MRI are the techniques which can be used to get structural and functional information about brain atrophy, connectivity, and network disruption. Positron emission tomography (PET) can be used to visualize the presence of molecular processes including amyloid-β and tau deposition, whereas computed tomography can be used to help assess the structure in clinical practice. The two modalities together can offer key biomarkers to make a diagnosis and monitor the disease (Lee et al., 2024; Wankhede et al., 2025).

3.2 Molecular and omics data provide mechanistic depth

Genomics (e.g., polygenic risk scores derived with the aid of GWAS), transcriptomics, proteomics, and metabolomics can be used to identify disease-related pathways and biomarkers in biofluids (blood and cerebrospinal fluid). These strategies enable at an early stage of detection and stratification of the patient, and in many instances, the pathological changes would be detected years prior to clinical manifestations.

3.3 Clinical and behavioral data provide phenotypic context

Cognitive measures (i.e., memory and executive functional tests) and motor measures (i.e., gait, tremor analysis) are still important in diagnosing and staging of the disease. These measures together with imaging are very useful in the classification of the disorders like AD and PD.

3.4 Wearable and digital biomarkers are a new frontier

Gait, activity, speech, and cognitive performance of a patient can be monitored on a continuous basis by using devices that include accelerometers, smartphones, and sensor-based systems. The technologies are able to pick up subtle changes in behavior and enhance prediction of NDs at an early age when combined with imaging and genetic data. Fig. 9.1 shows the GenAI pipeline in ND research.

4. Applications of generative AI in neurodegenerative diseases diagnosis

The diagnosis of NDs is rapidly being changed through the use of GenAI that allows the extraction, synthesis, and interpretation of the complex biomedical data. Compared to the standard machine learning method, generative models, including GANs, diffusion models, and VAEs, are capable of learning underlying data distribution and producing realistic samples and, thus, achieve the major benefits of addressing data scarcity, heterogeneity, and early disease detection (Jung et al., 2024; Wang et al., 2025).

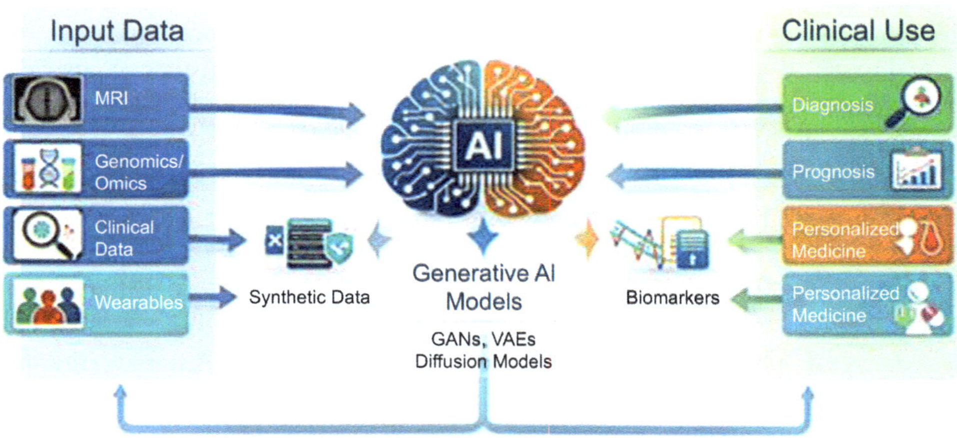

Figure 9.1 ***GenAI pipeline in neurodegenerative disease research.*** Fig. 9.1 shows the different input data and clinical uses of GenAI in neurodegenerative disease research.

4.1 Synthetic data generation

GenAI is important in the generation of synthetic data, specifically GANs and diffusion models. These models generate high fidelity neuroimaging data which supplements small datasets and enhances diagnostic model performance. Research indicates that GAN-based systems have a high diagnosing accuracy with AD as they augment the training datasets and increase the strength of classifications (Alam & Latifi, 2025). Moreover, synthetic data generation can deal with class imbalance and the limited number of rare datasets of NDs; as such a scenario is typical because of the expensive and ethical limitations of obtaining data. Generative models can be used to learn complex data distributions to provide balanced data and enhance generalization (Huynh & Deshpande, 2024).

4.2 Image reconstruction and enhancement

GenAI has a great capacity to improve neuroimaging quality by means of image reconstruction and restoration. The superresolution imaging is commonly performed with GANs and diffusion models, which can be used to convert low-resolution scans into high-resolution images with better anatomical detail. Also, these models are useful in noise removal and artifact correction, especially in MRI and PET imaging with low dose or acceleration scans, which tend to impose distortions (Wang et al., 2025). These advances increase the sensitivity of lesions and their diagnostic accuracy, and GenAI is inevitable in clinical neuroimaging processes.

4.3 Early disease detection

Early diagnosis of neurodegenerative conditions is one of the areas where GenAI has been promising. These models are able to detect subtle patterns of preclinical pattern which might not be noticeable in the normal analysis. GenAI corrects the progression of a disease and acquires latent representations, which can be used to identify the initial pathological alterations in diseases, including AD and PD (Basha et al., 2026). Also, multimodal data fusion, which combines imaging, genomic, and clinical data, can be enhanced with the help of generative models that will enhance the accuracy of the diagnosis and allow the characterization of the disease in a more comprehensive way (Liu et al., 2025).

4.4 Biomarker discovery

GenAI also advances biomarker discovery by extracting meaningful features from high-dimensional datasets. Models like VAEs can be used to discover latent space representations of disease-specific biomarkers. This method allows identifying new imaging and molecular biomarkers, which are essential in the initial diagnosis and targeted treatment. GenAI has also been applied to radiogenomics, connecting imaging phenotypes with genetic information to unravel the mechanisms of disease (Liu et al., 2025).

5. Applications in prognosis and progression of neurodegenerative diseases

Applications in prognosis and progression of NDs entail developing new approaches that will be used to understand, predict and make specific intervention to NDs like AD, PD, and other degenerative diseases.

5.1 Disease trajectory modeling

The goal of disease trajectory modeling is to model the time-varying dynamics of symptoms and biomarkers. Longitudinal cohort-based data-driven frameworks have the capacity to estimate the progression rates and disease stages by modeling continuous trajectories not discrete stages, which can be further used to characterize neurodegenerative timelines more finely (Oxtoby, 2023). These generative models combine longitudinal clinical and biomarker data to predict the course of the disease in individual patients, which allows heterogeneity among patients (Mekkes et al., 2024; Oxtoby, 2023). These methods facilitate the individual prognosis through the quantification of the changing cognitive and motor features with time.

5.2 Patient stratification

It is a method based on latent embeddings and unsupervised learning used to discover disease subtypes in heterogeneous populations. The subtype and stage inference are examples of computational methods that define individual subgroups of progression that share either phenotypic or biomarker distributions, which cannot be identified by standard clinical observation (Viani et al., 2025). Such latent embeddings have the potential to reveal the underlying mechanisms, and can be used to inform stratified patient groups to apply precision medicine, where interventions are to be applied to subtypes based on their progression and risk phenotypes (Viani et al., 2025). Multiomics, imaging, and clinical data integration can be useful in precision medicine applications to improve the stratification and treatment decision.

5.3 Outcome prediction

It aims at predicting the major endpoints that include cognitive impairment, motor impairment, and survival. Longitudinal datasets that have been trained with machine-learning models have demonstrated potential in the prediction of individual outcomes because they are able to capture temporal trends in clinical records and biomarkers that can be used to improve early detection and management strategies (Mekkes et al., 2024). In addition, these predictive systems are becoming a part of clinical decision support systems that enable physicians to have data-driven risk estimates and recommendations on specific treatment that enhance care planning and monitoring. With the evolution of computational methods, the combination of the trajectory models, latent stratification, and predictive analytics can be used to revolutionize the process of prognosis and progression management in NDs. Fig. 9.2 shows the applications of GenAI in the diagnosis and prognosis of NDs.

6. Case studies and disease-specific applications

GenAI has become a transformative tool in ND research because it can generate complex high dimensional data, realistic biomedical images, and aid in prognostic prediction (Seiler & Ritter, 2025). These functions are finding their use in NDs such as AD, PD, ALS, and others.

6.1 Alzheimer's disease

Generative models have been developed in AD to improve the imaging of the most important pathological hallmarks including amyloid-β plaques and tau tangles, which determine the disease course. Diffusion models and GANs have been used in generating high-resolution amyloid and tau PET images using lower resolution or other modalities to enhance the quality of diagnostic and decrease the need to perform invasive procedures (Alam & Latifi, 2025). As an example, latent diffusion models have been shown to have improved amyloid PET clarity and longitudinal consistency across various datasets, which can be used to detect and inter scan harmonization (Alam & Latifi, 2025).

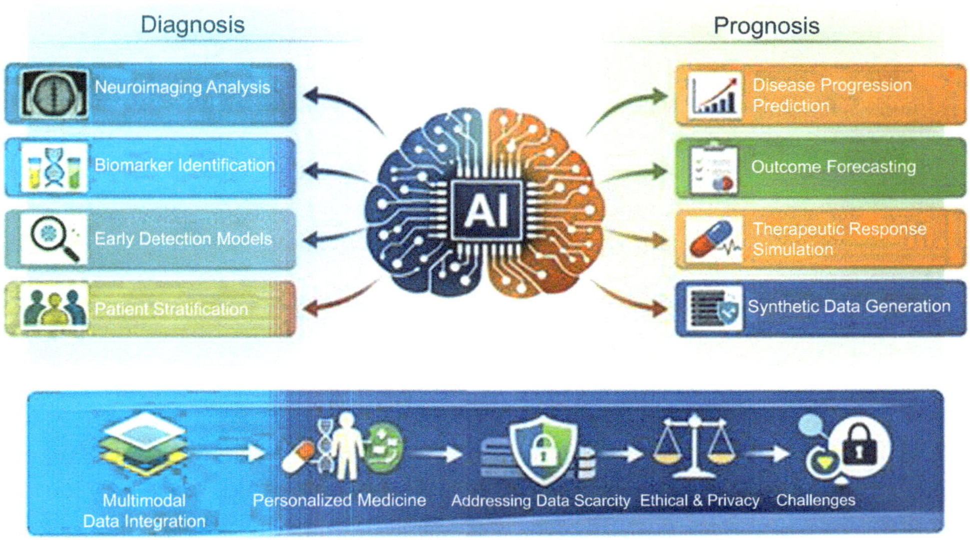

Figure 9.2 ***Applications of GenAI in the diagnosis and prognosis of the neurodegenerative diseases.*** Fig. 9.2 shows the different applications of GenAI in diagnosis and prognosis of neurodegenerative diseases.

Early diagnosis is still a priority in AD since symptoms of the disease are sometimes slow to manifest themselves in clinical terms. Patterns of progression based on the baseline biomarkers can be simulated by generative models and, therefore, mild cognitive impairment conversion to AD and future structural changes can be predicted (Das et al., 2026). This is reflected in attention aligned conditional diffusion auto encoders, which directly impose explicit alignment of multi modal clinical and imaging features that can be useful in improving anatomical accuracy in predicting progression (Das et al., 2026). The method enhances longitudinal modeling in comparison to the discriminative classifiers which are static, and this is useful in the early therapeutic intervention strategies.

6.2 Parkinson's disease

PD majorly involves dopaminergic neuron loss, making dopaminergic imaging vital for diagnosis and progression evaluation (Asuku & Olaniyan, 2026). Applications have been made to enhance the dopaminergic PET images using generative models and deep learning and to assist in cross modality synthesis, including the synthesis of PET to SPECT scans with CycleGANs, to solve the problem of data scarcity and improve downstream classification accuracy (Lopes et al., 2025). Generative architectures that are self-supervised and trained on dopamine transporter PET images have also been applied to produce realistic disease states and obtain latent representations that are predictive of the severity and progression of disease (Won et al., 2025).

In addition to image synthesis, GenAI can be used to predict motor progression by mapping multifaceted correlations between imaging features, clinical evaluations, as well as biomarkers profiles. Prognostic models based on early AI can produce projection images that can be visually interpreted to better understand the anticipated disease progression based on baseline imaging to help clinicians have more insight into future motor symptom projections (Sharma et al., 2026). These tools are consistent with the wider projects in the field of PD research to combine deep longitudinal data with machine learning to model individual progression (Shokrpour et al., 2025).

6.3 Amyotrophic lateral sclerosis

Prognostic modeling in ALS is an especially challenging problem as it is a rapidly developing disease and is heterogeneous in clinical presentation. Generative normative diffusion models that combine generative ability with normative reference distributions have been found to be useful in enhancing survival prediction by identifying subtle structural pattern signals in MRI associated with disease trajectory (Ijishakin et al., 2024). The combination of multiomics and clinical data, which will be facilitated by large AI-supported research consortia, will further increase the prognostic stratification and discovery of biomarkers in ALS (BGI Research, 2026).

6.4 Other disorders

In addition to AD, PD, or ALS, there are other neurodegenerative conditions to which GenAI can be applied, which include Huntington's disease, frontotemporal dementia, and multiple sclerosis. The conditions are advantageous to AI augmented imaging and multimodal analysis of the differentials and progression modeling especially in cases when the patterns of structural change differ dramatically across subtypes (Liu et al., 2025). With an ever-growing use of neurodegenerative research to combine imaging and genomics with clinical phenotypes through the use of generative models, opportunities to detect the disease earlier, provide more specific prognosis, and interventions continue to get larger.

7. Integration with multimodal and longitudinal data

GenAI has altered the ND research by allowing the synthetic and incorporation of complicated biomedical data, which is vital in the form of solid diagnosis and prognostic modeling. One of the directions that have been developed significantly is cross modal generation, especially between structural MRI and functional modalities like PET. More recently, constrained diffusion models have been trained to generate PET images by using MRI scans, enhancing access to functional information in PET, when it is not available because of cost or radiation issues (e.g., functional imaging constrained diffusion models generating high fidelity PET images from MRI data). These methods can utilize spatial correspondences and generative priors to recover missing modalities to improve the performance of multimodal diagnostics.

The incomplete modalities have remained to be an issue in clinical settings where the subject will not be fully multimodally acquired. It has been suggested that GANs can be used to fill the missing modalities without affecting clinically significant characteristics, hence enhancing the accuracy of the classification (Arockia Selvarathinam et al., 2026). Such imputation methods enable comprehensive multimodal analyses which enable models to utilize complementary information in imaging domains even in situations where there is partial missing data.

Besides the application of the static cross modal synthesis, the temporal generative models are used to capture the changes in the processes of NDs. Temporally aware diffusion architectures based on diffusion have also shown that it is possible to model temporal spatial changes in the brain and predict future morphological states based on longitudinal sequences of MRI (Puglisi et al., 2025). The models enable the description of the disease progression patterns, which are significant in the early prognosis and the planning of individual intervention.

GenAI complements multimodal combination and longitudinal studies of neurodegenerative studies. The cross-modal generation, modality imputation, and the temporally conscious models all aid in the provision of more precise and holistic diagnostic and prognostic models. Further methodological research and stringent validation of these abilities is necessary to translate them into normal clinical decision support.

8. Validation, interpretability, and clinical translation

The field of NDs diagnostics and prognostic modeling is changing at a rapid pace by using GenAI; however, rigorous validation, interpretability, and clinical translation are the key issues. To guarantee the strong performance of the model, the researchers will use model validation techniques, including internal cross validation and external data testing, to overcome the problem of overfitting and evaluate the model in the context of the different populations (Christodoulou et al., 2025). Nevertheless, much of the deep generative model literature, particularly that which is trained on neuroimaging or other multimodal clinical data, continues to show large drops in performance when externally validated, showing the importance of standardized validation frameworks (Christodoulou et al., 2025; Lin & Komperda, 2026). The second issue is explainability and interpretability of the generative models; those systems are often "black boxes," and the output produced by them may not be easily rationalized by the clinician. Explainable AI (XAI) system in SHAP or attention is being incorporated to offer an understanding of feature contributions to allow users to visualize why a model is predicting disease progression or risk (Hoghooghi Esfahani et al., 2025). This is vital to the clinical trust and ethical use. From regulatory and clinical implementation, diagnostic and prognostic GenAI systems have to pass complicated approval processes like other high risk medical devices and need to have well documented human control, constant monitoring, and risk control measures before implementation in practice. Lastly, unlike conventional machine learning techniques (e.g., support vector machines or random forests), generative models can generate realistic medical images and can learn with less labeled data, which would potentially enhance early detection and can increase feature learning. However, the traditional approaches tend to be more open and clinically more validable, and thus can be more readily implemented in the controlled environment. Overall, further external validation research, explainability models, and regulatory interface work will be essential to achieve the potential of GenAI in the care of ND.

9. Ethical, legal, and societal implications

GenAI can be used to transform ND diagnosis and prognosis, but it raises serious ethical, legal, and social challenges that should be tackled. In the first place, data privacy and security are crucial: generative models work with huge data volumes of sensitive health data, and any lack of strong measures, such as encryption and informed consent, means that patient data may be disclosed or abused, and legally guaranteed health rights will be violated.

A secure flow of data and governance models are needed to safeguard the personally identifiable information and preserve the autonomy of patients (Nasir et al., 2025).

Second, prejudice and equality is a significant problem. When trained on nonrepresentative data, models can be used to perpetuate or even increase differences in the accuracy of diagnosis and treatment across demographic groups, compromising equal care. Frequent auditing for fairness and inclusion of diverse populations in model training are very important mitigation measures (Tung et al., 2025).

Third, there is also the risk of the abusing use of synthetic data, which is normally suggested to ensure privacy. Synthetic data can also not be valid to the real world or will unintentionally encode biases, and this can be misleading to clinical research or decision making without proper validation and disclosure (Susser et al., 2024).

Lastly, there should be transparency and trust in a clinical setting. "The black box" characteristics of most generative models may undermine clinician and patient trust unless the explainability, transparent clinical supervision, and accountability model features are inherent in the AI implementation processes.

10. Challenges and limitations

GenAI has a potential transformative ability in the diagnosis and prognosis of NDs, although there are still major challenges. One of the main limitations is the quality and standardization of data; healthcare data are heterogeneous in terms of their acquisition protocols, imaging parameters, and clinical annotation, which compromises the model robustness and other-site consistency (Christodoulou et al., 2025). Furthermore, most of the researches use small or homogenous cohorts, which increases bias and reduce external validity. Clinical translation is also limited by computational cost and scalability since high performance AI models require significant processing power and infrastructure, which are usually not available in routine care environments. This restricts scalability particularly where the resource constrained clinics require real time inference and interface with the existing electronic health systems. The applicability to other populations also can be an issue: the models that have been trained on mainly Western or high-income samples often fail to work in ethnically diverse populations because of the lack of representation in the training data. Lastly, the adoption is slowed by the "black box" characteristics of most generative models, mistrust of clinicians, and lack of smooth interoperability with clinical decision support systems, which makes them difficult to integrate into healthcare workflows. These complex issues are important to solving the clinical promise of GenAI in the treatment of NDs.

11. Future directions

Recent developments show that GenAI will be of paramount use in revolutionizing the diagnosis and prognosis of NDs. The future work is probably to focus on the multimodal combination of the data provided by neuroimaging, genetic profiles, clinical history, and sensor signals to enhance the detection early and predictive accuracy. Application of bioinformatics and multiomics in conjunction with generative models can identify new molecular signatures and pathways in diseases such as AD and PD. Also, XAI, which will eliminate existing clinical adoption barriers by making model decisions readable to practitioners, will be required. Generative models with a clinician decision support framework would perhaps be able to standardize the collection and interpretation of data, increasing diagnostic confidence and simplifying the care pathway (Breithaupt et al., 2025). Model training on distributed patient data without interference with confidentiality will be possible through privacy preserving methods like federated learning, expanding the model generalizability to a wide range of populations. Finally, the ability of GenAI to generate artificial clinical samples and simulate the progression of a disease has the potential to advance prognostic models and speed up studies when the real data is limited (Breithaupt et al., 2025). The combination of these directions is expected to increase the accuracy of the diagnosis and preventive treatment of NDs.

12. Conclusion

GenAI is a groundbreaking solution in the prevention and diagnosis of neurodegenerative disorders, which provide new possibilities to increase the clinical decision-making process. Through the use of generative models that can create synthetic data, forecast disease progression, and detect subtle biomarkers, such technologies can overcome serious weaknesses of the existing diagnostic approaches, such as the lack of data and interpatient variability. Combination of GenAI with neuroimaging, genomics, and multimodal clinical data can reveal latent patterns that conventional analysis can miss, which can then be detected earlier and more accurately than other conditions like AD, PD, and ALS. Moreover, personalized treatment plans can be informed by predictive modeling with the help of AI, as well as clinical trials can be designed, and the final results can be improved. Nevertheless, to be translated into a regular clinical practice, it is necessary to be strictly validated, take into account ethical and privacy issues, and have a strong cooperation between computational scientists, clinicians, and regulatory agencies. As the field advances, the combination of human knowledge and AI-based information is bound to bring a paradigm shift in the management of NDs to the field of precision neuroscience, which is proactive, not reactive. Further study and interdisciplinary activity would be necessary to fulfill the potential of GenAI in enhancing the diagnoses, prognoses, and treatment of these disabling conditions.

References

Alam, M. M., & Latifi, S. (2025). Early detection of Alzheimer's disease using generative models: A review of GANs and diffusion models in medical imaging. *Algorithms, 18*(7). https://doi.org/10.3390/a18070434

Arockia Selvarathinam, A. L. X. R., Anbalagan, N., Srinivasu, P. N., Choi, J., & Ijaz, M. F. (2026). Parameter-optimized generative adversarial network framework for synthetic MRI generation: Fine-tuning critical variables for enhanced image fidelity. *Frontiers of Medicine, 12*. https://doi.org/10.3389/fmed.2025.1731370

Asuku, A., Adeyemo, M. B., Adeyemo, R. O., Agoro, S., Odugbesan, O., & Ayoola, S. (2022). Applications of stem cell therapy in neurodegenerative diseases. Available from: http://biosciencejournals.com/assets/archives/2022/vol10issue2/10-2-25-575.pdf.

Asuku, A. O., Ayinla, M. T., Ajibare, A. J., & Krishnamurthy, R. (2024). *Modulatory and neuroprotective roles of marine green spirulina on the brain and neurodegeneration marine greens: Environmental, agricultural, industrial and biomedical applications* (pp. 189–199). CRC Press. https://doi.org/10.1201/9781003369738-21

Basha, S., Pranavi, K. S., Chattopadhyay, A., Pai, A. R., & Mahato, K. K. (2026). Artificial intelligence and machine learning in neurodegenerative disease management: A 21st century paradigm. *Neurobiology of Disease, 220*, Article 107307. https://doi.org/10.1016/j.nbd.2026.107307

Breithaupt, A. G., Weiner, M., Tang, A., Possin, K. L., Sirota, M., Lah, J., Levey, A. I., Van Hentenryck, P., Zandehshahvar, R., Gorno-Tempini, M. L., Giorgio, J., Wang, J., Rauschecker, A. M., Rosen, H. J., Nosheny, R. L., Miller, B. L., & Pinheiro-Chagas, P. (2025). Integrating generative artificial intelligence in ADRD: A roadmap for streamlining diagnosis and care in neurodegenerative diseases. Available from: https://doi.org/10.48550/arXiv.2502.06842.

Chen, R. J., Lu, M. Y., Chen, T. Y., Williamson, D. F. K., & Mahmood, F. (2021). Synthetic data in machine learning for medicine and healthcare. *Nature Biomedical Engineering, 5*(6), 493–497. https://doi.org/10.1038/s41551-021-00751-8

Christodoulou, R. C., Woodward, A., Pitsillos, R., Ibrahim, R., & Georgiou, M. F. (2025). Artificial intelligence in Alzheimer's disease diagnosis and prognosis using PET-MRI: A narrative review of high-impact literature post-tauvid approval. *Journal of Clinical Medicine, 14*(16). https://doi.org/10.3390/jcm14165913

Das, A., Ram, K., Sivaprakasam, M. (2026). Align-cDAE: Alzheimer's disease progression modeling with attention-aligned conditional diffusion auto-encoder. Available from: https://doi.org/10.48550/arXiv.2603.01552.

Ganjizadeh, A., Wei, Y., & Erickson, B. (2024). From pixels to prognosis: AI-driven insights into neurodegenerative diseases. *Medical Research Archives, 12*(6). https://doi.org/10.18103/mra.v12i6.5512

Goodfellow, I., Bengio, Y., & Courville, A. (2016). *Deep learning. Deeplearningbook.org*. MIT Press. Available from: https://www.deeplearningbook.org/.

Ho, J., Jain, A., & Abbeel, P. (2020). Denoising diffusion probabilistic models. 2. Available from: https://arxiv.org/abs/2006.11239.

Hoghooghi Esfahani, H., Toyonaga, S., & Oyibo, K. (2025). The application of explainable artificial intelligence in the prediction, diagnoses, treatment, and management of chronic diseases: A systematic review. *Digital Health, 11*. https://doi.org/10.1177/20552076251355669

Huynh, N., & Deshpande, G. (2024). A review of the applications of generative adversarial networks to structural and functional MRI based diagnostic classification of brain disorders. *Frontiers in Neuroscience, 18*. https://doi.org/10.3389/fnins.2024.1333712

Ijishakin, A., Hadjasavilou, A., Abdulaal, A., Montana-Brown, N., Townend, F., Spinelli, E., Filippi, M., Agosta, F., Cole, J., Malaspina, A. (2024). Normative diffusion autoencoders: Application to amyotrophic lateral sclerosis. Available from: https://doi.org/10.48550/arXiv.2407.14191.

Jung, H. K., Kim, K., Park, J. E., & Kim, N. (2024). Image-based generative artificial intelligence in radiology: Comprehensive updates. *Korean Journal of Radiology, 25*(11), 959–981. https://doi.org/10.3348/kjr.2024.0392

Kingma, D. P., & Welling, M. (2014). Auto-encoding variational bayes. 11. Available from: http://arxiv.org/abs/1312.6114.

Lee, J., Renslo, J., Wong, K., Clifford, T. G., Beutler, B. D., Kim, P. E., & Gholamrezanezhad, A. (2024). Current trends and applications of PET/MRI hybrid imaging in neurodegenerative diseases and normal aging. *Diagnostics, 14*(6). https://doi.org/10.3390/diagnostics14060585

Lin, Y. J., & Komperda, I. (2026). Integrating multimodal neuroimaging for neurological disorders: A systematic framework for clinical translation. *SN Comprehensive Clinical Medicine, 8*(1). https://doi.org/10.1007/s42399-026-02330-x

Liu, H., Zhang, X., & Liu, Q. (2025). A review of AI-based radiogenomics in neurodegenerative disease. *Frontiers in Big Data, 8*. https://doi.org/10.3389/fdata.2025.1515341

Loni, M., Poursalim, F., Asadi, M., & Gharehbaghi, A. (2025). A review on generative AI models for synthetic medical text, time series, and longitudinal data. *Npj Digital Medicine, 8*(1). https://doi.org/10.1038/s41746-024-01409-w

Lopes, L., Jiao, F., Xue, S., Pyka, T., Krieger, K., Ge, J., Xu, Q., Fahmi, R., Spottiswoode, B., Soliman, A., Buchert, R., Brendel, M., Hong, J., Guan, Y., Bassetti, C. L. A., Rominger, A., Zuo, C., Shi, K., & Wu, P. (2025). Dopaminergic PET to SPECT domain adaptation: A cycle GAN translation approach. *European Journal of Nuclear Medicine and Molecular Imaging, 52*(3), 851–863. https://doi.org/10.1007/s00259-024-06961-x

Mekkes, N. J., Groot, M., Hoekstra, E., de Boer, A., Dagkesamanskaia, E., Bouwman, S., Wehrens, S. M. T., Herbert, M. K., Wever, D. D., Rozemuller, A., Eggen, B. J. L., Huitinga, I., & Holtman, I. R. (2024). Identification of clinical disease trajectories in neurodegenerative disorders with natural language processing. *Nature Medicine, 30*(4), 1143–1153. https://doi.org/10.1038/s41591-024-02843-9

Nag, A., Hassan, M. M., Karim, A., & Reddy, C. K. K. (2025). *Generative AI in neurodegenerative disorders*. River Publishers. https://doi.org/10.1201/9788743801764

Nasir, M., Siddiqui, K., & Ahmed, S. (2025). Ethical-legal implications of AI-powered healthcare in critical perspective. *Frontiers in Artificial Intelligence, 8*. https://doi.org/10.3389/frai.2025.1619463

Ng, A., & Jordan, M. (2001). On discriminative vs. generative classifiers: A comparison of logistic regression and naive Bayes. In T. Dietterich, S. Becker, & Z. Ghahramani (Eds.), *Advances in neural information processing systems, Vol. 14*. MIT Press. Available from: https://proceedings.neurips.cc/paper_files/paper/2001/file/7b7a53e239400a13bd6be6c91c4f6c4e-Paper.pdf.

Olufemi Asuku, A., & Tope Olugbemi, O. (2026). *Evolution and innovation in Parkinson's disease therapies in under-resourced settings*. IGI Global Scientific Publishing. https://doi.org/10.4018/979-8-3373-7437-6

Oxtoby, N. P. (2023). Data-driven disease progression modeling. In *Machine learning for brain disorders* (pp. 511–532). https://doi.org/10.1007/978-1-0716-3195-9_17

Puglisi, L., Alexander, D. C., & Ravì, D. (2025). Brain latent progression: Individual-based spatiotemporal disease progression on 3D brain MRIs via latent diffusion. *Medical Image Analysis, 106*, Article 103734. https://doi.org/10.1016/j.media.2025.103734

Rana, M. S., Longa, F. E. A., Riaz, A., & Sheu Alamu, O. (2025). Artificial intelligence in early diagnosis of neurodegenerative disorders: A systematic review of clinical applications and challenges. *International Journal of Medical and Health Research, 3*(3), 81–91. https://doi.org/10.61424/ijmhr.v3i3.382

Rao, H., Zeng, M., Zhao, X., & Miao, C. (2024). A survey of artificial intelligence in gait-based neurodegenerative disease diagnosis. Available from: https://doi.org/10.48550/arXiv.2405.13082.

Seiler, M., & Ritter, K. (2025). Pioneering new paths: The role of generative modelling in neurological disease research. *Pflugers Archiv European Journal of Physiology, 477*(4), 571–589. https://doi.org/10.1007/s00424-024-03016-w

Sharma, A., Agarwal, A., Wa Kalenga, M. K., Gupta, V., & Srivastava, V. (2026). AI in Parkinson's disease: A short review of machine learning approaches for diagnosis. *Processes, 14*(2), 199. https://doi.org/10.3390/pr14020199

Shokrpour, S., MoghadamFarid, A. M., Abkenar, S. B., Haghi Kashani, M., Akbari, M., & Sarvizadeh, M. (2025). Machine learning for Parkinson's disease: A comprehensive review of datasets, algorithms, and challenges. *Npj Parkinson's Disease, 11*(1). https://doi.org/10.1038/s41531-025-01025-9

Stefano, G. B. (2023). Artificial intelligence as a tool for the diagnosis and treatment of neurodegenerative diseases. *Republic Brain Sciences, 13*(6). https://doi.org/10.3390/brainsci13060938

Susser, D., Schiff, D. S., Gerke, S., Cabrera, L. Y., Cohen, I. G., Doerr, M., Harrod, J., Kostick-Quenet, K., McNealy, J., Meyer, M. N., Price, W. N., & Wagner, J. K. (2024). Synthetic health data: Real ethical promise and peril. *Hastings Center Report, 54*(5), 8–13. https://doi.org/10.1002/hast.4911

The future of ALS care: Multi-omics research and the "Know ALS" AI pilot. (2026). Bgi.com. Available from: https://www.bgi.com/global/news/multi-omics-research-and-the-know-als-ai-pilot.

Tung, T., Hasnaeen, S. M. N., & Zhao, X. (2025). Ethical and practical challenges of generative AI in healthcare and proposed solutions: A survey. *Frontiers in Digital Health, 7*. https://doi.org/10.3389/fdgth.2025.1692517

Vaswani, A., Shazeer, N., Parmar, N., Uszkoreit, J., Jones, L., Gomez, A.N., Kaiser, Ł., & Polosukhin, I (2017). Attention is all you need. Available from: https://doi.org/10.48550/arXiv.1706.03762.

Viani, A., Custo, A., d'Angremont, E., Garibotto, V., Frisoni, G. B., Gutman, B. A., & Lorenzi, M. (2025) Disease progression modeling and stratification for detecting sub-trajectories in the natural history of pathologies: Application to Alzheimer's disease trajectory modeling. *Imaging Neuroscience, 3*. https://doi.org/10.1162/IMAG.a.954

Wang, S., Zhou, X., Li, C., Wang, S., Li, Y., Tan, T., & Zheng, H. (2025). Generative artificial intelligence in medical imaging: Foundations, progress, and clinical translation. *Research: Ideas for Today's Investors, 8*. https://doi.org/10.34133/research.1029

Wankhede, N. L., Koppula, S., Umare, P. D., Umare, M. D., Lokhande, S. K., Taksande, B. G., Upaganlawar, A. B., Umekar, M. J., Kopalli, S. R., & Kale, M. B. (2025). Neuroimaging biomarkers: A critical review of their role in neurological disease diagnosis and management. *Brain Disorders, 18*. https://doi.org/10.1016/j.dscb.2025.100222

Won, J., Lee, G. Y., Jo, S., Lee, J., Lee, S., Kim, J. S., Sung, C., Oh, J. S., Kwon, K. Y., Park, S. B., Lee, J., Yum, J., Chung, S. J., & Kim, N. (2025). Enhancing 3D dopamine transporter imaging as a biomarker for Parkinson's disease via self-supervised learning with diffusion models. *Cell Reports Medicine, 6*(7). https://doi.org/10.1016/j.xcrm.2025.102207

Further reading

Yu, M., Wu, M., Yue, L., Bozoki, A., & Liu, M. (2026). Functional imaging constrained diffusion for brain PET synthesis from structural MRI. *Medical Image Analysis, 111*, Article 104044. https://doi.org/10.1016/j.media.2026.104044

CHAPTER 10

Generative AI for brain connectivity and network analysis: Understanding brain organization and its alterations in disease

L. Leo Prasanth[1], Godwyn Paulson Pitchaimani[2], Rachel Nallathamby[1], R. Sivaramakrisknan[1], P. Hemalatha[3] and I. Nandhini[1]

[1]Vel Tech Rangarajan Dr. Sagunthala R&D Institute of Science and Technology, Department of Computer Science and Engineering, Chennai, Tamil Nadu, India; [2]Escuela Superior de Ingeniería y Arquitectura (ESIA), Unidad Ticomán, Instituto Politécnico Nacional (IPN), Gustavo A. Madero, Mexico; [3]Rajalakshmi Engineering College, Department of Computer Science and Business Systems, Chennai, Tamil Nadu, India

1. Introduction

The human brain is one of the most complex biological systems, encompassing approximately 86 billion neurons interconnected through an intricate network of synapses. Understanding how these neural components interact to support cognition, emotion, and behavior has long been a central focus in neuroscience. This pursuit has led to the rise of network neuroscience, which models the brain as a graph comprising nodes (neurons or brain regions) and edges to decode its organizational principles and pathological disruptions (Gong et al., 2023). Data from neuroimaging modalities such as functional MRI (fMRI), electroencephalography (EEG), and diffusion tensor imaging (DTI) provide insights into brain connectivity but are often high-dimensional, temporally dynamic, and prone to noise, and are often incomplete (Wang & Chen, 2025). Traditional analytical approaches are limited by the complexity, paving the way to the adoption of generative AI (Gen AI), a powerful suite of machine learning (ML) models capable of learning underlying distributions and generating realistic, high-fidelity data (Bengesi et al., 2024). Recent advances in generative models such as VAEs, GANs, and diffusion models have catalyzed transformative progress in brain image reconstruction (Shaul et al., 2020), multimodal network integration (Zuo, Di et al., 2025), and simulation of disease-specific connectivity alterations (Zhou et al., 2025). For example, BG-GAN has been employed to model structural–functional connectivity in Alzheimer's disease, while BrainNetDiff demonstrates how multimodal diffusion can empower personalized network synthesis (Zong et al., 2024). These models not only reconstruct missing data and synthesize realistic brain networks (Barile et al., 2021) but also enable dynamic modeling of network evolution, providing prognostic insights into psychiatric and neurodegenerative conditions (Marzi et al., 2023).

Generative Artificial Intelligence for Neuroimaging
ISBN 978-0-443-45010-5
https://doi.org/10.1016/B978-0-443-45010-5.00013-7

In Clinical Psychological research, Gen AI is increasingly used to develop biomarkers that represent objective measures of a biological and molecular state of human biology, reflecting the state of cells or organisms at a specified point in time, thereby supporting early diagnosis and individualized treatment planning (Walter et al., 2019). Models such as ViT-GAN have even been proposed for identifying schizophrenia-associated structural and functional markers by linking magnetic resonance imaging (MRI) features with functional connectivity (Bi et al., 2024). Similarly, AI-driven synthesis of structural networks from functional data has proven effective in both healthy and diseased populations (Zhang et al., 2022). Beyond reconstruction and augmentation, generative modeling has been proposed as a framework to explore developmental brain changes and diversity (Akarca et al., 2021), emulate cognitive processes in robotic systems (Taniguchi et al., 2022), and understand brain–behavior relationships in conditions such as autism (Kundu et al., 2024). Foundational work by Friston & Price (2001) laid the theoretical groundwork for these applications by proposing dynamic, generative representations of brain function. Despite this promise, several challenges remain. Key concerns include model interpretability, limited sample sizes, individual variability, and the ethical implications of AI-driven neuroscience (Sai et al., 2024). Moreover, integrating multimodal data streams in a biologically plausible and computationally tractable manner continues to be a significant hurdle (Isallari & Rekik, 2021).

This chapter aims to synthesize these developments by first laying out the theoretical foundations of brain network analysis and Gen AI. It then explores recent applications in modeling healthy and pathological brain connectivity, with a focus on neurodegenerative, psychiatric, and developmental disorders. Finally, the challenges, ethical considerations, and future directions presented by this rapidly evolving field are examined. By bridging neuroscience with Gen AI, this chapter aspires to advance precision diagnostics, enhance clinical decision-making, and ultimately contribute to a deeper understanding of human brain organization and dysfunction.

1.1 Background: Generative AI for brain connectivity and network analysis

The human brain is an intricately organized and highly dynamic network, where complex patterns of structural and functional connectivity underlie cognition, behavior, and neurological health. Advances in neuroimaging, such as MRI, fMRI, and diffusion-weighted imaging (DWI), have enabled large-scale mapping of the brain's connectome. This has led to the emergence of **network neuroscience**, a discipline that models the brain as a graph composed of nodes (regions) and edges (connections). However, analyzing brain connectivity remains a significant challenge because of high-dimensional data, intersubject variability, and limitations in data acquisition techniques (Wang et al., 2023). **Gen AI** has emerged as a powerful technology to address these challenges. Generative models, particularly **VAEs**, **GANs**, and **hybrid VAE-GAN**

frameworks, are increasingly being used to synthesize, augment, and analyze brain connectivity representations (Laino et al., 2022). For example, VAE-GAN-based approaches provide superior performance in synthesizing realistic structural connectomes by learning latent representations across the two brain hemispheres (Zuo et al., 2023). GANs, specifically designed to handle high-dimensional imaging data, have been successfully applied in various brain MRI tasks, including modality translation, denoising, superresolution, and disease-specific image synthesis (Yao et al., 2024). These models also play a pivotal role in the **diagnosis and classification of neurological and psychiatric disorders**. GANs trained on functional network connectivity (FNC) features have demonstrated robust performance in distinguishing mental disorders such as schizophrenia and bipolar disorder (Zhao et al., 2020). Generative models have further been utilized to simulate abnormal brain connectivity patterns associated with schizophrenia, offering insights into disease progression and network disorganization (Zhang et al., 2021). The ability to realistically generate DWIs using deep generative models has also advanced the quality and availability of synthetic neuroimaging datasets (Hirte et al., 2021).

In addition, comprehensive reviews have highlighted the growing influence of Gen AI in both structural and functional neuroimaging, emphasizing its role in personalized diagnosis, data augmentation, and multimodal integration (Ali et al., 2022). Collectively, these developments mark a paradigm shift in how researchers understand and model brain networks.

i. Brain Connectivity and Network Neuroscience

Brain connectivity explores how different regions of the brain interact, communicate, and influence one another. These connections are broadly classified into three primary categories: structural, functional, and effective connectivity. Fig. 10.1 visually

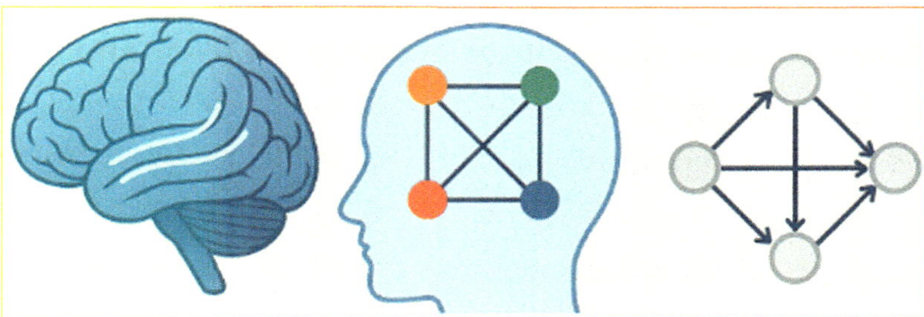

Structural Connectivity	**Functional Connectivity**	**Effective Connectivity**
anatomical connections via white matter tracts, often studied using diffusion MRI	Statistical dependencies among brain regions based on time-series signals	Directional influences among neural systems

Figure 10.1 Comparative overview of brain connectivity types.

presents a comparative overview of these connectivity types, each representing distinct but complementary perspectives on brain communication and coordination.

- **Structural Connectivity** reflects the physical architecture of the brain. It describes the anatomical pathways formed by white matter tracts, which connect various brain regions. These connections are typically assessed using diffusion MRI (dMRI), which maps the orientation and integrity of neural fibers. Structural connectivity provides the "wiring diagram" of the brain and serves as the underlying layer upon which other forms of connectivity are built.
- **Functional Connectivity** examines statistical dependencies between spatially distant brain regions based on temporal correlations in their activity. This type of connectivity is commonly derived from fMRI or EEG recordings. It identifies patterns of coactivation between brain regions, revealing functional networks that support cognitive processes such as memory, attention, and perception.
- **Effective Connectivity** goes a step further by capturing the directionality of neural connections and the causal influence of one neural system on another. It models how activity in one region directly affects another, often using methods such as dynamic causal modeling (DCM) or Granger causality analysis. Effective connectivity is crucial for understanding how information flows dynamically within brain circuits.

In network neuroscience, the brain is modeled as a graph, where nodes represent anatomical or functional regions of interest (ROI), and edges denote connections, whether structural, functional, or effective. Using graph theory, researchers can quantify properties such as degree (number of connections per node), modularity (network division into communities), efficiency (ease of information transfer), and centrality (influence of nodes within the network) (Gong et al., 2023). These network-level insights have significantly advanced our knowledge of cognitive functioning, brain development, and the neural basis of disorders such as schizophrenia and Alzheimer's disease.

2. Importance of generative AI for brain connectivity and network analysis

The human brain is a complex, dynamic system comprising intricate networks of structural and functional connections. Understanding the organization of these networks and how they deviate in neurological and psychiatric conditions is a central goal of contemporary neuroscience (Betzel & Bassett, 2017). In recent years, Gen AI has emerged as a transformative tool for advancing brain connectivity and network analysis, offering novel approaches to modeling, reconstructing, and interpreting the brain's connectivity architecture.

2.1 Overcoming limitations of traditional analytic approaches

Traditional statistical and ML methods often struggle to capture the nonlinear, high-dimensional, and stochastic nature of brain networks. They are particularly limited in handling missing data, small sample sizes, and complex dependencies across modalities and time. Gen AI models, such as VAEs and GANs, offer a powerful alternative. These models learn to approximate the underlying data distribution, enabling them to generate realistic synthetic data, infer missing connectivity patterns, and robustly generalize from limited or noisy observations (He et al., 2025).

2.2 Enhancing structural and functional connectivity modeling

Gen AI plays a critical role in modeling both structural connectivity (white matter pathways captured via diffusion MRI) and functional connectivity (correlations between brain regions based on fMRI or EEG). For example, recent works have used VAE-GAN frameworks to synthesize structural brain connectomes (a collection of neural pathways) with high fidelity, even reconstructing hemisphere-separated connectomes by aggregating cross-domain features (Vatansever et al., 2021). These methods enhance our ability to simulate, visualize, and analyze the brain's wiring, which is especially valuable when direct measurements are unreliable or unavailable.

Similarly, GAN-based models have been employed to synthesize functional connectivity patterns that preserve essential statistical properties of brain dynamics. This facilitates the exploration of how brain networks evolve across time, adapt during tasks, or change in disease.

2.3 Understanding disease-related alterations in brain networks

Generative models are especially valuable for studying neurological and psychiatric disorders, where alterations in brain connectivity are key biomarkers (Wright & Anticevic, 2024). By learning from healthy and diseased populations, these models can generate representative connectivity patterns that highlight abnormal pathways, network disruptions, or compensatory mechanisms (Zhang et al., 2021). For example, GANs have been used to distinguish schizophrenia-related changes in structural and functional networks by generating synthetic data that highlights and captures the disease-specific features (Zhao et al., 2020). Moreover, these models can simulate the progression of brain diseases over time by generating longitudinal connectivity maps that can aid in visual understanding of the stages of progression. This opens new avenues for early diagnosis, monitoring disease trajectories, and evaluating treatment responses with greater precision.

2.4 Facilitating multimodal and cross-population integration

Gen AI facilitates the integration of multimodal data, combining structural, functional, and metabolic imaging to build holistic brain network representations. These unified models help uncover hidden relationships across modalities and provide a more comprehensive understanding of the brain's organization. In addition, generative models are effective in domain adaptation and transfer learning, making them suitable for generalizing across different domains, demographics, scanner types, or imaging protocols (Shaheen & Melnik, 2025).

2.5 Improving data augmentation and simulation

One of the most practical benefits of generative models is their use in data augmentation. In clinical neuroscience, acquiring large datasets is often difficult due to cost, patient availability, and ethical constraints. Generative models can produce realistic synthetic brain data that augment small or imbalanced datasets, thus improving the training of discriminative models for diagnosis, prognosis, or classification tasks (Khalighi et al., 2024). Furthermore, simulated datasets can be used to benchmark algorithms, test hypotheses, or study rare neurological conditions under controlled variations, which may not be feasible with real-world data.

2.6 Enabling explainable and hypothesis-driven neuroscience

Beyond performance metrics, Gen AI contributes to interpretability and theory-building in neuroscience. By controlling latent variables or modifying input features, researchers can analyze how specific neural changes affect cognitive functions or disease phenotypes (Sengar et al., 2025). This causal modeling capability offers a new paradigm for exploring hypotheses about brain–behavior relationships and network disruptions in mental health. Despite these advancements, the extraction of stable and meaningful connectivity patterns remains a challenge because of the high dimensionality and noise inherent in brain imaging data. Ongoing research aims at developing more robust methodologies that can improve reliability and interpretability across all three connectivity domains.

3. Challenges in brain connectivity analysis

Despite significant advancements in neuroimaging technologies and analytical frameworks, brain connectivity analysis remains fraught with several inherent challenges. These limitations affect the reliability, reproducibility, and interpretability of network-based findings in both research and clinical settings. Fig. 10.2 illustrates the severity of five primary challenges in brain connectivity analysis: incomplete and noisy data (Zuo, Chen, et al., 2025), high intersubject variability (Ashburner & Klöppel,

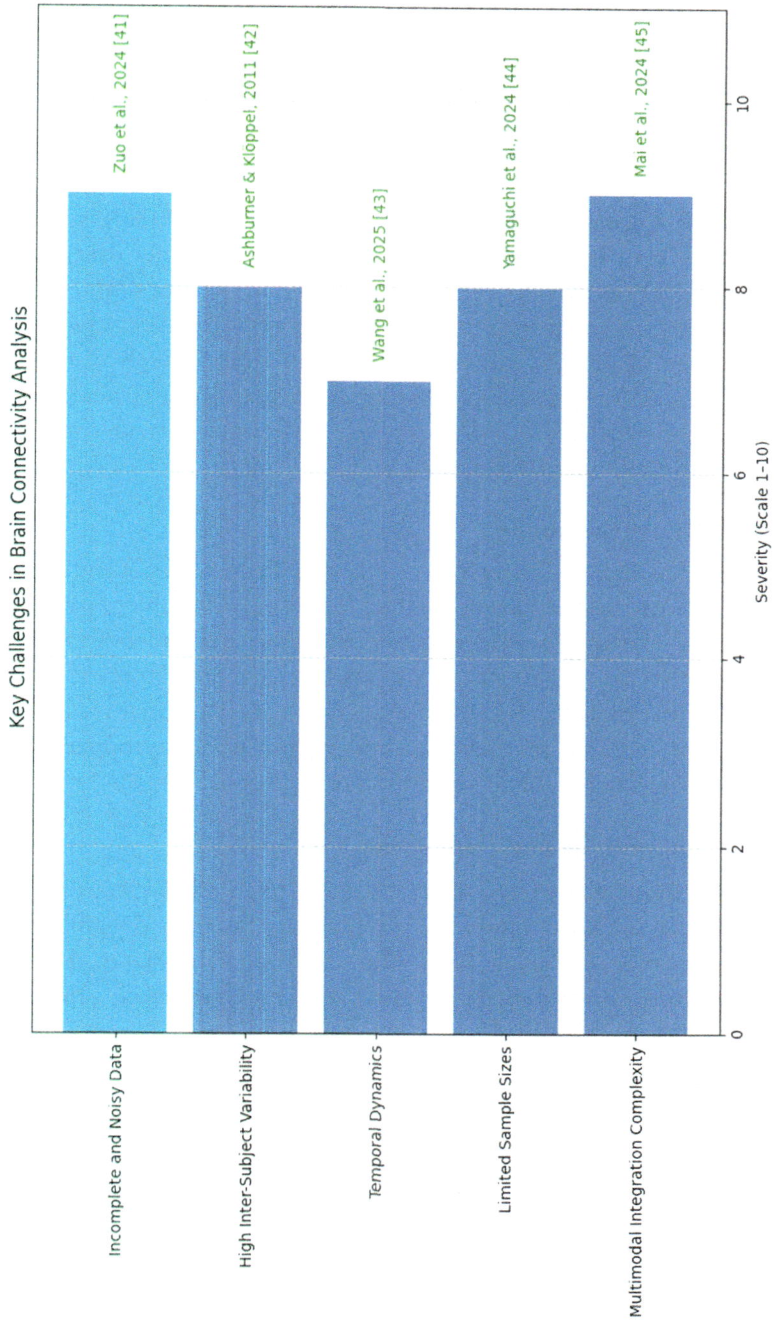

Figure 10.2 Severity of key challenges in brain connectivity analysis.

2011), temporal dynamics (Wang et al., 2025), limited sample sizes (Yamaguchi et al., 2024), and multimodal integration complexity (Mai et al., 2025). Each of these factors poses significant obstacles to reliable brain network modeling, highlighting the need for advanced Gen AI methods capable of mitigating these issues under data uncertainty. The most critical challenges are outlined in the following sections.

3.1 Incomplete and noisy data

Neuroimaging data are often contaminated with artifacts, including subject motion, physiological noise (e.g., respiration or cardiac pulsation), scanner drift, and environmental interference. These artifacts lead to low signal-to-noise ratios (SNRs) and missing data, especially in high-resolution or long-duration scans such as resting-state fMRI or diffusion MRI. Even small movements during scanning can severely distort the underlying connectivity measures, making it difficult to distinguish true neural signals from noise. This poses a fundamental obstacle to accurately constructing connectivity networks and extracting meaningful patterns.

3.2 High intersubject variability

Brain anatomy and functional organization vary significantly across individuals due to genetic, developmental, and environmental factors. This intersubject variability makes it difficult to identify robust group-level connectivity features, especially when studying heterogeneous populations. For instance, connectivity patterns observed in one cohort may not generalize to others because of differences in age, sex, cognitive ability, or clinical diagnosis. Consequently, findings may suffer from poor reproducibility, limiting their utility in personalized medicine or cross-site studies.

3.3 Temporal dynamics of brain networks

Most brain connectivity analyses are static, assuming fixed relationships among regions during the entire scan duration. However, the brain operates as a dynamic system, with connections continuously evolving across time and cognitive states. These temporal fluctuations are critical for understanding functional reorganization, task engagement, and disease progression. Modeling such dynamic functional connectivity (dFC) or time-varying effective connectivity poses additional challenges, including high computational complexity, handling increased data dimensionality, and the need for robust statistical frameworks that can capture short-term but meaningful network patterns and modified structures.

3.4 Limited sample sizes

Many brain connectivity studies, particularly those involving rare neurological or psychiatric disorders, suffer from small sample sizes due to data collection challenges, imaging costs, or ethical constraints. Small datasets increase the risk of overfitting, reduce statistical power, and limit the generalizability of findings. They also hinder the application of complex ML models that require large volumes of labeled data for effective training and validation.

3.5 Multimodal integration complexity

Modern neuroscience increasingly relies on multimodal imaging, combining data from different techniques such as structural MRI, diffusion MRI, fMRI, EEG, or Positron Emission Tomography (PET). While each modality provides complementary information, such as structural integrity, functional activity, and metabolic status, their integration is technically and conceptually challenging. Differences in spatial and temporal resolution, noise profiles, and preprocessing pipelines complicate data fusion, and inconsistencies between modalities can lead to conflicting interpretations of connectivity patterns.

4. Rise of generative AI in neuroimaging

Gen AI has emerged as a transformative force in neuroimaging, offering novel solutions to longstanding challenges in data acquisition, reconstruction, and interpretation. At its core, Gen AI employs advanced ML models capable of learning complex data distributions and synthesizing new data points that closely resemble the original training data (Sengar et al., 2025). These models go beyond simple replication; they capture the underlying generative processes, allowing for a wide range of applications in neuroscience and clinical neurology, such as simulating brain connectivity or reconstructing missing imaging data (Prasanth & Uma, 2025). Fig. 10.3 illustrates the hierarchical evolution of intelligent systems, demonstrating the field's progression from foundational AI to the frontier of Gen AI. The hierarchy begins with AI, which encompasses systems designed to emulate human cognitive processes, ranging from rule-based logic to

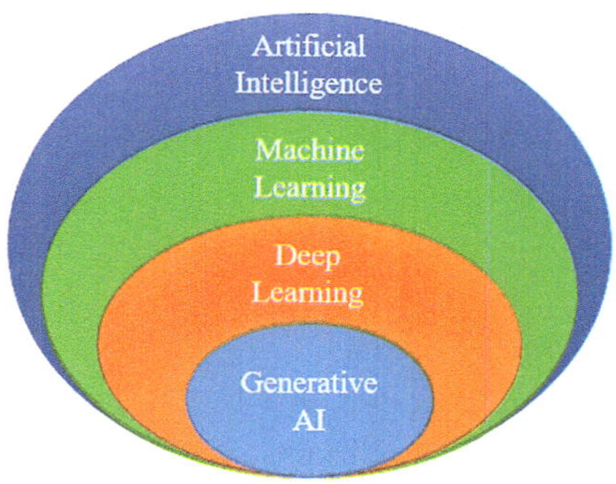

Figure 10.3 The hierarchical rise of generative AI.

adaptive algorithms (Catherine Storey, 2025). Nested within AI, ML enables systems to learn from data, improving performance through experience. Deep learning (DL), a specialized subset of ML, utilizes deep neural networks to capture complex features from high-dimensional data and has achieved remarkable success in tasks such as image recognition and natural language processing (Jain & Patient Care, 2024).

At the apex of this evolution is Gen AI, which distinguishes itself by creating new data instances that are statistically consistent with real-world inputs. It leverages sophisticated architectures such as generative adversarial networks (GANs), variational autoencoders (VAEs), diffusion models, and normalizing flows to model intricate data distributions and generate high-fidelity outputs (Catherine Storey, 2025). This paradigm shift from predictive analytics to creative synthesis is enabling Self-supervised learning to unveil brain dysfunctional signatures in neuropsychiatric disorders in personalized medicine and brain network modeling (Li et al., 2025).

In critical domains such as neuroimaging, Gen AI is revolutionizing practice by enabling the simulation of brain networks, reconstruction of missing or corrupted scans, and discovery of hidden patterns in brain connectivity. Its ability to move beyond analysis toward data generation and personalization marks a fundamental shift in how we approach scientific research, clinical diagnostics, and precision medicine.

The hierarchical evolution of intelligent systems, positioning Gen AI as an advanced subset of broader computational paradigms:

- AI encompasses all systems that mimic human intelligence, including rule-based systems and data-driven algorithms.
- ML lies within AI and focuses on systems that improve performance through experience, learning patterns from data.
- DL is a specialized subset of ML that uses deep neural networks to model complex, high-dimensional data representations, particularly effective in image, speech, and language processing.
- Gen AI, a cutting-edge branch within DL, specializes in creating new data instances that resemble real-world inputs. It includes powerful models such as GANs, VAEs, diffusion models, and normalizing flows, which are increasingly applied in domains like neuroimaging to simulate, reconstruct, and interpret brain data.

4.1 Foundational generative models

Several foundational generative models have played pivotal roles in the advancement of this field.

4.1.1 Variational autoencoders (VAEs)

VAEs are probabilistic generative models that learn to encode high-dimensional input data (such as brain scans) into a lower-dimensional latent space. This latent representation captures essential features of the data distribution. A decoder then reconstructs the input from this latent space, enabling both dimensionality reduction and the generation of new, realistic data samples. In neuroimaging, VAEs have been instrumental in modeling variability in brain structures and in identifying subtle patterns linked to neurological disorders (Zuo et al., 2023).

4.1.2 Generative adversarial networks (GANs)

GANs consist of two competing neural networks: a generator, which synthesizes new data samples, and a discriminator, which attempts to distinguish real samples from those generated (Zhao et al., 2020). Through this adversarial process, the generator improves until it produces highly realistic outputs. GANs are widely used in neuroimaging for tasks such as generating synthetic MRI or fMRI data, enhancing image resolution, and reconstructing missing or corrupted imaging regions with high fidelity.

4.1.3 Diffusion models

These models generate new data by learning the reverse of a gradual noising process. Starting from random noise, diffusion models iteratively denoise the signal to synthesize high-quality samples (Cao et al., 2024), and this can be very effective in augmenting the available data. Because of their stability and sample diversity, diffusion models are rapidly gaining popularity in the neuroimaging community, particularly for generating high-resolution anatomical or functional brain images.

4.1.4 Normalizing flows

Normalizing flows offer a reversible mapping between complex and simple distributions using a sequence of invertible transformations (Wilms et al., 2022). Unlike VAEs or GANs, they allow for exact likelihood estimation and efficient sampling. This makes them ideal for density estimation and uncertainty quantification in brain imaging studies, particularly where interpretability and tractability are essential.

Gen AI is increasingly utilized to address limitations in traditional neuroimaging analysis by enabling data-driven synthesis, reconstruction, and latent feature discovery. As summarized in Table 10.1, VAEs are effective in capturing latent features and identifying subtle disease-related alterations in brain networks. GANs are widely used to generate realistic neuroimaging data, reconstruct damaged or missing regions, and augment datasets for underrepresented neurological conditions. Diffusion models enable the creation of high-fidelity brain images through iterative denoising, facilitating better simulation of brain activity. Normalizing flows allow for exact likelihood estimation and flexible, probabilistic modeling of complex neural connectivity patterns. These generative techniques collectively enhance precision and scalability in simulating and interpreting brain networks.

The rise of Gen AI in neuroimaging marks a paradigm shift from static analysis of brain data to dynamic modeling and synthesis. By enabling robust reconstructions,

Table 10.1 Generative AI and their applications in brain connectivity analysis.

Generative AI technique	Applications in brain connectivity and network analysis
Variational autoencoders (VAEs)	- Learning latent representations of brain networks - Dimensionality reduction for connectivity patterns - Identifying subtle disease-related changes
Generative adversarial networks (GANs)	- Synthesizing realistic brain imaging data - Reconstructing missing or corrupted connectivity maps - Data augmentation for rare neurological disorders
Diffusion models	- High-fidelity generation of structural and functional brain images - Denoising neuroimaging data - Simulating brain activity patterns
Normalizing flows	- Exact likelihood estimation for brain connectivity distributions - Uncertainty quantification in network analysis - Flexible modeling of complex connectivity dynamics

realistic simulations, and interpretable latent embeddings, these models are poised to advance both our understanding of the brain and our ability to diagnose and treat neurological conditions in a personalized manner. As computational power and algorithmic sophistication continue to grow, Gen AI will undoubtedly remain at the forefront of innovation in neuroimaging.

5. Applications in neuroimaging and neuroscience

Gen AI models are now being leveraged to tackle a variety of challenges in neuroscience, bringing innovations in both research and clinical settings.

5.1 Reconstructing missing or corrupted brain imaging data

Medical imaging is often plagued by noise, motion artifacts, or missing slices due to patient movement or acquisition errors. Generative models, particularly GANs and diffusion models, can effectively reconstruct these corrupted areas, restoring data integrity and usability for downstream analysis.

5.2 Synthesizing realistic structural or functional networks

Generative models can simulate biologically plausible brain networks, aiding in the creation of synthetic datasets for algorithm testing or augmentation of limited datasets. These models help overcome data scarcity, especially in rare neurological conditions.

5.3 Learning latent representations of brain connectivity

By embedding complex brain data into a lower-dimensional latent space, models such as VAEs can reveal hidden patterns and associations in structural or functional connectivity. These latent features are often more robust to noise and can be used for clustering, classification, and regression tasks in neuroscience research.

5.4 Predicting disease-specific network alterations

Gen AI can model deviations from normative connectivity patterns, enabling the identification of disease-specific biomarkers. These capabilities are particularly relevant in disorders such as Alzheimer's disease, schizophrenia, and autism spectrum disorders, where subtle changes in brain networks can serve as early indicators.

5.5 Enabling personalized medicine through patient-specific simulations

Perhaps one of the most promising applications is the use of generative models for simulating individual brain dynamics. By tailoring generative models to patient-specific data, it becomes possible to predict treatment outcomes, simulate disease progression, and design personalized intervention strategies.

6. Understanding brain organization and its alterations in disease

Gen AI techniques, particularly GANs and VAEs, offer transformative advantages in the analysis of brain connectivity and network neuroscience. These models are capable of learning complex distributions from high-dimensional neuroimaging data and generating plausible synthetic representations that align with biological constraints. The key advantages include.

6.1 Data augmentation and imputation

Generative models can synthesize realistic brain connectivity matrices or neuroimaging samples to augment small datasets or impute missing data, thereby addressing challenges related to limited sample sizes and noisy measurements. This improves model generalizability and statistical robustness (Yamaguchi et al., 2024).

6.2 Cross-modal synthesis and integration

Gen AI enables the fusion and translation of brain imaging data across modalities (e.g., fMRI, DTI, and EEG), facilitating a more detailed understanding of brain structure-function relationships. This supports multimodal integration where traditional methods often fall short (Mai et al., 2025).

6.3 Capturing intersubject variability

Deep generative models can learn latent representations that account for individual differences in brain anatomy and connectivity. This supports personalized brain network modeling and subgroup discovery in heterogeneous clinical populations (Ashburner & Klöppel, 2011).

6.4 Simulation of disease progression

Generative models can simulate alterations in brain networks associated with neurological and psychiatric conditions, providing insights into potential biomarkers, disease staging, and progression dynamics (Zhang et al., 2021).

6.5 Functional and effective connectivity modeling

By learning spatiotemporal patterns in time-series data, Gen AI can infer both functional dependencies and causal influences between brain regions, enabling the reconstruction of functional and effective connectivity with higher fidelity (Wang et al., 2025).

6.6 Noise reduction and superresolution

Techniques such as GAN-based superresolution can enhance the quality of neuroimaging data by reducing artifacts, increasing resolution, and improving SNRs, especially useful in low-quality or clinical imaging settings (Hirte et al., 2021).

6.7 Exploratory hypothesis generation

Gen AI facilitates the creation of counterfactual brain networks and "what-if" scenarios, enabling researchers to explore the impact of structural or functional disruptions, which is valuable for cognitive neuroscience and therapeutic planning (Laino et al., 2022).

Gen AI enhances our capacity to explore and interpret the complex architecture of the brain, enabling robust and scalable approaches for understanding how connectivity patterns are altered in disease and across development. Its integration into brain network analysis represents a promising paradigm shift toward more interpretable, generalizable, and biologically grounded neuroscience.

7. Future directions and challenges

7.1 Data quality and availability

The performance of generative models is highly dependent on the quality and quantity of training data. However, high-resolution, annotated neuroimaging datasets are often limited due to privacy concerns, cost, and clinical variability. Future work should focus on creating open-access, standardized datasets and improving data-sharing frameworks.

7.2 Model interpretability and clinical trust

Most Gen AI models function as black boxes, making it difficult for clinicians to interpret the generated results. Developing interpretable generative models and explainability frameworks is critical to gaining trust in clinical settings.

7.3 Personalized neuroimaging simulation

Future research should aim at advancing patient-specific simulations to support personalized diagnosis and treatment planning. This includes modeling individual variations in brain structure and connectivity using generative techniques.

7.4 Integration with multimodal data

Combining generative models with multimodal data (e.g., fMRI, EEG, genetics, and clinical records) could enhance the richness of simulated brain networks. This direction could uncover deeper biomarkers and cross-domain correlations.

7.5 Robustness and generalization

Ensuring the generalization of generative models across diverse populations and scanners remains a challenge. Future work should focus on training robust models that perform consistently across various clinical environments.

7.6 Ethical and regulatory concerns

The synthetic generation of brain data raises ethical questions, including consent, data ownership, and potential misuse. Establishing ethical guidelines and regulatory standards will be essential for safe deployment in healthcare.

7.7 Real-time and low-resource applications

Making generative models efficient enough for real-time analysis or deployment in low-resource clinical settings remains an open challenge, requiring advances in model optimization and hardware integration.

The application of Gen AI in brain connectivity and network analysis presents immense promise, but several key challenges must be addressed to fully realize its clinical and scientific potential. Critical future directions include improving data quality and availability, enhancing model interpretability to build clinical trust, and developing patient-specific simulations for personalized medicine. Integrating multimodal data can enrich generative outputs, while improving model robustness and generalization is essential for broader applicability. Ethical and regulatory considerations around synthetic neuroimaging data must also be carefully managed. Lastly, optimizing models for real-time and resource-constrained environments will be vital for practical deployment in clinical settings.

8. Conclusion

Gen AI represents a transformative advancement in the field of brain connectivity and network analysis, enabling a shift from traditional descriptive modeling to data-driven synthesis and simulation. Techniques such as VVAEs, GANs, diffusion models, and normalizing flows offer powerful tools to reconstruct, simulate, and interpret complex neural networks with unprecedented fidelity. These models are particularly valuable in revealing latent representations of brain organization, simulating disease-specific alterations, and supporting precision medicine through personalized neuroimaging. In the context of neurological and psychiatric disorders, generative models provide novel insights into how brain connectivity patterns are disrupted, facilitating earlier diagnosis, better stratification, and tailored therapeutic interventions. Furthermore, by enabling synthetic data generation, these models address challenges related to limited data availability, enhancing the robustness of downstream analysis and ML applications.

Despite their promise, generative models must overcome key challenges, including data scarcity, interpretability, computational demands, and ethical considerations. Continued interdisciplinary collaboration, along with advances in multimodal integration and model transparency, will be essential to fully realize the clinical and scientific potential of Gen AI in neuroscience. The integration of Gen AI into brain connectivity research marks a paradigm shift, offering not only a deeper understanding of brain function and dysfunction but also paving the way for innovative diagnostic and therapeutic strategies in the era of personalized and precision neuroscience.

References

Akarca, D., Vértes, P. E., Bullmore, E. T., Baker, K., Gathercole, S. E., Holmes, J., Kievit, R. A., Manly, T., Bathelt, J., Bennett, M., Bignardi, G., Bishop, S., Bottacin, E., Bridge, L., Brkic, D., Bryant, A., Butterfield, S., Byrne, E. M., Crickmore, G., … Astle, D. E. (2021). A generative network model of neurodevelopmental diversity in structural brain organization. *Nature Communications, 12*(1). https://doi.org/10.1038/s41467-021-24430-z

Ali, H., Rafiul Biswas, Md, Mohsen, F., Shah, U., Alamgir, A., Mousa, O., & Shah, Z. (2022). The role of generative adversarial networks in brain MRI: A scoping review. *Insights Into Imaging, 13*(1). https://doi.org/10.1186/s13244-022-01237-0

Ashburner, J., & Klöppel, S. (2011). Multivariate models of inter-subject anatomical variability. *NeuroImage, 56*(2), 422–439. https://doi.org/10.1016/j.neuroimage.2010.03.059

Barile, B., Marzullo, A., Stamile, C., Durand-Dubief, F., & Sappey-Marinier, D. (2021). Data augmentation using generative adversarial neural networks on brain structural connectivity in multiple sclerosis. *Computer Methods and Programs in Biomedicine, 206,* Article 106113. https://doi.org/10.1016/j.cmpb.2021.106113

Bengesi, S., El-Sayed, H., Sarker, M. K., Houkpati, Y., Irungu, J., & Oladunni, T. (2024). Advancements in generative AI: A comprehensive review of GANs, GPT, autoencoders, diffusion model, and transformers. *IEEE Access, 12,* 69812–69837. https://doi.org/10.1109/ACCESS.2024.3397775

Betzel, R. F., & Bassett, D. S. (2017). Generative models for network neuroscience: Prospects and promise. *Journal of the Royal Society Interface, 14*(136). https://doi.org/10.1098/rsif.2017.0623

Bi, Y., Abrol, A., Jia, S., Sui, J., & Calhoun, V. D. (2024). Gray matters: ViT-GAN framework for identifying schizophrenia biomarkers linking structural MRI and functional network connectivity. *NeuroImage, 297,* Article 120674. https://doi.org/10.1016/j.neuroimage.2024.120674

Cao, H., Tan, C., Gao, Z., Xu, Y., Chen, G., Heng, P. A., & Li, S. Z. (2024). A survey on generative diffusion models. *IEEE Transactions on Knowledge and Data Engineering, 36*(7), 2814–2830. https://doi.org/10.1109/TKDE.2024.3361474

Catherine Storey, V. (2025). Knowledge management in a world of generative AI: Impact and implications. *ACM Transactions on Management Information Systems.* https://doi.org/10.1145/3719209

Friston, K. J., & Price, C. J. (2001). Dynamic representations and generative models of brain function. *Brain Research Bulletin, 54*(3), 275–285. https://doi.org/10.1016/S0361-9230(00)00436-6

Gong, C., Jing, C., Chen, X., Pun, C. M., Huang, G., Saha, A., Nieuwoudt, M., Li, H. X., Hu, Y., & Wang, S. (2023). Generative AI for brain image computing and brain network computing: A review. *Frontiers in Neuroscience, 17.* https://doi.org/10.3389/fnins.2023.1203104

He, R., Sarwal, V., Qiu, X., Zhuang, Y., Zhang, Le, Liu, Y., & Chiang, J. (2025). Generative AI models in time-varying biomedical data: Scoping review. *Journal of Medical Internet Research, 27,* Article e59792. https://doi.org/10.2196/59792

Hirte, A. U., Platscher, M., Joyce, T., Heit, J. J., Tranvinh, E., & Federau, C. (2021). Realistic generation of diffusion-weighted magnetic resonance brain images with deep generative models. *Magnetic Resonance Imaging, 81,* 60–66. https://doi.org/10.1016/j.mri.2021.06.001

Isallari, M., & Rekik, I. (2021). Brain graph super-resolution using adversarial graph neural network with application to functional brain connectivity. *Medical Image Analysis, 71,* Article 102084. https://doi.org/10.1016/j.media.2021.102084

Jain, N. S., & Patient Care, R. (2024). The impact of ChatGPT and generative AI in healthcare. *International Journal of Scientific and Research Publications, 14*(11), 24–38. https://doi.org/10.29322/ijsrp.14.11.2024.p15506

Khalighi, S., Reddy, K., Midya, A., Pandav, K. B., Madabhushi, A., & Abedalthagafi, M. (2024). Artificial intelligence in neuro-oncology: Advances and challenges in brain tumor diagnosis, prognosis, and precision treatment. *Npj Precision Oncology, 8*(1). https://doi.org/10.1038/s41698-024-00575-0

Kundu, S., Sair, H., Sherr, E. H., Mukherjee, P., & Rohde, G. K. (2024). Discovering the gene-brain-behavior link in autism via generative machine learning. *Science Advances, 10*(24). https://doi.org/10.1126/sciadv.adl5307

Laino, M. E., Cancian, P., Politi, L. S., Della Porta, M. G., Saba, L., & Savevski, V. (2022). Generative adversarial networks in brain imaging: A narrative review. *Journal of Imaging, 8*(4). https://doi.org/10.3390/jimaging8040083

Li, Y., Yang, Y., Chen, Y., Ye, C., & Ma, T. (2025). Self-supervised learning to unveil brain dysfunctional signatures in neuropsychiatric disorders: Methods and applications. *Health Data Science, 5,* Article 0282.

Mai, W., Zhang, J., Fang, P., & Zhang, Z. (2025). Brain-conditional multimodal synthesis: A survey and taxonomy. *IEEE Transactions on Artificial Intelligence, 6*(5), 1080–1099. https://doi.org/10.1109/tai.2024.3516698

Marzi, S. J., Schilder, B. M., Nott, A., Frigerio, C. S., Willaime-Morawek, S., Bucholc, M., Hanger, D. P., James, C., Lewis, P. A., Lourida, I., Noble, W., Rodriguez-Algarra, F., Sharif, J. A., Tsalenchuk, M., Winchester, L. M., Yaman, Ü., Yao, Z., Ranson, J. M., & Llewellyn, D. J. (2023). Artificial intelligence for neurodegenerative experimental models. *Alzheimer's and Dementia, 19*(12), 5970–5987. https://doi.org/10.1002/alz.13479

Prasanth, L. L., & Uma, E. (2025). Revolutionizing neurostimulator care: Enhancing remote health monitoring through SDN-cloud networks. *Telecommunication Systems, 88*(1). https://doi.org/10.1007/s11235-024-01255-x

Sai, S., Gaur, A., Sai, R., Chamola, V., Guizani, M., & Rodrigues, J. J. P. C. (2024). Generative AI for transformative healthcare: A comprehensive study of emerging models, applications, case studies, and limitations. *IEEE Access, 12*, 31078–31106. https://doi.org/10.1109/access.2024.3367715

Shaheen, H., & Melnik, R. (2025). Brain network dynamics and multiscale modeling of neurodegenerative disorders: A review. *IEEE Access, 13*, 33074–33100. https://doi.org/10.1109/access.2025.3542634

Shaul, R., David, I., Shitrit, O., & Riklin Raviv, T. (2020). Subsampled brain MRI reconstruction by generative adversarial neural networks. *Medical Image Analysis, 65*. https://doi.org/10.1016/j.media.2020.101747

Sengar, S. S., Bin Hasan, A., Kumar, S., & Carroll, F. (2025). Generative artificial intelligence: A systematic review and applications. *Multimedia Tools and Applications, 84*(21), 23661–23700. https://doi.org/10.1007/s11042-024-20016-1

Taniguchi, T., Yamakawa, H., Nagai, T., Doya, K., Sakagami, M., Suzuki, M., Nakamura, T., & Taniguchi, A. (2022). A whole brain probabilistic generative model: Toward realizing cognitive architectures for developmental robots. *Neural Networks, 150*, 293–312. https://doi.org/10.1016/j.neunet.2022.02.026

Vatansever, S., Schlessinger, A., Wacker, D., Kaniskan, H.Ü., Jin, J., Zhou, M. M., & Zhang, B. (2021). Artificial intelligence and machine learning-aided drug discovery in central nervous system diseases: State-of-the-arts and future directions. *Medicinal Research Reviews, 41*(3), 1427–1473. https://doi.org/10.1002/med.21764

Walter, M., Alizadeh, S., Jamalabadi, H., Lueken, U., Dannlowski, U., Walter, H., Olbrich, S., Colic, L., Kambeitz, J., Koutsouleris, N., Hahn, T., & Dwyer, D. B. (2019). Translational machine learning for psychiatric neuroimaging. *Progress in Neuro-Psychopharmacology and Biological Psychiatry, 91*, 113–121. https://doi.org/10.1016/j.pnpbp.2018.09.014

Wang, R., Bashyam, V., Yang, Z., Yu, F., Tassopoulou, V., Chintapalli, S. S., Skampardoni, I., Sreepada, L. P., Sahoo, D., Nikita, K., Abdulkadir, A., Wen, J., & Davatzikos, C. (2023). Applications of generative adversarial networks in neuroimaging and clinical neuroscience. *NeuroImage, 269*. https://doi.org/10.1016/j.neuroimage.2023.119898

Wang, R., & Chen, Z. S. (2025). Large-scale foundation models and generative AI for BigData neuroscience. *Neuroscience Research, 215*, 3–14. https://doi.org/10.1016/j.neures.2024.06.003

Wang, S., Zhou, T., Shen, Y., Li, Y., Huang, G., & Hu, Y. (2025). Generative AI enables EEG super-resolution via spatio-temporal adaptive diffusion learning. *IEEE Transactions on Consumer Electronics, 71*(1), 1034–1045. https://doi.org/10.1109/tce.2025.3528438

Wilms, M., Bannister, J. J., Mouches, P., MacDonald, M. E., Rajashekar, D., Langner, S., & Forkert, N. D. (2022). Invertible modeling of bidirectional relationships in neuroimaging with normalizing flows: Application to brain aging. *IEEE Transactions on Medical Imaging, 41*(9), 2331–2347. https://doi.org/10.1109/TMI.2022.3161947

Wright, S. N., & Anticevic, A. (2024). Generative AI for precision neuroimaging biomarker development in psychiatry. *Psychiatry Research, 339*. https://doi.org/10.1016/j.psychres.2024.115955

Yamaguchi, H., Sugihara, G., Shimizu, M., & Yamashita, Y. (2024). Generative artificial intelligence model for simulating structural brain changes in schizophrenia. *Frontiers in Psychiatry, 15*. https://doi.org/10.3389/fpsyt.2024.1437075

Yao, Z., Wang, Z., Xie, W., Zhan, Y., Wu, X., Dai, Y., Pei, Y., Wang, Z., & Zhang, G. (2024). Applications of generative artificial intelligence in brain MRI image analysis for brain disease diagnosis. *Neuropharmacology and Therapy, 1*(2). https://doi.org/10.15212/npt-2024-0007

Zhang, X., Braun, U., Harneit, A., Zang, Z., Geiger, L. S., Betzel, R. F., Chen, J., Schweiger, J. I., Schwarz, K., Reinwald, J. R., Fritze, S., Witt, S., Rietschel, M., Nöthen, M. M., Degenhardt, F., Schwarz, E., Hirjak, D., Meyer-Lindenberg, A., Bassett, D. S., & Tost, H. (2021). Generative network models of altered structural brain connectivity in schizophrenia. *NeuroImage, 225*. https://doi.org/10.1016/j.neuroimage.2020.117510

Zhang, L., Wang, L., & Zhu, D. (2022). Predicting brain structural network using functional connectivity. *Medical Image Analysis, 79*, Article 102463. https://doi.org/10.1016/j.media.2022.102463

Zhao, J., Huang, J., Zhi, D., Yan, W., Ma, X., Yang, X., Li, X., Ke, Q., Jiang, T., Calhoun, V. D., & Sui, J. (2020). Functional network connectivity (FNC)-based generative adversarial network (GAN) and its applications in classification of mental disorders. *Journal of Neuroscience Methods, 341*, Article 108756. https://doi.org/10.1016/j.jneumeth.2020.108756

Zhou, T., Ding, C., Jing, C., Liu, F., Hung, K., Pham, H., Mahmud, M., Lyu, Z., Qiao, S., Wang, S., & Tsang, K. F. (2025). BG-GAN: Generative AI enable representing brain structure-function connections for Alzheimer's disease. *IEEE Transactions on Consumer Electronics*. https://doi.org/10.1109/TCE.2025.3543943

Zong, Y., Jing, C., Chan, J. H., & Wang, S. (2024). BrainNetDiff: Generative AI empowers brain network construction via multimodal diffusion. In *Proceedings - International symposium on biomedical imaging*. https://doi.org/10.1109/ISBI56570.2024.10635395. http://ieeexplore.ieee.org/xpl/conferences.jsp

Zuo, Q., Chen, L., Shen, Y., Kwok-Po Ng, M., Lei, B., & Wang, S. (2025). BDHT: Generative AI enables causality analysis for mild cognitive impairment. *IEEE Transactions on Automation Science and Engineering, 22*, 5601–5613. https://doi.org/10.1109/TASE.2024.3425949

Zuo, Q., Di, Y., Ye, C., Shi, B., Chen, J., Wei, H., Li, R., & Lei, B. (2025). Bidirectional generative diffusion model with cascaded symmetric attention for brain connectivity-to-connectivity translation. *Biomedical Signal Processing and Control, 108*, Article 107900. https://doi.org/10.1016/j.bspc.2025.107900

Zuo, Q., Tian, H., Li, R., Guo, J., Hu, J., Tang, L., Di, Y., & Kong, H. (2023). Hemisphere-separated cross-connectome aggregating learning via VAE-GAN for brain structural connectivity synthesis. *IEEE Access, 11*, 48493–48505. https://doi.org/10.1109/ACCESS.2023.3276989

CHAPTER 11

Generative AI for simulating brain development and aging: Modeling normal and pathological processes

Ajay Sharma and Arun Malik
Department of Computer Science and Engineering, Lovely Professional University, Phagwara, Punjab, India

1. Introduction

The human brain is a distinctly dynamic and complicated system, present process profound adjustments from embryonic development through youth, adulthood, and into old age. Traditionally, the knowledge of this complex lifecycle has been shaped via methodologies, including neuroimaging, histopathological analyses, electrophysiology, and cognitive outcomes. While these techniques offer valuable insights, they are often confined by restricted temporal decisions, small pattern sizes, and ethical concerns related to information acquisition. In recent years, the emergence of generative synthetic artificial intelligence (AI) has opened transformative avenues in neuroscience research, particularly for simulating and providing information on the temporal evolution of the mind in each healthy and pathological state. The generative AI, comprising models that include variational autoencoders (VAEs), generative adversarial networks (GANs), and transformer-based architectures, permits the synthesis and modeling of high-dimensional data. These fashions are especially applicable to reconstruct brain pictures, simulate disorder trajectories, and generate artificial datasets that could complement scientific data. In the context of brain improvement and growing old, generative AI has the potential to provide granular, individualized insights into neurobiological mechanisms, from synaptic pruning in childhood to neurodegeneration in Alzheimer's disease (AD). This section works into how generative AI is being employed to model brain function and getting old, highlighting its theoretical foundations, current packages, and its software in simulating each ordinary and pathological trajectory. The creator has also explored the mathematical frameworks underlying these models and mentioned how artificial facts generated through AI can mitigate problems, which include fact scarcity and bias in neuroscience (Amoroso et al., 2018; Braak & Braak, 1998; Mei et al., 2021).

The exploration of brain development and aging has traditionally been anchored in methodologies that include neuroimaging, histological examinations, and longitudinal medical observations. These traditional procedures have supplied beneficial insights into structural and functional brain modifications throughout the human lifespan.

Generative Artificial Intelligence for Neuroimaging
ISBN 978-0-443-45010-5
https://doi.org/10.1016/B978-0-443-45010-5.00007-1

They are frequently constrained by factors such as restrained temporal decision, excessive statistics acquisition fees, and challenges in shooting dynamic or rare pathological activities. In recent years, the arrival of generative AI has added a transformative paradigm in neuroscience, providing powerful computational tools to simulate and understand the intricacies of mind maturation, cognitive decline, and neurodegenerative diseases. Generative AI, especially via deep learning frameworks such as VAEs and GANs, enables the advent and manipulation of complex, high-dimensional neural facts. These fashions can study the underlying styles of mind development and grow old from massive-scale datasets, and use this discovered illustration to generate new, realistic record samples. Such abilities allow researchers to honestly reconstruct age-related changes in the brain, expect ailment development, and predict the onset of neurodegenerative issues such as AD and Parkinson's disease (PD) (Fig. 11.1). The era of synthetic data via AI serves a crucial position in addressing boundaries of scientific research, together with information scarcity, privacy concerns, and population biases. Synthetic neuroimaging datasets may be used to teach and validate gadgets, gaining knowledge of algorithms, and consequently improving the robustness and generalizability of neuroscience studies. The modern work focuses on those rising applications, showcasing how generative AI is reshaping our method to analyzing the mind, now not only by augmenting empirical data but also by allowing simulations that display the trajectories of both healthy and pathological brain states (Chaddad et al., 2018; Ghaffar Nia et al., 2023; Helaly et al., 2022).

1.1 Understanding brain development and aging

1.1.1 Normal brain maturation from infancy to adulthood

The development of the human mind from infancy through adulthood is a highly orchestrated procedure regarding sequential and overlapping levels, which includes neuronal proliferation, synaptogenesis, synaptic pruning, and myelination. During the prenatal and early postnatal intervals, there may be an exponential boom in synaptic connections, organizing the foundational neural networks for sensory, motor, and cognitive functions. As improvement progresses, the mind undergoes widespread synaptic pruning, a process wherein weaker or redundant synapses are eliminated to improve the efficiency and specificity of neural communication. This pruning is especially outstanding in the course of youth and is vital for optimizing cognitive functions and behavioral regulation. One of the longest-lasting developmental phases occur inside the prefrontal cortex (PFC), a region associated with higher-order executive capabilities, including reasoning, decision-making, and impulse control. Structural maturation of the PFC continues into the third decade of life, underscoring the protracted timeline of human neurodevelopment.

The myelination, the procedure by which axons are insulated to facilitate faster neural transmission, regularly increases at some point in this era, improving the purposeful connectivity between brain areas. Generative AI fashions trained on excessive-decision, longitudinal neuroimaging datasets provide novel possibilities to simulate those developmental trajectories (Guleria et al., n.d.; Gupta et al., 2022). These fashions can seize and reproduce diffused, time-dependent changes in mind structure and function, making an allowance for the prediction of destiny cognitive consequences primarily based on early neural markers. Furthermore, by using environmental and genetic information, generative models can simulate the effect of external factors, including socioeconomic status, training, and youth pressure, on neuroplasticity and maturation. This facilitates a deeper expertise of how reviews form the developing mind and supports the improvement of customized interventions for at-risk populations (Brosch et al., 2013; Islam & Zhang, 2018).

1.1.2 Structural and functional changes in aging

The aging mind features a variety of anatomical and physiological modifications that affect cognitive and behavioral performance. Structurally, aging is associated with volumetric reductions in both gray and white matter, with prominent atrophy located within the hippocampus, PFC, and cerebellum. These adjustments are followed by a decline in synaptic density, dendritic complexity, and neurotransmitter performance, all of which contribute to diminished neuroplasticity and a gradual lack of cognitive features. Functionally, growing older is characterized by reduced processing pace, impaired executive function, and less efficient attention allocation. While some cognitive domain names, consisting of vocabulary and preferred information, continue to be stable or even enhance with age, others, especially the ones related to executive functioning and episodic reminiscence, show measurable decline. Compensatory mechanisms, consisting of multiplied bilateral mind activation and recruitment of changed neural pathways, are often observed in healthy older adults and are ideal to support cognitive resilience. AI-based techniques, especially generative fashions, allow the simulation of those age-associated changes by using data from multimodal datasets that encompass neuroimaging, behavioral assessments, and genetic profiles. These models can tune age-associated differences over time and perceive early deviations from normative aging styles. By reconstructing patterns of cortical thinning, ventricular enlargement, and connectivity disruptions, generative AI affords an effective framework for analyzing the heterogeneity of aging processes. AI fashions can be used in the discovery of biomarkers that distinguish between healthy and pathological aging, allowing early identification of people at risk for neurodegenerative illnesses. These insights are vital for the design of preventive strategies and the assessment of interventions aimed at keeping cognitive characteristics in older adults (Baur et al., 2019; Bernal et al., 2019).

1.1.3 Pathological processes in neurological disorders

Neurological and neuropsychiatric disorders frequently disrupt the everyday developmental and aging pathways of the mind, leading to structural and practical abnormalities that substantially impair cognitive and behavioral capabilities. Neurodegenerative diseases, which include AD and PD, represent outstanding examples of pathological aging. AD is often characterized by the buildup of amyloid-beta plaques, neurofibrillary tangles formed through hyperphosphorylated tau proteins, and large synaptic and neuronal loss (Lankadasu et al., 1 C E.a, 1 C E b; Sharma et al., 2021). These pathological features result in innovative memory impairment, language disorder, and govt deficits. In contrast, Parkinson's ailment is associated with the degeneration of dopaminergic neurons inside the substantia nigra, resulting in motor signs such as tremors, pressure, and bradykinesia, along with non-motor symptoms including cognitive decline and temper disturbances. Both illnesses showcase wonderful spatiotemporal patterns of progression that may be difficult to seize the use of traditional analytical techniques. Generative AI models have proven a high-quality promise in simulating disease progression through learning from comprehensive medical, genetic, and neuroimaging datasets. These fashions can expect the onset and trajectory of signs, perceive excessive-risk people, and simulate responses to healing interventions. For instance, AI can generate artificial magnetic resonance imaging (MRI) scans at diverse ailment ranges, supplying a visual and quantitative understanding of disorder evolution. This functionality is especially useful for medical trials, wherein synthetic data can increase constrained affected person samples and enhance the robustness of outcome measures. Beyond neurodegeneration, generative models were used to explore neurodevelopmental disorders, together with autism spectrum disorder (ASD). Individuals with ASD regularly show early brain overgrowth, altered white matter development, and unusual connectivity patterns. By modeling those functions, generative AI contributes to the understanding of the neurobiological substrates of ASD and helps the improvement of diagnostic gear and focused interventions. The potential of generative models to simulate each regular and pathological mind state gives a powerful supplement to empirical research, allowing a more complete knowledge of the mechanisms underlying brain fitness and disorder (Hinrichs et al., 2011; Jack et al., 2010; Nordberg et al., 2010).

1.2 Generative AI: Foundations and techniques

1.2.1 Introduction to generative AI

Generative AI represents a transformative class of deep learning methodologies capable of creating new data samples that carefully resemble genuine, actual-world observations. In the context of neuroscience, generative AI has emerged as a pivotal device for addressing long-standing challenges associated with data shortage, patient privacy, and the complexity of modeling dynamic organic systems. These models, which consist of architectures consisting of GANs, VAEs, and extra current diffusion models and transformer-based models, are especially properly appropriate for simulating the elaborate structural and functional styles found in neuroimaging data. Unlike traditional AI fashions, which are commonly designed for tasks that include classification, regression, or segmentation, generative fashions are uniquely placed to synthesize new record samples, making them beneficial for several applications, which include ailment trajectory modeling, augmentation of scientific datasets, and the imputation of missing or corrupted imaging records. In particular, they provide a noninvasive and highly powerful method to simulate the development of neurological issues, along with AD, PD, and a couple of sclerosis, by using learning from high-dimensional datasets and producing plausible destiny states of mind pathology (Kumar & Sharma, 2023; Sharma et al., 2024; Sharma & Kumar, 2022). The generative AI enables the reconstruction of incomplete or artifact-weighted-down scans, thereby improving the reliability of scientific diagnostics and permitting more sturdy downstream analyses. These models additionally guide the speculation era and testing through allowing researchers to genuinely manipulate brain variables (e.g., age, genetic risk, and disease status) and take a look at their artificial effects in silico. This capacity to emulate complicated neurobiological strategies presents a unique computational lens through which neuroscientists can explore the mechanisms of mind development, aging, and pathology. The use of artificial information generated via AI mitigates problems of patient confidentiality and regulatory restrictions, allowing for wider sharing and collaboration across research institutions. As these technologies continue to evolve, they preserve a giant promise for reshaping how facts are generated, interpreted, and applied in neuroscience, ultimately accelerating the improvement of customized and precision-based neurological care (Bengio, 2009; Hinton et al., 2006; Jack et al., 2008; Nordberg et al., 2010).

- **Key Techniques Used in Neuroscience**
- **Variational Autoencoders (VAEs):** VAEs (Fig. 11.1) are a class of generative models rooted in probabilistic reasoning that have ended up being an increasingly valued tool in neuroscience for modeling high-dimensional brain imaging information. VAEs are especially adept at mastering compressed, low-dimensional representations called "latent variables" from complex inputs such as structural MRI, fMRI, or diffusion tensor imaging (DTI). These latent variables seize the underlying generative methods of the brain and are instrumental in responsibilities that include reconstructing missing data, simulating neurodevelopmental patterns, and forecasting neurodegenerative development. The mathematical basis of a VAE lies in optimizing the "Evidence Lower Bound (ELBO)," which balances two goals: maximizing the fidelity of reconstructed facts while ensuring the latent space stays regularized to observe a known distribution (usually a multivariate Gaussian). This method permits the model to analyze a significant latent area from which new, possible samples can be drawn. Conceptually, a VAE includes neural networks: an encoder that maps brain images to a latent area distribution, and a decoder that reconstructs brain images from this latent representation. Importantly, the probabilistic nature of VAEs allows for the quantification of uncertainty, a valuable feature in medical applications where ambiguity in diagnosis or analysis is not unusual. In modeling mind

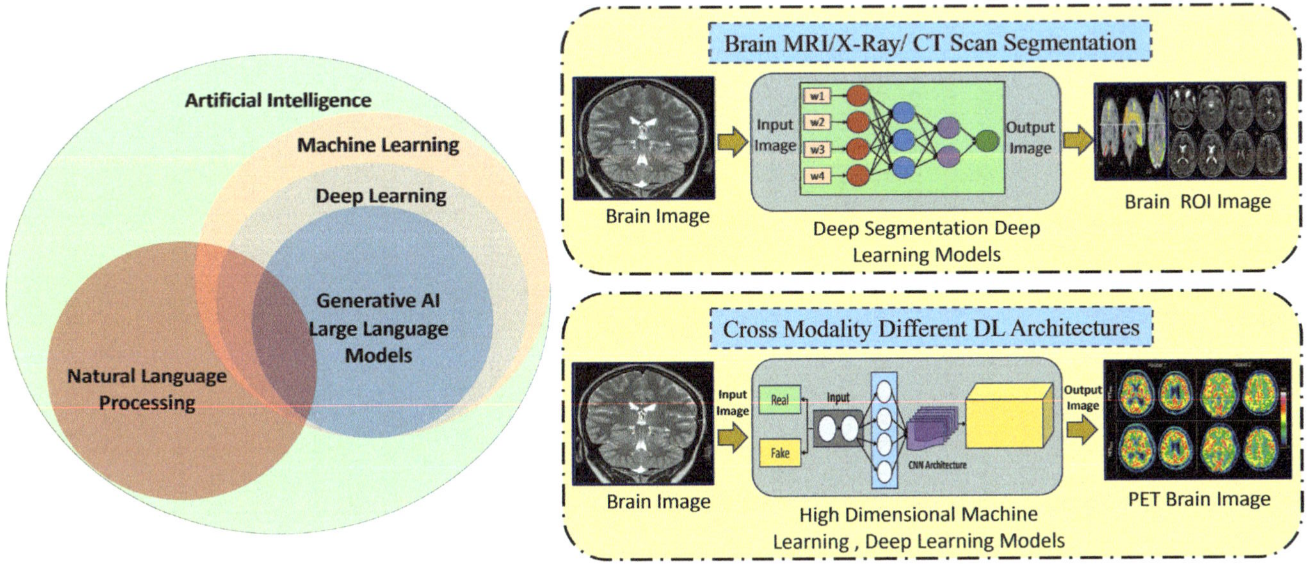

Figure 11.1 The artificial intelligence and deep learning-based architecture for MRI/X-Ray/CT scan image segmentation.

improvement, VAEs were used to capture the nonlinear trajectories of neuroanatomical maturation, together with cortical thinning, white matter growth, and changes in purposeful connectivity. By training on longitudinal imaging datasets, including the adolescent brain cognitive development (ABCD) and the Human Connectome Project, VAEs can identify latent factors associated with age-related neural differences. These fashions are capable of producing artificial MRI scans at one-of-a-kind developmental tiers, thereby facilitating the examination of early biomarkers of extraordinary development, consisting of conditions such as ASD or ADHD (Varoquaux & Cheplygina, 2022; Zhao et al., 2019).

For getting old research, VAEs are employed to map the continuum from wholesome aging to pathological states, which include slight cognitive impairment (MCI) or AD. The capacity to interpolate among latent representations of healthy and diseased brains permits visualization of disease development in a patient-specific way. Moreover, VAEs can assist with perceiving early deviations from normative aging patterns by using learning from numerous populations, providing new insights into personalized neuroscience. Another critical utility lies in anomaly detection. By getting to know the distribution of healthful mind structures, VAEs can flag deviations indicative of neuropathology, even at preclinical levels. This supports early prognosis and intervention techniques. Furthermore, in situations of incomplete or noisy records, a commonplace challenge in neuroimaging VAEs can efficiently reconstruct plausible brain pictures, ensuring greater robust downstream evaluation (Kingma & Welling, 2014, 2019; Kwon et al., 2019; Laino et al., 2022; Rosca et al., 2017; Varoquaux & Cheplygina, 2022; Zhao et al., 2019).

VAEs (Table 11.1) constitute a powerful tool in neuroscience for simulating the spatiotemporal dynamics of the brain throughout the lifespan. Their probabilistic nature, capacity for unsupervised getting to know, and potential to generate biologically viable synthetic information make them vital in advancing our knowledge of normal and pathological brain functions.

- **Generative Adversarial Networks (GANs):** GANs (Fig. 11.1) have emerged as an innovative deep getting to know framework capable of synthesizing noticeably practical records, particularly in the visual domain, such as neuroimaging. In neuroscience, GANs are increasingly utilized for generating synthetic brain scans, modeling neurodegenerative disease trajectories, and augmenting datasets for robust machine learning education. The center architecture of a GAN entails neural networks: a "Generator," which creates synthetic information, and a "Discriminator," which evaluates whether or not a pattern is actual (from schooling records) or fake (from the generator). These networks are educated adversarial, with the generator enhancing its output fine over the years to idiot the discriminator. The education objective of GANs is primarily based on a minimax game, where the generator goals to limit the discriminator's potential to distinguish fake from actual, even as the discriminator seeks to maximize it. This opposed setup encourages the generator

Table 11.1 Table showing the variational autoencoder with mathematical formulation and application.

Variational autoencoders (VAEs)	Mathematical formulation	Application in neuroscience
VAEs are probabilistic generative models that encode input data (e.g., brain MRI scans) into a lower-dimensional latent space and then reconstruct it back, while also learning a distribution over this latent space.	- $x$denote the input neuroimaging data (e.g., brain scan), - $z$denote the latent variable, - $q\_\varphi(z\|x)$ be the encoder (approximate posterior), - $p\_\theta(x\|z)$ be the decoder (likelihood), - $p(z)$ be the prior (usually standard normal N(0, I)). Und (ELBO) on the marginal likelihood log $p(x)$: $L_{\text{VAE}\ (\theta,\varphi;x)} = E_{\{q_\varphi(z\|x)\ \}}[\log p_\theta(x\|z)]$ $- D_{KL}\left(q_{(\varphi(z\|x)\|p(z)\ )}\ \right)$	Application in neuroscience: - x Can be structural or functional brain images. - Latent representations $z$capture Developmental or degenerative trends. - Sampling from $p(z)$ allows the generation of synthetic but realistic brain scans.

to study the true distribution of the statistics. In neuroimaging programs, GANs have been used to generate artificial structural MRIs, fMRIs, or even connectome information, which might be visually and statistically indistinguishable from actual scans. Such synthetic statistics are valuable for schooling diagnostic algorithms, particularly within the context of a statistics shortage or imbalance throughout age groups and disease categories (Lankadasu et al., 2024; Satti et al., 2024).

One of the most compelling uses of GANs in neuroscience is the simulation of ailment development. For example, GANs have been trained on cross-sectional datasets of AD patients to generate artificial brain scans representing intermediate stages of sickness. Conditional GANs (GANs), which include auxiliary facts, such as age or clinical analysis, allow the best manipulation over the era system. This capability is important for knowledge of the trajectory of neurodegeneration and evaluating hypothetical interventions in silico. GANs additionally help the challenge of mind photo awesome-decision, wherein low-resolution or artifact-weighted down scans are enhanced the use of the discovered priorities from amazing pics. This is specifically beneficial in pediatric and geriatric populations, wherein motion artifacts or scanner obstacles often degrade images exceptionally. Cycle GANs and style-transfer GANs similarly allow an area model, allowing the transformation of pixels throughout modalities (e.g., T1-weighted to T2-weighted MRI) or between exceptional imaging websites, consequently lowering variability and harmonizing information across studies. GAN-generated information is used to mitigate privacy concerns. Since artificial snapshots do not correspond to real people, they can be freely shared among researchers, facilitating broader collaboration and replication studies. This is valuable in rare instances wherein the affected person's information is confined and sensitive. Despite their big ability, GANs come with challenges, including schooling instability and the risk of mode disintegration, where the generator produces a limited form of outputs. Recent improvements, consisting of Wasserstein GANs and modern growing GANs, deal with these boundaries and have proven improved performance in neuroimaging responsibilities (Ali et al., 2022; Hinton & Salakhutdinov, 2006; Laino et al., 2022; Lin et al., 2018; Mori et al., 2013).

The GANs (Table 11.2) provide an adaptable and mighty framework for simulating complex brain changes across improvement and aging. Their ability to generate sensible and numerous neuroimaging statistics positions them as a cornerstone within the subsequent technology of computational neuroscience tools.

- **Transformer-Based Models:** Transformer-based models (Table 12.3) (Fig. 11.1), at the beginning designed for natural language processing (NLP), are rapidly gaining interest in neuroscience for their ability to model sequential and high-dimensional data. At their core, transformers rely upon self-interest mechanisms that dynamically weigh the importance of various input additives, allowing the model to capture long-term dependencies and complicated temporal styles. These homes make transformers mainly applicable for analyzing longitudinal brain records, in which

Table 11.2 Table showing the generative adversarial network, mathematical formulation and the GAN's application.

Generative adversarial networks (GANs)	Mathematical formulation	Application in neuroscience
GANs use a game-theoretic approach between two neural networks: - Generator $G(z;\theta\_g)$ Generates fake brain data from latent input $z \sim p(z)$, - Discriminator $D(x;\theta\_d)$ Classifies real versus synthetic data	The GAN training objective is a minimax optimization problem: $\min_G \max_D E_{\{x\sim p_{\text{data}(x)}\}}[\log D(x)] + E_{\{z\sim p(z)\}}[\log(1-D(G(z)))]$	- G synthesizes realistic *MRIs*,*fMRI* Activation maps, or disease-state transitions - D Ensures realism and fidelity of generated scans - Conditional *GANs* can be used: $G(z\|y)$ where y is a label (e.g., age and diagnosis)

temporal correlations and interlocal interactions are essential. The application of transformers in neuroscience has been facilitated with the aid of their scalability and versatility in managing variable-period inputs. For example, longitudinal neuroimaging facts, cognitive scores, and scientific assessments amassed throughout more than one timepoints can be represented as sequences fed into a transformer model. The version can then learn how to predict destiny brain states, sickness progression, or treatment response. Unlike recurrent neural networks (RNNs) or long-term quick-term period reminiscence (LSTM) models, transformers do not suffer from vanishing gradient problems and may be effectively parallelized, making them more suitable for large-scale neurodevelopmental and getting old studies. Transformer models, along with bidirectional encoder representations from transformers (BERT) and vision transformers (ViTs), were adapted to system spatial-temporal brain information, along with 3D voxel-wise functions from MRI scans. These models may be best tuned to pick out developmental trajectories, locate early markers of cognitive decline, or classify neuropsychiatric issues. Specifically, transformers have proven promise in predicting brain age, a critical biomarker in getting old studies, by learning diffused anatomical patterns throughout the cortex and subcortical systems (Cong et al., 2024a, 2024b; Hayat & Aramvith, 2024).

- A growing area of hobby is using transformer encoders and decoders for multimodal integration. For example, combining structural MRI, diffusion imaging, EEG time-collection, and behavioral statistics permits a holistic view of mind features. Cross-attention mechanisms in transformers enable the fusion of these modalities, crucial to the progression of prediction of neurological outcomes and a more comprehensive understanding of the brain's development. From a generative attitude, transformer decoders have been utilized in autoregressive generative responsibilities

Table 11.3 Table showing the transformer-based models, their mathematical expression, and application in neuroscience.

Transformer-based models	Mathematical formulation	Application in neuroscience
Transformers model sequential data using self-attention mechanisms, making them suitable for longitudinal or multimodal neuroimaging data	Given a sequence of inputs $X = [x1, x2, ., xT]$, The scaled Dot-Product attention for query Q, key K, and value V Matrices are computed as: Attention $(Q, K, V) = \text{softmax}\left(\frac{QK^T}{} sqrt(d_k)\right) V$ where: - $Q = XW^Q, K = XW^K, V = XW^V$ - W^Q, W^K, W^V are learned projection matrices, - d_k is the dimensionality of the key vectors (for normalization) Transformer encoder block (simplified): Output $=$ LayerNo $+$ FeedForward(X)	- Inputs xt Time-series data (e.g., sequential MRI volumes, EEG signals, cognitive scores) - Predict future brain states, diagnose progressive disorders - Self-attention captures temporal or spatial dependencies between brain regions or timepoints. - Cross-attention in multi-modal transformers integrates MRI + genetic + behavioral data.

where future brain scans are synthesized primarily based on preceding imaging records. This supports situation modeling, which includes estimating the neuroanatomical impact of a genetic mutation or simulating mind modifications under distinct environmental conditions. Moreover, transformers educated on large, numerous datasets can generalize properly throughout populations, increasing the equity and inclusivity of AI-pushed neuroscience (Child et al., 2019; Cong et al., 2024a; Ghorbel et al., 2023; Hayat & Aramvith, 2024).

The transformer-based models preserve monstrous capability in neuroscience for modeling temporal dynamics, integrating heterogeneous information, and generating predictive insights into brain development and aging. Their enduring model from NLP to neuroinformatics signifies a first-rate breakthrough in understanding, the evolving architecture and features of the human mind.

1.2.2 Data sources for training AI models

Generative AI requires large datasets to teach robust fashions. These datasets consist of multimodal neuroimaging (MRI, fMRI, PET), electrophysiological recordings, genetic records, and scientific assessments. The integration of various information resources enhances the accuracy and generalizability of AI-pushed simulations of brain characteristics and pathology. Generative AI models in neuroscience demand rich, diverse, and large-scale datasets to ensure correct simulation of brain improvement, aging, and pathology. These datasets generally include neuroimaging modalities (e.g., MRI, fMRI, and PET), electrophysiological recordings (e.g., electroencephalogram (or electroencephalography) (EEG) and magnetoencephalography (MEG), genetic information (e.g., single nucleotide polymorphism (SNPs) and gene expression profiles), and comprehensive clinical checks. Integration of such multimodal information complements the version's capability to capture complex interactions throughout biological systems and enhance predictions of cognitive outcomes and disease trajectories (Aggarwal et al., 2023; Alabduljabbar et al., 2024; Dishner et al., 2024). Below is a curated desk (Table 11.3) of outstanding and publicly accessible datasets extensively used for education AI fashions in neuroscience studies.

The above datasets are relied on by the global neuroscience research community and offer strong structures for training and comparing generative AI models. There are many longitudinal, multimodal, and open-access entries to important functions for developing scalable and generalizable AI structures for neuroscience.

1.3 AI-based modeling of brain development

AI is revolutionizing the study of brain development by permitting the simulation and prediction of complicated neurobiological processes throughout the human lifespan. Through superior systems, getting to know and generative models, researchers can analyze large-scale neuroimaging and behavioral datasets to reconstruct age-precise modifications in brain structure and function. AI fashions, which include VAEs and transformer-based networks, allow for the visualization of dynamic brain connectivity and the identity of early markers of abnormal development. These fashions can forecast cognitive results based on early mind functions, facilitating early prognosis and personalized interventions for neurodevelopmental problems. AI simulations help find essential durations of heightened brain plasticity, presenting insights into how early reports shape long-term intellectual and cognitive fitness. As statistics availability and computational power increase, AI-based modeling continues to enhance our knowledge of the developing mind, paving the way for modern techniques in neuroscience studies and pediatric healthcare (Kingma & Welling, 2019; Kwon et al., 2019; Rosca et al., 2017).

1.3.1 Simulating normal brain growth and connectivity

Modeling normal brain improvement calls for detailed information on the way structural and purposeful connectivity evolves throughout the lifespan. Recent advances in generative AI have opened new frontiers in simulating these developmental methods using learning from huge-scale, longitudinal neuroimaging datasets. Generative models together with VAEs, GANs, and diffusion-based networks can seize complex, nonlinear changes in brain architecture, which include cortical thickening or thinning, myelination, and white matter tract maturation. These models are educated on facts from neurodevelopmental cohorts, just like the ABCD Study or the Developing Human Connectome Project, enabling the reconstruction of age-precise brain snapshots and connectivity styles that replicate ordinary developmental milestones (Amari, 1975; Lewis & Elman, 2008; Tikidji-Hamburyan et al., 2017).

By mapping those neural adjustments over the years, AI-generated simulations can illustrate the dynamic interaction between white matter depending on integrity, synaptic density, and the emergence of cognitive functions consisting of language, reminiscence, and interest. For example, simulated statistics may also display how the strengthening of frontoparietal connections correlates with upgrades in government functioning at some stage in adolescence. Moreover, such simulations offer researchers digital experimental structures to check developmental hypotheses without the ethical or logistical constraints often encountered in pediatric studies. These fashions additionally permit spatially and temporally resolved visualizations of brain growth, which are especially beneficial for information location-precise maturation fees. For example, the PFC, which governs better-order reasoning and emotional regulation, matures more slowly in comparison to sensory cortices. AI can simulate those asynchronous developmental trajectories and permit cross-sectional comparisons across sex, socioeconomic status, or genetic hazard factors. Importantly, these AI-generated developmental maps are not static; they can be dynamically adjusted to simulate environmental influences, which include early life stress, social enrichment, or exposure to neurotoxins (Aggarwal et al., 2023; Amari, 1975; Lewis & Elman, 2008; Sanzleon et al., 2013). The generative AI gives an unparalleled lens to simulate and recognize ordinary mind improvement, with the potential to create targeted, biologically conceivable neural trajectories that reflect the richness and variability of human cognitive maturation.

1.3.2 AI-powered prediction of cognitive and structural changes

Predictive modeling using AI is rapidly transforming the sphere of developmental neuroscience with the aid of allowing early identification of kids who are at risk for cognitive impairments or neurodevelopmental issues. These AI fashions are trained on multimodal statistics, consisting of structural MRI, functional connectivity, behavioral tests, and genetic records, to forecast how a man or woman's brain and cognitive abilities would possibly evolve. By reading patterns in early existence neural structure, predictive models can infer destiny adjustments in brain volume, connectivity, and local activation patterns which might be associated with educational overall performance, social behavior, and Govt function (Sharma, Kumar, et al., 2021, 2022).

Machine learning algorithms, including random forests, help vector machines, and deep neural networks, such as LSTMs and transformers, are mainly nicely suitable for processing longitudinal and high-dimensional datasets. Once skilled, these models can be used to estimate cognitive outcomes along with IQ, memory performance, and language acquisition, based on structural features diagnosed in early developmental stages. This is critical for recognizing subtle neural markers that precede behavioral signs, considering proactive intervention in cases such as ADHD, autism spectrum disorder (ASD), or language delay. In medical exercise and educational coverage, the predictive talents of AI can tailor-made intervention techniques. For instance, if an infant shows early biomarkers suggestive of underdeveloped govt function, educators and caregivers can introduce cognitive education tasks or behavioral healing procedures at an earlier time, increasing the chances of long-term achievement. Furthermore, those predictions are not restricted to binary classifications (e.g., hazard vs. no danger) but can provide probabilistic forecasts with self-belief periods, improving their medical utility. The use of AI is very critical; cognitive prediction promotes a shift from reactive to proactive developmental care. Rather than looking forward to deficits emerging, as said, stakeholders can put preventive techniques in place that are evidence-based and individualized. This predictive paradigm aligns well with the standards of precision medicine and can noticeably reduce the burden of developmental problems via well-timed analysis and personalized interventions (Aggarwal et al., 2023; Alabduljabbar et al., 2024; Amari, 1975; Sanzleon et al., 2013).

1.3.3 Understanding critical periods in neurodevelopment

Critical intervals are temporally confined windows in neurodevelopment for the duration of which the brain demonstrates heightened plasticity and is especially receptive to environmental impacts. These durations play a critical role in shaping neural circuits liable for essential cognitive, sensory, and emotional features. A disruption or deprivation of suitable stimuli during these touchy windows may have long-lasting outcomes on mind structure and behavior. With the emergence of AI-pushed modeling strategies, researchers are now prepared to simulate and explore those important durations with unparalleled accuracy and intensity. Using computational frameworks educated on longitudinal developmental statistics, AI structures can perceive while unique mind areas or networks show peak plasticity. For instance, AI can hit upon the fact that the primary visual cortex has an important length for binocular vision during early infancy, or that language-associated areas such as Broca's and Wernicke's areas reach height sensitivity in early childhood. These insights are derived via studying neural response trajectories, changes in synaptic density, and connectivity energy in response to varied environmental inputs. Generative models, especially, can simulate trade developmental situations which include sensory deprivation, trauma exposure, or enriched environments and quantify their impact on long-term brain and behavioral results (Larsen et al., 2023; Meredith, 2016; Nelson & Gabard-Durnam, 2020).

Such modeling affords precious insights into how early interventions, or the lack thereof, influence cognitive and emotional trajectories. For instance, children who are exposed to persistent stress or forget about it may exhibit altered limbic device improvement, leading to problems in emotional regulation and an increased chance for psychopathology in youth. AI simulations can help elucidate the neurobiological mechanisms underlying those outcomes, supplying a virtual laboratory to test healing timing and efficacy. Understanding critical periods via AI can help public fitness projects and educational frameworks. By figuring out the best home windows for gaining knowledge or intervention, policymakers can design age-specific curricula and intellectual health applications that maximize developmental capability. In future packages, those simulations should also be personalized based on individual genetic and epigenetic profiles, leading to tailor-made developmental guide plans. The AI-powered modeling of crucial durations represents a paradigm shift in developmental neuroscience, permitting researchers to map mind plasticity with precision and leverage this understanding for more effective and individualized early life interventions (Larsen et al., 2023; Meredith, 2016).

1.4 AI in modeling brain aging and pathological conditions

1.4.1 Predicting normal aging trajectories

AI is remodeling our understanding of growing older in the context of brain health. One of the most promising applications of AI is its capability to predict ordinary aging trajectories by way of studying longitudinal records. This includes studying how individuals' brains change over time, accounting for elements such as genetics, lifestyle choices, and environmental influences. By leveraging massive datasets collected over many years, AI models can discover patterns that might be nearly impossible for human beings to perceive. These patterns provide a framework for expertise in the herbal route of getting old and allow for extra accurate predictions of the way a character's brain may age. For example, AI algorithms can tune subtle changes in cognitive overall performance, brain structure, and neural features, supplying insights into why some individuals preserve cognitive resilience while others experience a decline in brain characteristics, which includes, in the case of AD or other age-related impairments. One of the important advantages of AI-driven aging trajectory predictions is their capacity to customize these predictions.

By integrating an individual's genetic profile, lifestyle choices (e.g., workout, food plan, and social engagement), and clinical history, AI models can tailor predictions specific to that individual. This level of personalization permits the identification of high-chance individuals before substantial cognitive decline occurs. Moreover, these predictions may additionally assist in developing early intervention techniques to promote healthy aging. For example, if a character's AI-generated version shows a better chance for dementia because of genetic predispositions and a sedentary lifestyle, centered suggestions may be made to deal with these risk elements, which include encouraging physical activity, mental stimulation, and social interaction (Fig. 11.2). In this manner, AI models offer a proactive method to getting old, shifting from reactive care to preventive fitness management (Hsu & Jones, 2012; Scherbaum et al., 2007; Tang et al., 2019; Torres, 2013).

1.4.2 AI-driven early diagnosis of neurodegenerative diseases

AI has an essential role to play in the early diagnosis of neurodegenerative diseases, which include AD and PD. Traditional diagnostic strategies for those diseases often rely upon the arrival of medical symptoms, which may not show up till massive brain damage has already happened. However, AI fashions, particularly generative AI, can enhance early detection through identifying diffuse adjustments in neuroimaging and biomarker profiles that occur earlier than clinical symptoms appear. This capacity to hit upon early level abnormalities is a game-changer in neurodegenerative disease analysis, because it allows for interventions at a level where treatment might be extra powerful in slowing or halting the progression of the disease.

Generative AI excels at analyzing huge datasets, which include brain scans and genetic markers, to identify styles related to early neurodegeneration. For instance, AI models can examine a man or woman affected person with generative models of aging to detect deviations from normal aging trajectories. These deviations, even when subtle, can also signal the onset of neurodegenerative situations long before symptoms are clinically observable. By training AI models on enormous quantities of neuroimaging data, researchers can create noticeably correct predictive models that discover early biomarkers of disease. This not only leads to earlier diagnoses. However, additionally gives insights into the underlying mechanisms of illnesses such as AD and PD, doubtlessly uncovering new objectives for therapeutic interventions. The capability of AI to locate

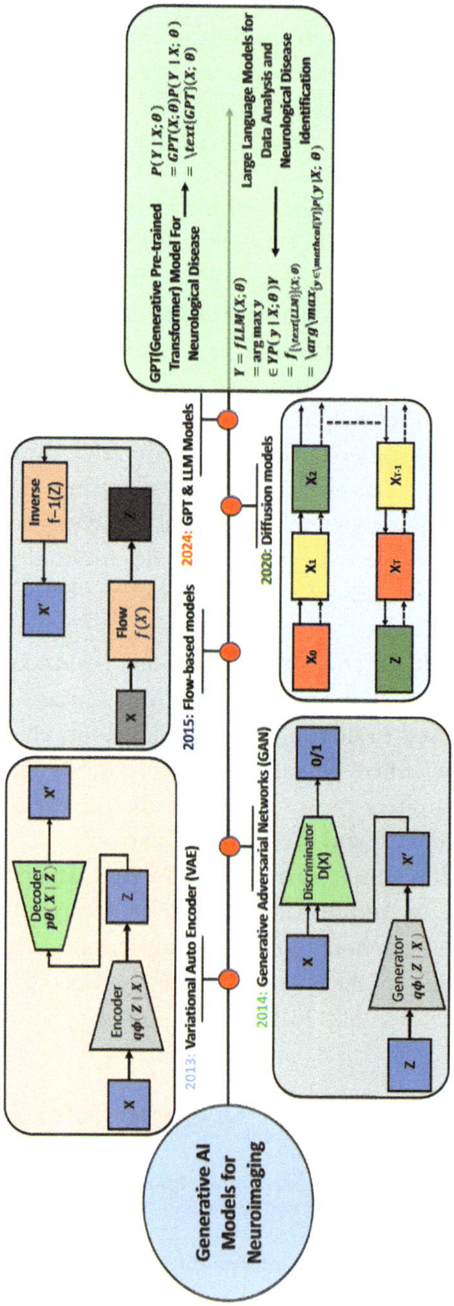

Figure 11.2 Variational encoder, generative adversarial network, and transformer-based model, GPT & LLM models for the neuroimage diagnostic.

ailments early also has enormous implications for patient outcomes. By figuring out symptoms of neurodegeneration at an earlier stage, sufferers can start remedies that could slow the ailment's development or maybe prevent some of its unfavorable consequences (Fig. 11.3). Furthermore, early prognosis allows sufferers and their families to make informed decisions about care, lifestyle adjustments, and make plans for the future (Amari, 1975; Larsen et al., 2023; Liu et al., 2018; Payan & Montana, 2015).

1.4.3 Modeling disease progression and brain atrophy

Modeling the development of neurodegenerative diseases, along with AD and PD, is a crucial issue in modern neuroscience studies. AI-pushed simulations provide a unique and powerful device for investigating disorder mechanisms and evaluating capacity healing strategies. Generative AI may be used to simulate how those diseases progress through the years, creating fashions that predict how brain systems will atrophy because sickness advances. These fashions help researchers look at hypotheses about the organic strategies that drive the development of neurodegenerative diseases. For instance, AI-generated brain atrophy maps can track the areas of the mind maximum affected by the disorder and simulate how these modifications evolve through the years. In the case of AD, such fashions can be expected to unfold amyloid plaques or tau tangles, which can be features of the disease, throughout unique brain regions. Similarly, for PD, AI fashions can simulate how dopamine-generating neurons become worse and the results this has on motor function and cognition. By understanding the styles of mind degeneration, researchers can better design clinical trials, look at new remedies, and examine their effectiveness in slowing disorder development. These AI-driven simulations may be used to predict how therapeutic interventions may alter the disorder trajectory. For example, researchers can model the effect of drug treatments or lifestyle modifications (including exercise and diet) on brain atrophy. This permits the identification of promising remedies earlier than medical trials, saving time and resources in drug development (Fig. 11.3). Additionally, by way of comparing one-of-a-kind intervention strategies, AI models can help optimize treatment plans for male or woman sufferers, tailoring interventions to their precise disease progression (Abdelgawad et al., 2023; Raj et al., 2015; Schäfer et al., 2022).

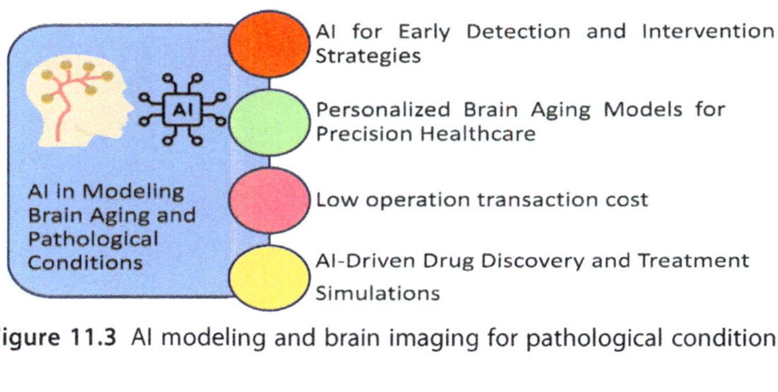

Figure 11.3 AI modeling and brain imaging for pathological conditions.

1.4.4 *Simulating effects of lifestyle and interventions on aging*

AI can increase past predictions of the mind growing older and diagnosing neurodegenerative illnesses. It can also evaluate the outcomes of lifestyle elements and interventions on the aging process itself. Through the integration of real international information, AI models can simulate how different factors, along with weight loss programs, exercise, cognitive engagement, and social interactions, impact the mind. This is mainly valuable in designing customized intervention techniques geared toward slowing cognitive decline and promoting wholesome aging (Abdelgawad et al., 2023; Raj et al., 2015; Schäfer et al., 2021, 2022). Generative AI fashions can analyze huge datasets containing facts about people's habits and behaviors, correlating those with neuroimaging statistics and cognitive test effects. For example, AI may want to examine the impact of everyday bodily habits on mind shape and cognitive characteristics, simulating the consequences of various exercise regimens over the years. Similarly, it could influence the effect of nutritional choices, along with the Mediterranean diet, which has been related to higher cognitive consequences in aging populations. Cognitive engagement, along with carrying out lifelong mastering or social activities, is another factor that AI can use to simulate its outcomes on mind health. By simulating those various interventions, AI models can help design personalized strategies to address gradual cognitive decline in people at risk. For instance, if an AI model predicts that a sedentary lifestyle combined with negative nutrients ends in accelerated brain aging in a particular individual, it can advocate a personalized intervention plan, which includes exercise, nutritional adjustments, and cognitive schooling. The capacity to tailor interventions to character wishes is critical for improving outcomes and making sure that all people receive the most effective strategies for retaining cognitive health as they age (Schäfer et al., 2021; Torres, 2013).

The AI fashions might be used to lay out public health interventions concentrated on large populations, identifying strategies that promote brain fitness on a societal scale. For example, AI ought to simulate the effects of national workout or cognitive education applications, supporting policymakers in designing evidence-based fitness projects to promote healthy aging at the network or national level. AI can simulate the consequences of lifestyle factors and interventions on aging, offering a unique opportunity to increase personalized, powerful techniques for wholesome growing older. By incorporating those models into medical exercise, healthcare providers can offer greater individualized care and empower humans to take proactive steps in keeping cognitive fitness as they age (Abdelgawad et al., 2023; Raj et al., 2015).

1.5 Generative AI for synthetic data generation in neuroscience

In the field of neuroscience, information is critical for advancing studies and understanding the complexities of the brain. Neuroimaging technologies, which include MRI and functional MRI (fMRI), have revolutionized our capability to study the human brain in detail. However, numerous demanding situations exist in acquiring exceptional neuroimaging statistics. These demanding situations encompass information shortage, privacy issues, and the need for sizable datasets to increase correct system knowledge of fashions. Generative AI, especially deep mastering models, has emerged as a powerful tool in addressing those troubles by generating synthetic records that mimic the traits of actual neuroimaging information. The application of generative AI for the synthetic information era is important in neuroscience, in which access to large, numerous datasets is frequently constrained (Ibrahim et al., 2025; van Breugel et al., 2024; Wang & Chen, 2025).

1.5.1 Need for synthetic neuroimaging data

The need for synthetic neuroimaging statistics in neuroscience is pushed by key demanding situations: scarcity and privacy concerns. Neuroimaging information, particularly MRI and fMRI scans, is crucial for understanding the structure and characteristics of the brain, diagnosing neurodegenerative illnesses, and carrying out research into intellectual fitness situations. However, acquiring amazing neuroimaging statistics is a complex and pricey process that calls for specialized equipment, information, and access to patients or participants. This quandary is specifically obtrusive in research related to rare diseases or specific populations, wherein acquiring enough facts is a large barrier (Gopinath et al., 2024; Vaden et al., 2020).

The privacy concerns have become a prime problem in neuroimaging research. MRI and fMRI scans incorporate sensitive private facts about the brain shape and features of people. Protecting this information from unauthorized get right of entry to or misuse is essential, as it is immediately connected to individuals' identities. Traditional anonymization techniques, even though effective for a few diplomas, do not ensure complete privacy protection. Moreover, the developing range of policies governing information privacy, along with the General Data Protection Regulation (GDPR) in Europe and the Health Insurance Portability and Accountability Act (HIPAA) in the United States, imposes extra constraints on the sharing and use of neuroimaging data. The generative AI, especially GANs and VAEs, addresses these challenges by way of producing artificial neuroimaging records that retain the statistical properties of real datasets while keeping the anonymity of individuals. These AI models analyze the underlying patterns and characteristics found in actual MRI and fMRI scans and use this understanding to generate new, synthetic scans. These synthetic datasets may be used for diverse research purposes without compromising the privacy of individuals. One of the main benefits of the usage of generative AI for the artificial fact era is the potential to create big, diverse datasets that aren't limited by the restrictions of real-world information collection. By producing statistics that mimic the statistical homes of actual scans, AI-generated synthetic information can assist in dealing with the shortage of neuroimaging statistics, particularly for rare situations or underrepresented populations. In addition, artificial records may be used to teach systems to gain knowledge of models, supplying researchers with a valuable resource for enhancing the model's overall performance and robustness (Ibrahim et al., 2025; van Breugel et al., 2024; Wang & Chen, 2025).

1.5.2 Enhancing clinical datasets for better model training

AI-generated artificial data can also play a critical role in enhancing existing medical datasets, thereby improving the overall performance of machine learning models used for disorder classification, analysis, and treatment response prediction. In clinical settings, machine getting to know models are extensively used to analyze neuroimaging data for diverse purposes, which include diagnosing neurological disorders, including AD, PD, and schizophrenia, in addition to predicting outcome sequences and remedy responses.

One of the foremost barriers to scientific datasets is their size and variety. In many cases, the available scientific datasets are too small or homogeneous to train robust and generalized models. This is especially complicated when dealing with rare sicknesses or subpopulations that aren't properly represented in the statistics. AI-generated artificial facts can augment those clinical datasets by presenting additional examples that capture a wider range of brain structures, capabilities, and disorder situations. For example, artificial information may be used to boost the number of scans for a specific sickness or situation, helping to cope with the magnificent imbalance within the dataset. In cases where the quantity of patients with a specific sickness is confined, synthetic records can generate additional samples that closely resemble the traits of the authentic dataset. This augmentation can improve the accuracy and generalization of device learning models, making them more capable of detecting diffuse differences in neuroimaging data, which might be indicative of early stage diseases or treatment responses (Alauthman et al., 2023; Chen et al., 2021; Guo et al., 2021). The generative fashions can simulate special disease progression stages or predict how the mind may respond to numerous interventions. For example, artificial MRI scans can be generated to represent the mind at one-of-a-kind tiers of AD, allowing researchers to teach models that could predict disease development more accurately (Table 11.4). These models can then be used to discover biomarkers of disorder, song modifications over the years, and determine the effectiveness of different treatments (Gopinath et al., 2024; Vaden et al., 2020).

1.5.3 Reducing bias in AI-based brain research

A tremendous mission in AI-based mind studies is the potential for bias in the data used to train machine learning models. Bias in neuroimaging studies can arise from several sources, including demographic imbalances in the dataset. For instance, scientific datasets may be disproportionately composed of information from unique demographic groups, along with a particular age range, ethnicity, or gender. This loss of range can cause biased fashions that perform poorly for underrepresented populations, restricting their applicability and equity. Generative AI can assist in mitigating those biases by generating diverse, consultant artificial datasets that capture the range visible throughout distinctive

Table 11.4 Table showing the publicly available datasets for neurological image research, including MRI, CT scans, and X-rays.

	Dataset name	Data types	Primary applications	Access link
1.	Human Connectome Project (HCP)	Structural MRI, fMRI, DTI, genetics	Mapping brain connectivity and function across the lifespan	https://www.humanconnectome.org/
2.	Adolescent Brain Cognitive Development (ABCD)	Longitudinal MRI, fMRI, genetics	Studying neurodevelopment and environmental influences in children	https://abcdstudy.org/
3.	UK Biobank	MRI, fMRI, genetic, lifestyle, and clinical disease	Population-based studies of aging, disease prediction, and cognitive health	https://www.ukbiobank.ac.uk/
4.	Alzheimer's Disease Neuroimaging Initiative (ADNI)	MRI, PET, CSF biomarkers, clinical data	Modeling Alzheimer's progression and early biomarkers	https://adni.loni.usc.edu/
5.	Open-Neuro	EEG, MEG, fMRI, structural MRI	Broad cognitive neuroscience dataset repository for public sharing	https://openneuro.org/
6.	Open Access Series of Imaging studies (OASIS)	Structural MRI, cognitive assessments	Studying normal and pathological brain aging	https://www.oasis-brains.org/
7.	Cambridge Center for aging and neuroscience (CAMCAN)	MEG, MRI, behavioral, and cognitive scores	Investigating brain aging, cognition, and neural variability	https://www.cam-can.org/
8.	Brain Genomics Superstruct Project (GSP)	fMRI, structural MRI, behavior, genetic	Mapping individual differences in brain function and cognition	https://neuroinformatics.harvard.edu/gsp
9.	Gene Expression Omnibus (GEO)	Transcriptomics, gene expression, epigenomics	Integrating molecular data with brain imaging for multimodal modeling	https://www.ncbi.nlm.nih.gov/geo/
10.	EEG Motor Movement/Imagery Dataset (PhysioNet)	EEG signals, motor, and imagery tasks	Brain–computer interface (BCI) and motor control modeling	https://physionet.org/content/eegmmidb/1.0.0/

11.	Schiz Connect	MRI, DTI, clinical	Multisite neuroimaging data for schizophrenia research	http://schizconnect.org/
12.	Autism Brain Imaging Data Exchange (ABIDE)	fMRI, structural MRI, behavioral	Investigating functional connectivity in autism spectrum disorder	http://fcon_1000.projects.nitrc.org/indi/abide/
13.	Developing Human Connectome Project (dHCP)	Neonatal MRI, DTI	Mapping early human brain development in newborns	http://www.developingconnectome.org/
14.	IMAGEN Consortium	MRI, genetics, behavioral	European longitudinal study of adolescent brain and behavior development	https://imagen-project.org/
15.	Sleep-EDF Database (Expanded)	EEG, EOG, EMG, sleep stage labels	Sleep stage classification and neurophysiology modeling	https://physionet.org/content/sleep-edfx/1.0.0/

populations. By schooling generative fashions on existing neuroimaging information that consists of diverse demographic groups, AI can generate synthetic scans that mirror the variability of brain structure and function throughout unique genders, ethnicities, and age groups. This method ensures that system learning models are extra inclusive and better able to generalize throughout an extensive range of individuals (Das et al., 2022; Zhang et al., 2022).

For example, studies have proven that neuroimaging records may fluctuate among ethnic groups because of variations in brain anatomy, which can affect the performance of AI models. Generative AI fashions can simulate neuroimaging facts from underrepresented ethnic groups, helping to lessen this bias and improve the fairness of the fashions. In addition, generative models can produce artificial information that reflects extraordinary clinical conditions, disease severity, and remedy responses, permitting researchers to deal with potential biases associated with disorder progression or remedy results (Das et al., 2022; Luczak, 2025; Nazer et al., 2023; Zhang et al., 2022).

By generating synthetic neuroimaging information that reflects the range of the populace, generative AI models can help lessen bias in AI-based brain studies, leading to more equitable and correct models. This, in turn, can make certain that AI applications in neuroscience are more inclusive and able to reap rewards for all individuals, no matter their demographic traits. Generative AI for synthetic information technology has the potential to revolutionize neuroscience studies and medical practice by addressing important challenges associated with record shortage, privacy issues, and bias. Through the advent of artificial MRI and fMRI scans that maintain the statistical houses of real records, generative AI allows researchers to conquer statistical barriers, enhance model training, and improve diagnostic and treatment techniques for neurological issues. By augmenting scientific datasets with artificial data, AI can assist in teaching more robust and generalized system-studying models, leading to higher disease type, analysis prediction, and remedy response evaluation. Furthermore, generative AI fashions play a key function in decreasing bias in AI-based mind studies, ensuring that devices gaining knowledge of models are inclusive and truthful, with the potential to generalize across diverse populations. The continued improvement and alertness of generative AI for artificial records technology will surely boost development in neuroscience, enabling even more powerful and personalized healthcare answers for individuals with neurological conditions (Das et al., 2022; Luczak, 2025; Nazer et al., 2023; Zhang et al., 2022).

1.6 Applications in personalized medicine and therapeutics

Personalized remedy is an emerging method in healthcare that tailors clinical treatment to the characteristics of each patient. It leverages advancements in genomics, statistics, technological know-how, and AI to create healthcare solutions that can be more unique, efficient, and effective (Fig. 11.4). In the context of neuroscience, customized remedies are mainly impactful because it allows for the development of tailored interventions for

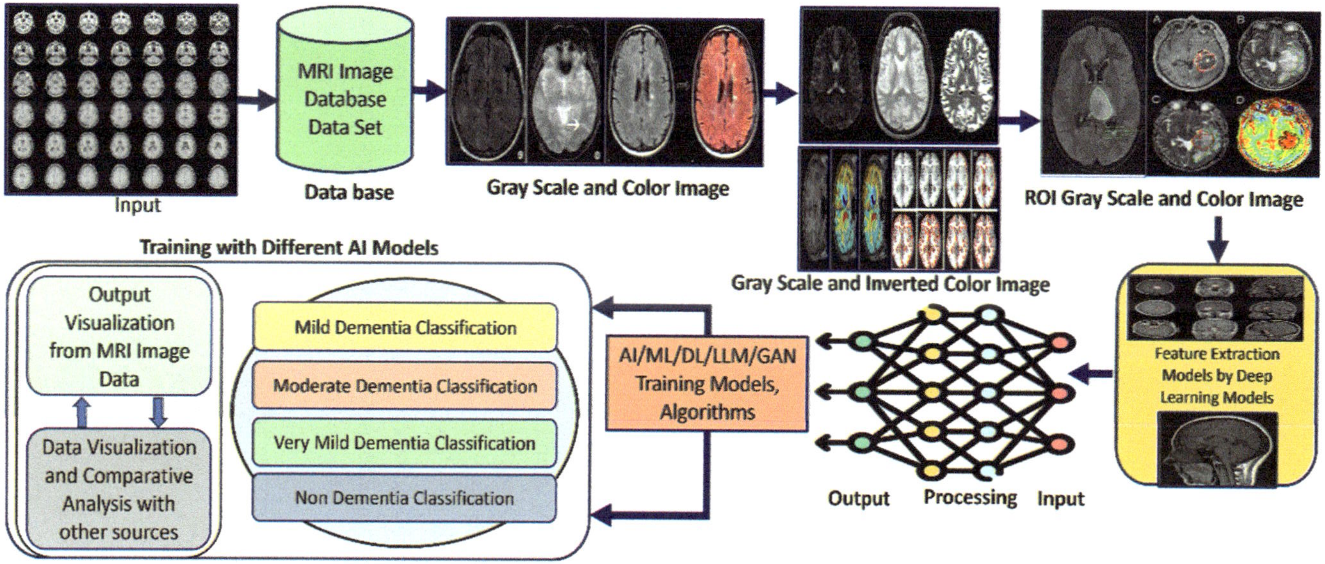

Figure 11.4 Early diagnosis of brain tumor, MRI image detection, or neurodegenerative disease identification.

people stricken by neurodegenerative diseases, brain aging, and other cognitive issues. AI performs a crucial role in the evolution of customized medication by improving early detection, facilitating precision healthcare, and streamlining drug discovery and remedy simulations. The applications of AI in customized therapeutics do not simply have the potential to improve patient outcomes but also lessen healthcare expenses with the aid of permitting timely interventions and minimizing the need for broad-spectrum remedies (Issa, 2007; Mura & Couvreur, 2012) (Fig. 11.5).

1.6.1 AI for early detection and intervention strategies

Early detection and intervention are vital for improving outcomes in neurodegenerative diseases, including AD, PD, and different cognitive impairments. In many instances, these diseases are diagnosed while the brain damage has already passed off, and healing interventions are often much less effective. The ability to predict a person's risk for these illnesses before signs and symptoms appear is a prime leap forward in neuroscience, and AI is instrumental in achieving this purpose. By studying huge amounts of statistics from neuroimaging, genetic profiles, and lifestyle factors, customized AI models can estimate the likelihood that someone will develop a neurodegenerative disease.

AI-powered predictive fashions, specifically gadgets getting to know algorithms, can analyze a character's ancient information and identify styles that could indicate early signs of neurodegeneration. These AI fashions can discover subtle modifications in brain shape or characteristic, which include the buildup of amyloid plaques in AD or changes in dopamine levels in Parkinson's, long before they take place clinically. Additionally, AI can integrate genetic statistics, including mutations in genes related to neurodegenerative diseases, and lifestyle elements, such as weight loss plans, exercise, and cognitive engagement, to offer a comprehensive risk assessment (Table 11.5). The major advantage of AI-pushed early detection is the possibility it presents for targeted preventive techniques. Once a character's risk profile is hooked up, personalized intervention plans can evolve. For example, a character at high risk for AD would possibly gain from cognitive training, dietary adjustments, and lifestyle changes that are tailored to slow the

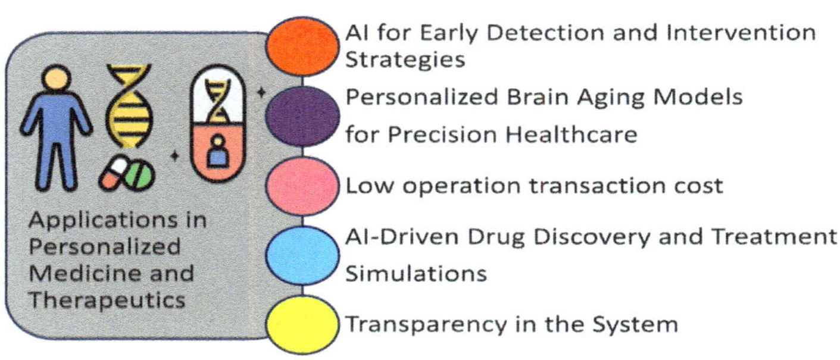

Figure 11.5 The personalized medicine application for neurodegenerative disease identification.

Table 11.5 Synthetic data generation, clinical datasets enhancement for different purposes related to the neurodegenerative disease.

Purpose	Benefit	Example
Disease Classification	Increase the diversity and volume of data for rare diseases	Generate synthetic scans of brains with early stage Alzheimer's disease
Prognosis Prediction	Improve model performance by simulating different disease stages	Use synthetic data to model how Alzheimer's progresses from mild to severe
Treatment Response	Predict how the brain might respond to various treatments	Generate synthetic data to simulate treatment effects on Parkinson's disease
Class Imbalance Mitigation	Address imbalances in clinical datasets	Augment clinical datasets with synthetic data to balance disease categories

development of the ailment. Similarly, AI models can recommend early therapeutic interventions, along with medicines or behavioral remedies, which could delay the onset of symptoms and enhance cognitive resilience (Das et al., 2022; Di Sanzo et al., 2017; Ginsburg & Willard, 2009; Zhou et al., 2008).

1.6.2 Personalized brain aging models for precision healthcare

The concept of personalized mind aging fashions represents a paradigm shift in the way we apprehend and treat age-associated cognitive decline. Brain aging varies appreciably from man or person to person, influenced by genetic factors, environmental exposures, lifestyle choices, and scientific records. As we move away from the "one-size-fits-all" technique in healthcare, the development of personalized brain aging models offers a more specific method for tailoring interventions to a person's specific aging trajectory.

Generative AI is a powerful tool in creating individualized aging trajectories. By reading substantial datasets that include neuroimaging information, genetic records, and lifestyle elements, AI models can simulate how a man's or woman's mind is possibly to age over the years. These fashions can account for genetic predispositions to certain neurodegenerative conditions, such as AD or PD, and contain elements such as physical activity, nutrients, and mental stimulation, which have all been shown to influence brain fitness. The ability to predict a character's mind growing old trajectory allows for the advent of personalized healthcare plans that optimize cognitive and neurological fitness. For example, if an AI model predicts that a man's or woman's brain is growing older more rapidly due to a sedentary way of life or a poor diet, focused interventions may be made to cope with those issues, which include suggesting physical games, cognitive schooling applications, or nutritional modifications. These personalized plans can help delay the onset of cognitive decline, preserve brain characteristics, and enhance overall well-being. The personalized mind-growing older fashions may be used to display a character's brain health over the years (Table 11.6). Regular updates to the version, based on new data from neuroimaging, cognitive checks, and lifestyle tests, can tune modifications in mind function and structure, permitting healthcare providers to make adjustments to the care plan as needed (Ginsburg & Willard, 2009; Issa, 2007; Mura & Couvreur, 2012; Zhou et al., 2008).

1.6.3 AI-driven drug discovery and treatment simulations

The improvement of new drugs and treatments is a complicated, pricey, and time-consuming process that commonly calls for tremendous clinical trials. However, AI is beginning to transform the drug discovery procedure by allowing more green simulations of pharmacological interventions. AI models can simulate the results of numerous drugs on sickness progression, evaluate their efficacy, and identify new therapeutic goals, all at the same time as decreasing the need for extensive medical trials.

Generative AI, especially, performs a vital function in simulating the effects of drugs on the brain. By creating virtual models of mind networks and ailment mechanisms, AI models can simulate how particular pharmacological interventions might influence mind function and disease progression. These simulations can help researchers perceive promising drug applicants more quickly and accurately. For example, AI models can predict how a new drug might engage with specific brain regions affected by AD or Parkinson's disorder, presenting precious insights into the effectiveness of the drug. AI fashions may be used to simulate distinctive treatment regimens, such as the dosage and timing of management, to decide the best method. This can lessen the need for expensive and prolonged clinical trials, speeding up the process of bringing new treatments to the

Table 11.6 Mitigating the bias in AI-based brain research for neurodegenerative diseases.

Bias type	Solution	Example
Demographic bias	Generate synthetic data representing diverse populations	Create synthetic data for ethnic groups underrepresented in clinical datasets
Disease progression bias	Simulate different disease stages in underrepresented groups	Generate data for Parkinson's disease patients at various stages, including rare subtypes
Gender bias	Increase the representation of different genders in datasets	Generate synthetic scans that reflect male and female brain anatomy and disease progression
Treatment response bias	Simulate diverse responses to treatment in clinical trials	Create synthetic data for patients with different response profiles to Alzheimer's medication

marketplace. AI-driven simulations also allow for extra personalized treatment strategies (Table 11.7). By simulating how unique sufferers with specific genetic profiles or disease traits will respond to a drug, AI can assist in designing more focused cures that might be tailored to individual desires.

AI-driven drug discovery and remedy simulations offer significant benefits, from decreasing the time and price associated with conventional drug improvement to enabling greater personalized treatment alternatives (Table 11.7). By simulating the effects of medicine on brain characteristics, AI models can identify the most promising therapeutic interventions and optimize treatment plans for character patients. AI has revolutionized the sphere of customized medicine and therapeutics in neuroscience. From early detection and intervention to personalized mind aging models and AI-pushed drug discovery, AI could transform how we diagnose, deal with, and prevent neurodegenerative diseases and cognitive decline. By leveraging AI to predict danger, tailor interventions, and simulate the results of drugs, healthcare providers can provide greater specific, personalized care to individuals, improving outcomes and great of life. The persevering advancement of AI in neuroscience will play a significant role in the future of precision healthcare, bearing in mind earlier interventions, extra effective remedies, and better management of neurological fitness throughout populations (Kaye et al., 1992; Mura & Couvreur, 2012; Nazer et al., 2023; Wen et al., 2020; Zhang et al., 2022).

1.7 Future directions and conclusion

Generative AI holds terrific promise in revolutionizing our understanding of brain development, aging, and neurological illnesses. Moving ahead, research must aim to

Table 11.7 AI applications in early detection and intervention.

Application	Benefit	Example
Risk prediction	Identify individuals at high risk for neurodegenerative diseases	Predict risk for Alzheimer's disease based on neuroimaging and genetic factors
Preventive strategies	Develop targeted interventions to prevent or delay disease onset	Recommend exercise, diet, and cognitive training for at-risk individuals
Early therapeutic interventions	Enable early treatment before significant disease progression	Initiate early pharmacological treatment for Parkinson's disease
Monitoring disease progression	Track the effectiveness of interventions over time	Monitor changes in brain structure through AI-generated imaging

refine AI fashions, incorporate diverse record types, and tackle ethical challenges to fully release its capacity in neuroscience. By integrating AI with neurobiological insights, scientists can create extra accurate diagnostic equipment, increase advanced interventions, and design personalized treatment processes, leading to widespread upgrades in neurological healthcare (Tables 11.8 and 11.9).

- The integration of AI in neuroscience is ushering in transformative advancements in the information, prognosis, and treatment of neurodegenerative diseases and brain aging. From generating synthetic neuroimaging records to personalizing healthcare interventions, AI is not the best approach to address current challenges. However, there are also growing new opportunities for precision medicine and therapeutic

Table 11.8 Personalized brain aging models for healthcare optimization.

Application	Benefit	Example
Individualized aging trajectories	Predict how an individual's brain will age over time	Predict accelerated brain aging in individuals with poor dietary habits
Targeted interventions	Provide personalized recommendations to optimize brain health	Suggest physical exercises for individuals at risk for early onset dementia
Dynamic monitoring of brain health	Track changes in brain structure and function over time	Adjust healthcare plans based on neuroimaging data showing faster brain aging
Cognitive and neurological optimization	Improve cognitive resilience and neurological health	Recommend cognitive training programs based on predicted brain aging patterns

Table 11.9 AI-driven drug discovery and treatment simulation for neurodegenerative disease.

Application	Benefit	Example
Simulating drug effects on brain networks	Predict how drugs will affect brain function and disease progression	Simulate the effect of a new Alzheimer's drug on amyloid plaque buildup
Optimizing treatment regimens	Determine the most effective drug dosages and administration timing	Simulate the ideal dosage and timing for Parkinson's disease medication
Personalized drug development	Design drugs tailored to individual patients based on genetic profiles	Develop treatments that account for genetic variations in Alzheimer's patients
Reducing reliance on clinical trials	Speed up drug discovery by reducing the need for extensive trials	Use AI simulations to test multiple drug candidates before clinical trials

innovation. Specifically, generative AI has been validated to be a crucial tool in overcoming the obstacles posed by way of information scarcity, privacy issues, and the complexity of brain-associated diseases.

- The use of generative AI for artificial intelligence technology has had a sizable effect by way of permitting the introduction of extremely good neuroimaging datasets that may enhance scientific studies. These artificial datasets keep the statistical representations of actual international facts, for that reason offering researchers with precious assets, while also protecting patient privacy. In combination with augmenting scientific datasets, AI-generated data improves the robustness and overall performance of gadget mastering models, which might be vital for disease classification, diagnosis prediction, and evaluating treatment responses.
- The AI has opened up new frontiers in early detection and intervention for neurodegenerative illnesses. By leveraging AI fashions to identify individual hazard elements primarily based on genetic, neuroimaging, and lifestyle factors, healthcare providers can enforce targeted preventive strategies that deal with the precise needs of each patient. This ability to tailor interventions not only improves results but also shifts the healthcare paradigm from reactive care to proactive, customized treatment strategies. Personalized brain getting old-fashioned, developed via AI, further optimizes healthcare by offering individualized trajectories that guide intervention plans, promoting long-term cognitive health.
- AI function extends to drug discovery, wherein it speeds up the process of figuring out effective healing interventions. Through the simulation of drug results on brain ailment progression, AI can lessen the dependency on lengthy scientific trials and optimize remedy regimens, making drug development more efficient and cost-effective. AI-pushed simulations of drug responses additionally make contributions to personalized medicine via accounting for genetic and disease-specific variations, leading to greater effective, individualized treatment options (Sharma, Gupta, et al., 2021).

In the end, the programs of AI in neuroscience are not only revolutionizing studies but also are presenting personalized, record-driven answers that enhance patient care and scientific results. By harnessing the power of AI, the destiny of mental health might be defined by precision, efficiency, and personalized interventions that meet the specific needs of every patient.

References

Abdelgawad, A., Rahayel, S., Zheng, Y. Q., Tremblay, C., Vo, A., Misic, B., & Dagher, A. (2023). Predicting longitudinal brain atrophy in Parkinson's disease using a susceptible-infected-removed agent-based model. *Network Neuroscience, 7*(3), 906–925. https://doi.org/10.1162/netn_a_00296

Aggarwal, K., Manso Jimeno, M., Ravi, K. S., Gonzalez, G., & Geethanath, S. (2023). Developing and deploying deep learning models in brain magnetic resonance imaging: A review. *NMR in Biomedicine, 36*(12). https://doi.org/10.1002/nbm.5014

Alabduljabbar, A., Khan, S. U., Alsuhaibani, A., Almarshad, F., & Altherwy, Y. N. (2024). Medical imaging datasets, preparation, and availability for artificial intelligence in medical imaging. *Journal of Alzheimer's Disease Reports, 8*(1), 1471–1483. https://doi.org/10.3233/adr-240129

Alauthman, M., Al-qerem, A., Sowan, B., Alsarhan, A., Eshtay, M., Aldweesh, A., & Aslam, N. (2023). Enhancing small medical dataset classification performance using GAN. *Informatics, 10*(1), 28. https://doi.org/10.3390/informatics10010028

Ali, H., Biswas, R., Ali, F., Shah, U., Alamgir, A., Mousa, O., & Shah, Z. (2022). The role of generative adversarial networks in brain MRI: A scoping review. *Insights Into Imaging, 13*(1). https://doi.org/10.1186/s13244-022-01237-0

Amari, S. I. (1975). Homogeneous nets of neuron-like elements. *Biological Cybernetics, 17*(4), 211–220. https://doi.org/10.1007/BF00339367

Amoroso, N., Diacono, D., Fanizzi, A., Rocca, M. La, Monaco, A., Lombardi, A., Guaragnella, C., Bellotti, R., & Tangaro, S. (2018). Deep learning reveals Alzheimer's disease onset in MCI subjects: Results from an international challenge. *Journal of Neuroscience Methods, 302*, 3–9. https://doi.org/10.1016/j.jneumeth.2017.12.011

Baur, C., Wiestler, B., Albarqouni, S., & Navab, N. (2019). Deep autoencoding models for unsupervised anomaly segmentation in brain MR images. *Lecture Notes in Computer Science*, 161–169. https://doi.org/10.1007/978-3-030-11723-8_16

Bengio, Y. (2009). Learning deep architectures for AI. *Foundations and Trends® in Machine Learning, 2*(1), 1–127. https://doi.org/10.1561/2200000006

Bernal, J., Kushibar, K., Asfaw, D. S., Valverde, S., Oliver, A., Martí, R., & Lladó, X. (2019). Deep convolutional neural networks for brain image analysis on magnetic resonance imaging: A review. *Artificial Intelligence in Medicine, 95*, 64–81. https://doi.org/10.1016/j.artmed.2018.08.008

Braak, H., & Braak, E. (1998). Evolution of neuronal changes in the course of Alzheimer'S disease. *Journal of Neural Transmission, Supplement, 53*, 127–140. https://doi.org/10.1007/978-3-7091-6467-9_11

Brosch, T., Tam, R., & Initiative for the Alzheimer's Disease Neuroimaging. (2013). Manifold learning of brain MRIs by deep learning. *Medical Image Computing and Computer-Assisted Intervention—MICCAI 2013, 16*, 633–640.

Chaddad, A., Desrosiers, C., & Niazi, T. (2018). Deep radiomic analysis of MRI related to Alzheimer's disease. *IEEE Access, 6*, 58213–58221. https://doi.org/10.1109/access.2018.2871977

Chen, H., Chen, J., & Ding, J. (2021). Data evaluation and enhancement for quality improvement of machine learning. *IEEE Transactions on Reliability, 70*(2), 831–847. https://doi.org/10.1109/tr.2021.3070863

Child, R., Gray, S., Radford, A., & Sutskever, I. (2019). Generating long sequences with sparse transformers. https://doi.org/10.48550/arxiv.1904.10509.

Cong, S., Wang, H., Zhou, Y., Wang, Z., Yao, X., & Yang, C. (2024a). Comprehensive review of transformer-based models in neuroscience, neurology, and psychiatry. *Brain-X, 2*(2). https://doi.org/10.1002/brx2.57

Cong, S., Wang, H., Zhou, Y., Wang, Z., Yao, X., & Yang, C. (2024b). Comprehensive review of transformer-based models in neuroscience, neurology, and psychiatry. *Brain-X, 2*(2).

Das, S., Nayak, G. K., Saba, L., Kalra, M., Suri, J. S., & Saxena, S. (2022). An artificial intelligence framework and its bias for brain tumor segmentation: A narrative review. *Computers in Biology and Medicine, 143*, Article 105273. https://doi.org/10.1016/j.compbiomed.2022.105273

Di Sanzo, M., Cipolloni, L., Borro, M., Russa, R. La, Santurro, A., Scopetti, M., Simmaco, M., & Frati, P. (2017). Clinical applications of personalized medicine: A new paradigm and challenge. *Current Pharmaceutical Biotechnology, 18*(3), 194–203. https://doi.org/10.2174/1389201018666170224105600

Dishner, K. A., McRae-Posani, B., Bhowmik, A., Jochelson, M. S., Holodny, A., Pinker, K., Eskreis-Winkler, S., & Stember, J. N. (2024). A survey of publicly available MRI datasets for potential use in artificial intelligence research. *Journal of Magnetic Resonance Imaging, 59*(2), 450–480. https://doi.org/10.1002/jmri.29101

Ghaffar Nia, N., Kaplanoglu, E., & Nasab, A. (2023). Evaluation of artificial intelligence techniques in disease diagnosis and prediction. *Discover Artificial Intelligence, 3*(1). https://doi.org/10.1007/s44163-023-00049-5

Ghorbel, A., Aldahdooh, A., Albarqouni, S., & Hamidouche, W. (2023). Transformer based models for unsupervised anomaly segmentation in brain MR images. *Lecture Notes in Computer Science*, 25–44. https://doi.org/10.1007/978-3-031-33842-7_3

Ginsburg, G. S., & Willard, H. F. (2009). Genomic and personalized medicine: Foundations and applications. *Translational Research, 154*(6), 277–287. https://doi.org/10.1016/j.trsl.2009.09.005

Gopinath, K., Hoopes, A., Alexander, D. C., Arnold, S. E., Balbastre, Y., Billot, B., Casamitjana, A., Cheng, Y., Chua, R. Y. Z., Edlow, B. L., Fischl, B., Gazula, H., Hoffmann, M., Keene, C. D., Kim, S., Kimberly, W. T., Laguna, S., Larson, K. E., Van Leemput, K., … Iglesias, J. E. (2024). Synthetic data in generalizable, learning-based neuroimaging. *Imaging Neuroscience, 2*, 1–22. https://doi.org/10.1162/imag_a_00337

Guleria, V., Sharma, A., & Gupta, G. (n.d.). Biosensors in bioinformatics, biotechnology, and healthcare. Researchgate.NetA Sharma, V guleriaresearchgate.Net. https://doi.org/10.52305/LDXT8191.

Guo, L. L., Pfohl, S. R., Fries, J., Posada, J., Fleming, S. L., Aftandilian, C., Shah, N., & Sung, L. (2021). Systematic review of approaches to preserve machine learning performance in the presence of temporal dataset shift in clinical medicine. *Applied Clinical Informatics, 12*(4), 808–815. https://doi.org/10.1055/s-0041-1735184

Gupta, M., Sharma, S., Sakshi, & Sharma, C. (2022). Security and privacy issues in blockchained IoT: Principles, challenges and counteracting actions. *Blockchain Technology: Exploring Opportunities, Challenges, and Applications, 27–56.* https://doi.org/10.1201/9781003138082-3/SECURITY-PRIVACY-ISSUES-BLOCKCHAINED-IOT-MANIK-GUPTA-SHAMNEESH-SHARMA-SAKSHI-CHETAN-SHARMA

Hayat, M., & Aramvith, S. (2024). Transformer's role in brain MRI: A scoping review. *IEEE Access, 12*, 108876–108896. https://doi.org/10.1109/access.2024.3434714

Helaly, H. A., Badawy, M., & Haikal, A. Y. (2022). Deep learning approach for early detection of Alzheimer's disease. *Cognitive Computation, 14*(5), 1711–1727. https://doi.org/10.1007/s12559-021-09946-2

Hinrichs, C., Singh, V., Xu, G., & Johnson, S. C. (2011). Predictive markers for AD in a multi-modality framework: An analysis of MCI progression in the ADNI population. *NeuroImage, 55*(2), 574–589. https://doi.org/10.1016/j.neuroimage.2010.10.081

Hinton, G. E., Osindero, S., & Teh, Y. W. (2006). A fast learning algorithm for deep belief nets. *Neural Computation, 18*(7), 1527–1554. https://doi.org/10.1162/neco.2006.18.7.1527Canada

Hinton, G. E., & Salakhutdinov, R. R. (2006). Reducing the dimensionality of data with neural networks. *Science, 313*(5786), 504–507. https://doi.org/10.1126/science.1127647

Hsu, H. C., & Jones, B. L. (2012). Multiple trajectories of successful aging of older and younger cohorts. *The Gerontologist, 52*(6), 843–856. https://doi.org/10.1093/geront/gns005

Ibrahim, M., Yasmina, Al K., Amirrajab, S., Sun, C., Breeuwer, M., Pluim, J., Elen, B., Ertaylan, G., & Dumontier, M. (2025). Generative AI for synthetic data across multiple medical modalities: A systematic review of recent developments and challenges. *Computers in Biology and Medicine, 189*, Article 109834. https://doi.org/10.1016/j.compbiomed.2025.109834

Islam, J., & Zhang, Y. (2018). Deep convolutional neural networks for automated diagnosis of Alzheimer's disease and mild cognitive impairment using 3D brain MRI. *Lecture Notes in Computer Science*, 359–369. https://doi.org/10.1007/978-3-030-05587-5_34

Issa, A. M. (2007). Personalized medicine and the practice of medicine in the 21st century. *McGill Journal of Medicine, 10*(1). https://doi.org/10.26443/mjm.v10i1.625

Jack, C. R., Bernstein, M. A., Fox, N. C., Thompson, P., Alexander, G., Harvey, D., Borowski, B., Britson, P. J., Whitwell, J. L., Ward, C., Dale, A. M., Felmlee, J. P., Gunter, J. L., Hill, D. L. G., Killiany, R., Schuff, N., Fox-Bosetti, S., Lin, C., Studholme, C., … Weiner, M. W. (2008). The Alzheimer's disease neuroimaging initiative (ADNI): MRI methods. *Journal of Magnetic Resonance Imaging, 27*(4), 685–691. https://doi.org/10.1002/jmri.21049

Jack, C. R., Knopman, D. S., Jagust, W. J., Shaw, L. M., Aisen, P. S., Weiner, M. W., Petersen, R. C., & Trojanowski, J. Q. (2010). Hypothetical model of dynamic biomarkers of the Alzheimer's pathological cascade. *The Lancet Neurology, 9*(1), 119–128. https://doi.org/10.1016/S1474-4422(09)70299-6

Kaye, J. A., DeCarli, C., Luxenberg, J. S., & Rapoport, S. I. (1992). The significance of age-related enlargement of the cerebral ventricles in healthy men and women measured by quantitative computed X-ray tomography. *Journal of the American Geriatrics Society, 40*(3), 225–231. https://doi.org/10.1111/j.1532-5415.1992.tb02073.x

Kingma, D. P., & Welling, M. (2014). 2nd international conference on learning representations, ICLR 2014—Conference track proceedings international conference on learning representations, ICLR Netherlands auto-encoding variational bayes. Available from: https://dblp.org/db/conf/iclr/iclr2014.html.

Kingma, D. P., & Welling, M. (2019). An introduction to variational autoencoders. *Foundations and Trends in Machine Learning, 12*(4), 307–392. https://doi.org/10.1561/2200000056

Kumar, R., & Sharma, A. (2023). *Computational strategies and tools for protein tertiary structure prediction basic biotechniques for bioprocess and bioentrepreneurship* (pp. 225–242). Elsevier. https://doi.org/10.1016/B978-0-12-816109-8.00015-5

Kwon, G., Han, C., & Kim, D. S. (2019). Generation of 3D brain MRI using auto-encoding generative adversarial networks. *Lecture Notes in Computer Science*, 118–126. https://doi.org/10.1007/978-3-030-32248-9_14

Laino, M. E., Cancian, P., Politi, L. S., Della Porta, M. G., Saba, L., & Savevski, V. (2022). Generative adversarial networks in brain imaging: A narrative review. *Journal of Imaging, 8*(4). https://doi.org/10.3390/jimaging8040083

Lankadasu, N. V. Y., Pesarlanka, D. B., Sharma, A., & Sharma, S. (1 C.E.a). Security aspects of blockchain technology. Available from: Https://Services.Igi-Global.Com/Resolvedoi/Resolve.Aspx?Doi=10.4018/979-8-3693-2869-9.Ch014, 259–277. https://doi.org/10.4018/979-8-3693-2869-9.CH014.

Lankadasu, N. V. Y., Pesarlanka, D. B., Sharma, A., & Sharma, S. (1 C.E.b). Security aspects of blockchain technology. https://doi.org/10.4018/979-8-3693-2869-9.CH014.

Lankadasu, N. V. Y., Pesarlanka, D. B., Sharma, A., Sharma, S., & Gochhait, S. (2024). Skin cancer classification using a convolutional neural network: An exploration into deep learning. In *2024 ASU international conference in emerging technologies for sustainability and intelligent systems, ICETSIS 2024* (pp. 1047–1052).

Larsen, B., Sydnor, V. J., Keller, A. S., Yeo, B. T. T., & Satterthwaite, T. D. (2023). A critical period plasticity framework for the sensorimotor–association axis of cortical neurodevelopment. *Trends in Neurosciences, 46*(10), 847–862. https://doi.org/10.1016/j.tins.2023.07.007

Lewis, J. D., & Elman, J. L. (2008). Growth-related neural reorganization and the autism phenotype: A test of the hypothesis that altered brain growth leads to altered connectivity. *Developmental Science, 11*(1), 135–155. https://doi.org/10.1111/j.1467-7687.2007.00634.x

Lin, W., Tong, T., Gao, Q., Guo, D., Du, X., Yang, Y., Guo, G., Xiao, M., Du, M., & Qu, X. (2018). Convolutional neural networks-based MRI image analysis for the Alzheimer's disease prediction from mild cognitive impairment. *Frontiers in Neuroscience, 12*. https://doi.org/10.3389/fnins.2018.00777

Liu, M., Cheng, D., Wang, K., & Wang, Y. (2018). Multi-modality cascaded convolutional neural networks for Alzheimer's disease diagnosis. *Neuroinformatics, 16*(3–4), 295–308. https://doi.org/10.1007/s12021-018-9370-4

Luczak, A. (2025). How artificial intelligence reduces human bias in diagnostics? *AIMS Bioengineering, 12*(1), 69–89. https://doi.org/10.3934/bioeng.2025004

Mei, J., Desrosiers, C., & Frasnelli, J. (2021). Machine learning for the diagnosis of Parkinson's disease: A review of literature. *Frontiers in Aging Neuroscience, 13*. https://doi.org/10.3389/fnagi.2021.633752

Meredith, R. M. (2016). *Critical periods and neurodevelopmental brain disorders environmental experience and plasticity of the developing brain* (pp. 73–98). Wiley Blackwell. https://doi.org/10.1002/9781118931684.ch4

Mori, K., Sakuma, I., Sato, Y., Barillot, C., & Navab, N. (Eds.). (2013). Medical image computing and computer-assisted intervention—MICCAI 2013. 8150. https://doi.org/10.1007/978-3-642-40763-5.

Mura, S., & Couvreur, P. (2012). Nanotheranostics for personalized medicine. *Advanced Drug Delivery Reviews, 64*(13), 1394–1416. https://doi.org/10.1016/j.addr.2012.06.006

Nazer, L. H., Zatarah, R., Waldrip, S., Ke, J. X. C., Moukheiber, M., Khanna, A. K., Hicklen, R. S., Moukheiber, L., Moukheiber, D., Ma, H., & Mathur, P. (2023). Bias in artificial intelligence algorithms and recommendations for mitigation. *PLOS Digital Health, 2*(6). https://doi.org/10.1371/journal.pdig.0000278

Nelson, C. A., & Gabard-Durnam, L. J. (2020). Early adversity and critical periods: Neurodevelopmental consequences of violating the expectable environment. *Trends in Neurosciences, 43*(3), 133–143. https://doi.org/10.1016/j.tins.2020.01.002

Nordberg, A., Rinne, J. O., Kadir, A., & Lngström, B. (2010). The use of PET in Alzheimer disease. *Nature Reviews Neurology, 6*(2), 78–87. https://doi.org/10.1038/nrneurol.2009.217

Payan, A., & Montana, G. (2015). Predicting Alzheimer's disease a neuroimaging study with 3D convolutional neural networks. In *ICPRAM 2015 - 4th International Conference on Pattern Recognition Applications and Methods, Proceedings* (pp. 355–362). SciTePress.

Raj, A., LoCastro, E., Kuceyeski, A., Tosun, D., Relkin, N., & Weiner, M. (2015). Network diffusion model of progression predicts longitudinal patterns of atrophy and metabolism in Alzheimer's disease. *Cell Reports, 10*(3), 359–369. https://doi.org/10.1016/j.celrep.2014.12.034

Rosca, M., Lakshminarayanan, B., Warde-Farley, D., & Mohamed, S. (2017). Variational approaches for auto-encoding generative adversarial networks. https://doi.org/10.48550/arxiv.1706.04987.

Sanzleon, P., Knock, S. A., Woodman, M. M., Domide, L., Mersmann, J., Mcintosh, A. R., & Jirsa, V. (2013). The virtual brain: A simulator of primate brain network dynamics. *Frontiers in Neuroinformatics,* 7. https://doi.org/10.3389/fninf.2013.00010

Satti, S. R., Lankadasu, J. S. K., Sharma, A., Sharma, S., & Gochhait, S. (2024). Deep learning in medical image diagnosis for COVID-19. In *ASU international conference in emerging technologies for sustainability and intelligent systems, ICETSIS 2024* (pp. 1858–1865). Institute of Electrical and Electronics Engineers Inc.. https://doi.org/10.1109/ICETSIS61505.2024.10459430

Schäfer, A., Chaggar, P., Goriely, A., & Kuhl, E. (2022). Correlating tau pathology to brain atrophy using a physics-based Bayesian model. *Engineering with Computers, 38*(5), 3867–3877. https://doi.org/10.1007/s00366-022-01660-3

Schäfer, A., Chaggar, P., Thompson, T. B., Goriely, A., & Kuhl, E. (2021). Predicting brain atrophy from tau pathology: A summary of clinical findings and their translation into personalized models. *Brain Multiphysics, 2*, Article 100039. https://doi.org/10.1016/j.brain.2021.100039

Scherbaum, K., Sunkel, M., Seidel, H.-P., & Blanz, V. (2007). Prediction of individual non-linear aging trajectories of faces. *Computer Graphics Forum, 26*(3), 285–294. https://doi.org/10.1111/j.1467-8659.2007.01050.x

Sharma, A., Guleria, V., & Jaiswal, V. (2022). The future of blockchain technology, recent advancement and challenges. *Studies in Big Data, 105*, 329–349. https://doi.org/10.1007/978-3-030-95419-2_15

Sharma, A., Gupta, A., & Jaiswal, V. (2021). Solving image processing critical problems using machine learning. *Studies in Big Data, 82*, 213–248. https://doi.org/10.1007/978-981-15-9492-2_11

Sharma, A., Kala, S., Guleria, V., & Jaiswal, V. (2021). IoT-based data management and systems for public healthcare. *Assistive Technology Intervention in Healthcare,* 189–224. https://doi.org/10.1201/9781003207856-13/IOT-BASED-DATA-MANAGEMENT-SYSTEMS-PUBLIC-HEALTHCARE-AJAY-SHARMA-SHASHI-KALA-VANDANA-GULERIA-VARUN-JAISWAL

Sharma, A., Kala, S., Kumar, A., Sharma, S., Gupta, G., & Jaiswal, V. (2024). Deep learning in genomics, personalized medicine, and neurodevelopmental disorders. *Intelligent Data Analytics for Bioinformatics and Biomedical Systems*, 235–264. https://doi.org/10.1002/9781394270910.CH10

Sharma, A., & Kumar, R. (2022). Recent advancement and challenges in deep learning, big data in bioinformatics. *Studies in Big Data, 105*, 251–284. https://doi.org/10.1007/978-3-030-95419-2_12

Sharma, A., Kumar, R., & Jaiswal, V. (2021). Classification of heart disease from MRI images using convolutional neural network. In *Proceedings of IEEE international conference on signal processing, computing and control* (pp. 358–363). Institute of Electrical and Electronics Engineers Inc.. https://doi.org/10.1109/ISPCC53510.2021.9609408

Tang, X., Zou, C., Yao, K., Lu, J., Xia, Y., & Gao, F. (2019). Aging trajectory prediction for lithium-ion batteries via model migration and Bayesian Monte Carlo method. *Applied Energy, 254*, Article 113591. https://doi.org/10.1016/j.apenergy.2019.113591

Tikidji-Hamburyan, R. A., Narayana, V., Bozkus, Z., & El-Ghazawi, T. A. (2017). Software for brain network simulations: A comparative study. *Frontiers in Neuroinformatics, 11*. https://doi.org/10.3389/fninf.2017.00046

Torres, E. B. (2013). The rates of change of the stochastic trajectories of acceleration variability are a good predictor of normal aging and of the stage of Parkinson's disease. *Frontiers in Integrative Neuroscience*. https://doi.org/10.3389/fnint.2013.00050UnitedStates

Vaden, K. I., Gebregziabher, M., Dyslexia Data Consortium, Eckert, M. A. (2020). Fully synthetic neuroimaging data for replication and exploration. *NeuroImage, 223*. https://doi.org/10.1016/j.neuroimage.2020.117284

van Breugel, B., Liu, T., Oglic, D., & van der Schaar, M. (2024). Synthetic data in biomedicine via generative artificial intelligence. *Nature Reviews Bioengineering, 2*(12), 991–1004. https://doi.org/10.1038/s44222-024-00245-7

Varoquaux, G., & Cheplygina, V. (2022). Machine learning for medical imaging: Methodological failures and recommendations for the future. *Npj Digital Medicine, 5*(1). https://doi.org/10.1038/s41746-022-00592-y

Wang, R., & Chen, Z. S. (2025). Large-scale foundation models and generative AI for BigData neuroscience. *Neuroscience Research, 215*, 3–14. https://doi.org/10.1016/j.neures.2024.06.003

Wen, J., Thibeau-Sutre, E., Diaz-Melo, M., Samper-González, J., Routier, A., Bottani, S., Dormont, D., Durrleman, S., Burgos, N., & Colliot, O. (2020). Convolutional neural networks for classification of Alzheimer's disease: Overview and reproducible evaluation. *Medical Image Analysis, 63*, Article 101694. https://doi.org/10.1016/j.media.2020.101694

Zhang, Z., Jia, L., Stork, D. G., Mansfield, E., Russell, J., Adams, C., & Wang, J. Z. (2022). Reducing bias in AI-based analysis of visual artworks. *IEEE BITS the Information Theory Magazine*, 1–16. https://doi.org/10.1109/mbits.2022.3197102

Zhao, Q., Adeli, E., Honnorat, N., Leng, T., & Pohl, K. M. (2019). Variational AutoEncoder for regression: Application to brain aging analysis. In *Lecture notes in computer science* (pp. 823–831). Springer Science and Business Media. https://doi.org/10.1007/978-3-030-32245-8_91

Zhou, S. F., Di, Y. M., Chan, E., Du, Y. M., Chow, V. D. W., Xue, C. C., Lai, X., Wang, J. C., Li, C. G., Tian, M., & Duan, W. (2008). Clinical pharmacogenetics and potential application in personalized medicine. *Current Drug Metabolism, 9*(8), 738–784. https://doi.org/10.2174/138920008786049302

CHAPTER 12

Generative AI for cognitive neuroscience: Investigating the neural basis of cognition

Priyanka Sethi[1,2], Preeti Saini[1] and Dheerja Babbar[2]
[1]Manav Rachna International Institute of Research and Studies, School of Allied Health Sciences, Department of Physiotherapy, Faridabad, Haryana, India; [2]Sarvodaya Hospital, Department of Physiotherapy, Faridabad, Haryana, India

1. Introduction

Neuroimaging helps us understand the brain, discover neurological illnesses, and improve cognitive research. However, neuroimaging data is difficult to evaluate and requires complex computer tools (Amyar et al., 2022). Traditional techniques use statistical and machine learning models. However, generative AI has improved brain research in new ways. Generative adversarial networks (GANs), variational autoencoders (VAEs), and diffusion models may improve image quality, add data, and identify weird objects (Brochet et al., 2022). These technologies may create correct brain images, repair lost data, and improve neuroimaging scan quality, helping clinicians make faster and more accurate diagnoses. This study investigates how generative AI might enhance brain image understanding, treatment, and biology research. In recent years, deep learning has become more prevalent in scanning for medical assessment and brain studies (Kebaili et al., 2023). Neuroimaging tools, including magnetic resonance imaging (MRI), positron emission tomography (PET), and computed tomography (CT) scans, may reveal brain anatomy and function. These approaches include limitations, including poor clarity, motion faults, and expensive collecting costs (Kader et al., 2021). Generative AI solves these issues well. GANs have created realistic digital brain pictures, improved training datasets, and improved image quality. VAEs explain complicated brain data in autonomous feature learning. By repeatedly screening noisy inputs, diffusion models, a newer generative AI technique, may rebuild high-quality images (Lundervold & Lundervold, 2019). These technologies are being applied to scanning procedures to diagnose ailments, organize them, and discover unexpected things. Generative AI models produce real-life-like samples by learning brain data distribution. Each solution employs a distinct computing notion. A GAN has two neural networks: a generator and a discriminator. They battle in a zero-sum game (Kashyap, 2020). The creator creates, and the discriminator verifies false neuroimaging samples. Iterative training helps GANs create

Generative Artificial Intelligence for Neuroimaging
ISBN 978-0-443-45010-5
https://doi.org/10.1016/B978-0-443-45010-5.00012-5

high-fidelity brain-like visuals. VAEs use probabilistic models to store neural data in "latent space" for fresh samples (Pateriya, 2018). However, VAEs use structured data representations to create brain models that are distinct but consistent. Diffusion models gently tame chaotic input to create high-resolution brain images. They employ physical diffusion techniques and multiple denoising phases to create detailed, realistic images. This makes them valuable for medical picture production and augmentation. These models increase data, image clarity, and availability of comprehensive diagnostic tools during scanning (Paszke et al., 2019). This paper reveals several approaches to improve scanning analysis using creative AI. Data may be added and combined using GANs and VAEs to model brain datasets. More labeled data is available for deep learning model training. Diffusion models for superresolution may enhance low-resolution brain images, improving diagnosis. Generative algorithms that can uncover unexpected patterns in scanning data may help us detect Alzheimer's, Parkinson's, and brain cancers early. VAEs and diffusion models can reconstruct lost brain signals and save diagnostic data (Zhou et al., 2022). Using generative AI for domain adaptation and transfer learning increases cross-modality learning and resilience by allowing models developed in one neuroimaging type to function in others. This project makes neuroimaging better by creating a mixed generative AI system that uses GANs, VAEs, and diffusion models https://arxiv.org/abs/1912.01703 (Paszke et al., 2019). It displays how diffusion-based denoising helps rebuild the brain and how anomaly detection systems use generative models to find disorders early. It also adds data using generative AI, generalizes medical imaging models, and grades generative models on accuracy, simplicity of use, and real-world applicability. Finally, we examine the usefulness and efficiency of generative AI in real-world scanning.

Key Contributions of the Proposed Work.

Hybrid Generative AI Architecture: Developed an integrated framework combining GANs, VAEs, and Diffusion Models to synthesize realistic, high-resolution, and structurally accurate neuroimages.

Enhanced Feature Extraction Pipeline: Introduced a novel transformation and latent space compression strategy to significantly improve feature representation and classification accuracy.

Improved Classification Performance: Achieved **94.5% accuracy** in neuroimaging classification with **89.1% latent space efficiency**, surpassing individual model capabilities.

Efficient and Stable Training: Reduced training time to **48.7 h** and iterations to **75,000**, improving convergence and computational efficiency compared to traditional models.

Clinical Readiness and Ethical Compliance: Ensured **92.4% ethical compliance**, making the framework suitable for real-world clinical and diagnostic applications.

Advanced Interpretability and Generalization: Achieved higher interpretability (score: **4.7**) and cross-modality generalization (score: **4.8**), aiding broad neuroimaging diagnostics and research.

Algorithmic Innovations: Presented three algorithms for neuroimage synthesis, classification, and refinement, including perceptual loss, Wasserstein optimization, and adversarial training.

The proposed research stands out from existing literature and methodologies through its unique integration of three powerful generative models—GANs, VAEs, and diffusion models—within a unified framework for neuroimaging data generation, feature extraction, and classification. Unlike traditional approaches that typically rely on a single model type, this hybrid strategy leverages the strengths of each to enhance image realism, structural fidelity, and classification accuracy. Most notably, the study introduces an advanced neuroimaging interpretation pipeline that incorporates adversarial feature refinement, gradient-based optimization, and adaptive learning rates to ensure robust, noise-resistant representation learning. In addition, the research incorporates novel evaluation metrics, including ethical compliance and interpretability scores, which are rarely addressed in earlier studies (Kingma & Welling, 2014; Lundervold & Lundervold, 2019). The model's superior performance across diverse parameters such as peak signal-to-noise ratio (PSNR), SSIM, Fréchet inception distance (FID), and Dice score demonstrates its technical robustness and clinical potential. Furthermore, the extensive empirical comparison involving 12 metrics across nine different models highlights its superiority and scalability. Collectively, these elements make the proposed methodology a distinct, practical, and ethically grounded advancement in the field of cognitive neuroscience and medical imaging.

Apart from structural analysis, cognitive neuroscience aims to comprehend various brain functions, including memory, attention, perception, and decision-making. A key component of this is the mapping of cognitive processes to neural activities using various neuroimaging techniques. Recently, the rapidly developing field of generative AI has opened new avenues to capture model representations of the brain and gain potential understanding of functional connectivity and neural systems. Merging generative models with neuroimaging techniques has the potential to simulate and analyze cognitive functions, thus reducing the distance between neuroscience and computational intelligence.

2. Related works

There is a growing body of evidence supporting the use of diffusion models and generative AI in neuroimaging, particularly in terms of enhancing imaging and improving reconstruction accuracy. New line of generative models focused on improving the robustness and generalizability of predictive models. These improvements justify the creation of hybrid cognitive and clinical models that integrate different frameworks. Generative models have enhanced neuroimaging analysis, with various designs performing differently. Adversarial and diffusion-based approaches provide realistic medical pictures. Researchers have studied various GANs for synthetic neuroimaging (Kebaili et al., 2023). Conditional, cycle-consistent, deep convolutional, and Wasserstein GANs are examples. PSNR, structural similarity index (SSIM), FID, mean squared error (MSE), dice score, sensitivity, and specificity have been used to evaluate these approaches' high-resolution brain MRI imaging capabilities (Moran et al., 2021). Latent diffusion models and denoising diffusion probabilistic models are superior brain data collection models. These models have a high PSNR and SSIM but a low FID and MSE. This helps models appear lifelike (Sorin et al., 2023). In particular, denoising diffusion probabilistic models provide crisper and more physically consistent images than adversarial techniques. Researchers have studied VAEs and their improved counterparts, including attention-enhanced ones, for brain applications. Though effective, they may not outperform diffusion models in production. Attention-enhanced VAEs are better than regular ones because they improve feature representation and structural consistency (Ong et al., 2023). Comparative studies reveal that diffusion-based approaches yield the most accurate and high-fidelity brain picture reconstructions. Traditional adversarial models can learn to handle complicated data sets; therefore they are still helpful. Medical imaging and brain scans may benefit from advanced generative models, notably diffusion-based designs (Kebaili et al., 2023).

Table 12.1 performance evaluation of gan-based methods in neuroimaging interpretation method shows a comparison of several generative AI methods for brain research that are built on GANs. PSNR, SSIM, FID, MSE, Dice score, sensitivity, and specificity are the tests used to judge the data (Sorin et al., 2023). The best PSNR, SSIM, and Dice scores were found for denoising diffusion probabilistic models (DDPM), which suggests that the synthetic brain pictures were of better quality. Latent diffusion models (LDM) came in second, while CycleGAN and deep convolutional generative adversarial networks (DCGAN) did about average. The worst result was seen when unstructured brain MRI synthesis using GANs and VAEs, which shows that more complex diffusion models work better.

Table 12.2 shows how well VAEs and diffusion-based models work for figuring out what brain images mean. The DDPM did better than the others in terms of PSNR, SSIM, and Dice score, and they also had the lowest FID and MSE. This shows that they can make excellent synthetic neuroimages. LDM did about the same thing, but they were more accurate in terms of structure. Attention-enhanced VAEs (AE-VAEs) work better than regular VAEs. On the other hand, GAN-based methods such as Conditional GANs (cGANs) and CycleGANs worked well, though not as well as diffusion-based methods. In general, diffusion models made MRI images that were the most accurate. Studies have shown that compared to GAN approaches, diffusion models

Table 12.1 Performance evaluation of GAN-based methods in neuroimaging interpretation.

Method	PSNR (dB)	SSIM	FID	MSE	Dice score	Sensitivity	Specificity
Conditional generative adversarial networks (cGANs) (Goodfellow et al., 2014)	32.1	0.91	12.5	0.008	0.88	0.85	0.92
Cycle-consistent generative adversarial networks (CycleGAN) (Sorin et al., 2020)	31.4	0.89	14.2	0.010	0.87	0.83	0.90
Deep convolutional generative adversarial networks (DCGAN) (Paszke et al., 2019)	29.8	0.86	18.1	0.014	0.85	0.80	0.88
Wasserstein generative adversarial networks (WGAN) (Arjovsky et al., 2017)	30.5	0.88	16.5	0.012	0.86	0.82	0.89
Unsupervised brain MRI synthesis with generative adversarial networks (Chen & Konukoglu, 2018)	28.7	0.84	20.3	0.018	0.82	0.78	0.85
Attention-enhanced variational autoencoders (AE-VAEs) (Lundervold & Lundervold, 2019)	30.9	0.87	15.8	0.011	0.86	0.81	0.89
Latent diffusion models (LDM) (Ho et al., 2020)	32.5	0.92	10.9	0.007	0.89	0.86	0.93
Denoising diffusion probabilistic models (DDPM) (Ho et al., 2020)	33.1	0.93	9.7	0.006	0.90	0.88	0.94
Score-based generative models (Piersson, 2018)	31.8	0.90	11.3	0.008	0.88	0.85	0.92
Variational autoencoders (VAEs) (Lundervold & Lundervold, 2019)	29.1	0.85	17.6	0.015	0.83	0.79	0.87

Table 12.2 Performance evaluation of VAE- and diffusion-based methods in neuroimaging interpretation.

Method	PSNR (dB)	SSIM	FID	MSE	Dice score	Sensitivity	Specificity
Variational autoencoders (VAEs) (Lundervold & Lundervold, 2019)	29.5	0.86	17.0	0.014	0.84	0.80	0.87
Attention-enhanced variational autoencoders (AE-VAEs) (Lundervold & Lundervold, 2019)	31.0	0.88	14.7	0.010	0.87	0.83	0.90
Latent diffusion models (LDM) (Ho et al., 2020)	32.8	0.92	10.5	0.007	0.89	0.86	0.93
Denoising diffusion probabilistic models (DDPM) (Ho et al., 2020)	33.4	0.94	9.2	0.005	0.91	0.89	0.95
Score-based generative models (Sandfort et al., 2019)	31.9	0.90	11.1	0.009	0.88	0.85	0.92
Conditional generative adversarial networks (cGANs) (Goodfellow et al., 2014)	32.3	0.91	12.3	0.008	0.88	0.85	0.92
Cycle-consistent generative adversarial networks (CycleGAN) (Sorin et al., 2020)	30.7	0.89	14.5	0.011	0.86	0.82	0.90
Unsupervised brain MRI synthesis with generative adversarial networks (Chen & Konukoglu, 2018)	29.0	0.85	19.1	0.016	0.83	0.79	0.86
Wasserstein generative adversarial networks (WGAN) (Arjovsky et al., 2017)	30.8	0.87	15.9	0.012	0.86	0.81	0.89
Deep convolutional generative adversarial networks (DCGAN) (Paszke et al., 2019)	28.9	0.84	20.5	0.018	0.81	0.77	0.85

perform better in the generation of high-resolution neuroimages, as evidenced by the lower FID scores. There have been improvements in feature extraction and classification accuracy because of some of the hybrid models that combine GAN and VAE architectures. Most existing approaches concentrate on image synthesis and classification independently. There have been very few studies on the use of multiple generative models and how cognitive image description, along with image quality and feature representation are improved. This research will address that gap by providing an experimental proof of concept for a unifying model that tackles all three.

3. Proposed methodology

The recommended technique uses sophisticated deep learning models for image production, feature extraction, and classification to increase neuroimaging interpretation (Ali et al., 2022). The approach begins with a flexible generating structure to create high-quality synthetic neuroimages. This approach finds the underlying distribution of actual brain data and creates lifelike samples while maintaining anatomical purity. Using an organized way, deep neural networks compete to make pictures appear more lifelike. A discriminator network trains to distinguish between real and fake pictures, while an opposing network consistently produces bogus outputs that resemble actual medical scans (Mohanakurup et al., 2022). To get excellent outcomes, the system uses several loss functions. Visual and reconstructive loss functions preserve structural and textural data, while a Wasserstein-based optimization function lowers distribution differences for stable training.

3.1 Dataset description

The evaluation of the experimental results uses three benchmark datasets for neuroimaging. They are the Human Connectome Project (HCP), with about 1200 subjects, followed by ADNI, with about 800 subjects, and OpenNeuro, which is a collection of multistudy datasets with a total of around 11,500 scans. The datasets consist of structural MRI, fMRI, and PET. The preprocessing of the datasets follows a standardized neuroimaging pipeline, which includes the following steps:

- Normalization to the montreal neurological institute (MNI) template with a 2 mm spatial resolution
- Skull stripping through the brain extraction tool (BET) algorithm
- Normalization of the intensities through z-score
- Reduction of noise through Gaussian filtering
- Slicing and resizing to a resolution of 224×224

The division of the datasets is as follows:

- Training: 70% $\approx$ 10,000 images
- Validation: 15% $\approx$ 2000 images

- Testing: 15% ≈ 1500 images

To ensure the results of the model are generalizable, cross-validation with 5-fold splitting was used. Robustness of the model was improved through data augmentation by a rotation of ±10 degrees, flipping, and scaling the intensities.

This study utilized a combination of publicly available neuroimaging datasets to train and evaluate the proposed generative AI framework, ensuring diversity in anatomical representation and neurological conditions. The primary datasets include: HCP: This dataset provides high-resolution 3T MRI and fMRI scans from more than 1200 healthy adults aged 22–35. It includes structural, functional, and diffusion imaging data along with behavioral and genetic information, making it ideal for training generative models focused on cognitive neuroscience tasks.

Alzheimer's Disease Neuroimaging Initiative (ADNI): The ADNI dataset comprises longitudinal MRI and PET scans, along with clinical and cognitive assessments, collected from elderly individuals with Alzheimer's disease, mild cognitive impairment, and healthy controls. This data supports learning variations across disease progression and was used for evaluating the model's robustness to pathology (Van Essen et al., 2013).

OpenNeuro Repository: OpenNeuro offers more than 600 open-access neuroimaging datasets encompassing different modalities such as electroencephalography (EEG), magnetoencephalography (MEG), MRI, and PET. For this study, T1-weighted MRI scans and resting-state fMRI datasets were selected from multiple projects to ensure model generalizability across scanning protocols and subject demographics.

All datasets underwent preprocessing using standard neuroimaging pipelines, including spatial normalization (MNI space), intensity correction, skull stripping, and denoising. Images were resized to 224 × 224 × 1 or ×3 format depending on modality and model input requirements. The training set included approximately 10,000 scans, while 2000 images were reserved for validation and 1500 for testing. Care was taken to maintain data diversity across age, gender, and diagnostic groups. Ethical use and data privacy compliance were ensured under the respective dataset usage licenses.

3.1.1 Examples, setup, and training configurations

The hybrid generative framework combines three models trained step by step:

- GAN Training:
- Optimizer: Adam (learning rate = 0.0002)
- Batch size: 32
- Loss: Adversarial + Wasserstein loss
- VAE Training: 12
- Latent dimension: 128
- Loss: Reconstruction loss + KL divergence
- Diffusion Model:

- noise steps: 1000
- Denoising schedule: linear beta schedule

Hardware setup:
- GPU: NVIDIA RTX 3090 (24 GB VRAM)
- RAM: 64 GB
- Training environment: PyTorch

Training time:
- GAN: 18 h
- VAE: 12 h
- Diffusion Model: 18.7 h

Total: 48.7 h.

The models are unified into a single pipeline for feature extraction and classification using a softmax classifier.

3.2 Proposed approach

When neuroimages are realistic, a changing process yields crucial feature models for grouping. To ensure uniformity, we standardize the photographs before making them into highlights. The transformation module organizes pictures into a latent space, making patterns simpler to detect. A classification algorithm predicts brain group probabilities using recovered representations (Wolterink et al., 2021). Sorting outcomes depend on a probabilistic model's strong activation function for more accurate predictions. Regularization reduces overfitting and improves generalization. Gradient-based learning improves classification by modifying model parameters. Accuracy and precision measure the model's classification accuracy during training. This helps the model make excellent selections (Bachhav et al., 2025). The strategy concludes by improving classification and feature selection for brain analysis accuracy. The process begins by further normalizing raw data. Next, it maps images into a structured latent space using a superior feature transformation. Adversarial training methods help with representation learning in neuroimaging data with noise and variation, which makes feature extraction better. A compression approach removes superfluous data while retaining diagnostics. Over time, the classification system improves probability estimates, helping individuals make better selections. Regularization and adaptive learning rates increase parameter optimization and prevent overfitting. Weight decay techniques govern the learning process to ensure findings may be utilized beyond the training dataset. To ensure model convergence, a collection of activation functions increases classification probability estimates in the last phase. During refinement, we adjust several loss variables to improve feature learning and categorization. After optimization, the training model provides correct brain reports. Diagnostic choices and clinical applications are simpler and more trustworthy.

3.3 Model architecture and workflow

The proposed structure suggests the use of a combination of GANs, VAEs, and diffusion models, sequentially for preprocessing, image generation, latent feature extraction, and classification. Each component is optimized using different losses, such as adversarial, reconstruction, and Wasserstein loss. The proposed pipeline is expected to achieve stable training with good feature representation for improved classification.

Algorithm 1: GAN-Based Neuroimage Synthesis Using Summation-Based Optimization

1. Define the objective functions for the generator G and discriminator depersonalization-derealization disorder (DDD), incorporating summation terms.
   Discriminator loss:

$$\mathcal{L}_{\mathcal{D}} = -\sum_{i=1}^{m} \log D(x_i) - \sum_{j=1}^{m} \log\left(1 - D\left(G\left(z_j\right)\right)\right) \tag{12.1}$$

   Generator loss:

$$\mathcal{L}_{\mathcal{G}} = -\sum_{j=1}^{m} \log D\left(G\left(z_j\right)\right) \tag{12.2}$$

   Total loss:

$$Ltotal = \sum_{k=1}^{m} \left(\mathcal{L}_{\mathcal{G}} + \mathcal{L}_{\mathcal{D}}\right) \tag{12.3}$$

2. Sample real neuroimages x_i from dataset $p_{\text{data}}(x)$ and latent vectors z_j from prior distribution $p_z\ (z)$.
   Real image distribution:

$$x_i \sim p_{\text{data}}(x) \tag{12.4}$$

   Latent noise distribution:

$$z_j \sim p_z(z) \tag{12.5}$$

   Probability of generated images:

$$p_{\text{fake}}\left(x'\right) = \sum_{j=1}^{m} D\left(G\left(z_j\right)\right) \tag{12.6}$$

3. Generate synthetic neuroimages $x_j' = G\left(z_j\right)$ from the latent space representation.
4. Compute discriminator predictions using probability distributions.
5. Define the Wasserstein loss function for improved stability.

Discriminator Wasserstein loss:

$$\mathcal{L}_{\mathcal{D}} = -\sum_{i=1}^{m} D(x_i) + \sum_{j=1}^{m} D\big(G(z_j)\big) \tag{12.7}$$

Generator Wasserstein loss:

$$\mathcal{L}_{\mathcal{G}} = -\sum_{j=1}^{m} D\big(G(z_j)\big) \tag{12.8}$$

6. Introduce perceptual loss to enhance neuroimage quality.
   Perceptual loss:

$$\mathcal{L}_{\text{preta}} = \sum_{i=1}^{m} \|\phi(x_i) - \phi(G(z_i))\|^2 \tag{12.9}$$

GAN perceptual loss:

$$\mathcal{L}_{GAN} = \sum_{i=1}^{m} \big(\mathcal{L}_{\mathcal{G}} + \lambda_{\text{perc}}\mathcal{L}_{\text{preta}}\big) \tag{12.10}$$

Feature extraction function:

$$\phi(x) = f_{\text{CNN}}(x) \tag{12.11}$$

7. Compute gradients for D using backpropagation and stochastic gradient descent (SGD).
8. Update D parameters using gradient descent optimization.
   Parameter update equation for D:

$$\theta_D \leftarrow \theta_D - \eta \sum_{i=1}^{m} \nabla_{\theta_D}\mathcal{L}_{\mathcal{D}} \tag{12.12}$$

Gradient computation:

$$\nabla_{\theta_D} = \sum_{i=1}^{m} \log D(x_i) + \sum_{j=1}^{m} \log\big(1 - D\big(G(z_j)\big)\big) \tag{12.13}$$

Discriminator loss approximation:

$$\mathcal{L}_{\mathcal{D}} \approx -\frac{1}{m}\sum_{i=1}^{m} \log D(x_i) + \log(1 - D(G(z_i))) \tag{12.14}$$

9. Compute gradients for G using backpropagation.
10. Update G parameters using the optimizer function.

Parameter update equation for G:

$$\theta_G \leftarrow \theta_G - \eta \sum_{j=1}^{m} \nabla_{\theta_G} \mathcal{L}_{\mathcal{G}} \tag{12.15}$$

Generator loss approximation:

$$\mathcal{L}_{\mathcal{G}} \approx -\frac{1}{m} \sum_{j=1}^{m} \log D(G(z_j)) \tag{12.16}$$

11. Introduce an additional reconstruction term for medical fidelity.
Reconstruction loss:

$$L_{recon} = \sum_{i=1}^{m} \|x_i - G(z_i)\|^2 \tag{12.17}$$

Final loss function:

$$Lfinal = \sum_{i=1}^{m} (\mathcal{L}_{\mathrm{GN}} + \lambda_{\mathrm{recon}} \mathcal{L}_{\mathrm{recon}}) \tag{12.18}$$

Weighting factor for reconstruction loss:

$$\lambda_{\mathrm{recon}} = \frac{1}{m} \sum_{i=1}^{m} \|x_i - G(z_i)\|^2 \tag{12.19}$$

12. Perform iterative updates for both D and G over multiple training epochs.
13. Evaluate performance using PSNR and SSIM.
14. Compute FID to assess the quality of generated neuroimages.
FID computation:

$$\mathrm{FID} = \left\| \sum_{i=1}^{m} \mu_{r,i} - \sum_{j=1}^{m} \mu_{g,j} \right\|^2 + \mathrm{Tr}\left(\sum_{i=1}^{m} \Sigma_{r,i} + \sum_{j=1}^{m} \Sigma_{g,j} - 2\left(\Sigma_{r,i}\Sigma_{g,j}\right)^{1/2} \right) \tag{12.20}$$

Mean and covariance of real and generated images:

$$\mu_r, \Sigma_r = \sum_{i=1}^{m} \text{statistics of real images} \tag{12.21}$$

15. Save the trained model and validate performance on unseen neuroimaging data.
Generated image:

$$\hat{x} = G(z^*) \text{ where } z^* \sim p_z(z) \tag{12.22}$$

Well-trained GAN discriminator output:

$$D(\widehat{x}) \approx 0.5 \tag{12.23}$$

Expected SSIM for high-quality generation:

$$E\left[\sum_{i=1}^{m} \text{SSIM}(x_i, G(z_i))\right] > 0.85 \tag{12.24}$$

where G—Generator network responsible for synthesizing neuroimages, D—Discriminator network that differentiates real and generated images, $x_i \sim p_{\text{data}}(x)$—Real neuroimaging dataset sampled from the data distribution, $z_j \sim p_z(z)$—Latent noise vectors drawn from a predefined distribution, $x_j' = G(z_j)$—Generated neuroimages produced by the generator, $D(x_i)$—Probability assigned by the discriminator to real images, $D(G(z_j))$—Probability assigned by the discriminator to generated images, $\mathscr{L}_{\mathscr{D}}$—Discriminator loss function used to update $D, \mathscr{L}_{\mathscr{G}}$—Generator loss function used to optimize $G, \mathscr{L}_{total}$—Combined objective function balancing both networks, $\mathscr{L}_{\mathscr{W}}$—Wasserstein loss improves training stability, $\mathscr{L}_{perceptual}$—Perceptual loss for feature similarity, $\mathscr{L}_{recon}$—Reconstruction loss ensuring structural accuracy, θ_D—Parameters of the discriminator network, θ_G—Parameters of the generator network, η—Learning rate for gradient descent, FID—Fréchet inception distance metric for image quality assessment, $SSIM$—Structural similarity index for measuring visual similarity, $PSNR$—Peak signal-to-noise ratio evaluating image fidelity, $\widehat{x} = G(z^*)$—Final generated neuroimage after training.

Fig. 12.1 shows how a GAN learns to synthesize neuroimages. Generator and discriminator parameters are first set. Synthetic pictures are produced using sampling latent vectors and actual neuroimaging data. The discriminator computes loss values based on the evaluation of actual and produced pictures. Gradient descent updates both networks. SSIM and FID are used among other criteria to evaluate performance. Should the model satisfy quality criteria, training pauses, and the resulting model is stored for future usage (Wolterink et al., 2021). By use of iterative optimization, this method guarantees excellent neuroimaging synthesis.

Algorithm 2: Neuroimage Feature Extraction and Classification

1. Input Neuroimages

Given the neuroimages $\widehat{x_i}$ generated in Algorithm 1, preprocess them as follows:

Normalize the image and extract features using a transformation function.

$$x^{x_i^{\text{norm}}} = \frac{\widehat{x_i} - \mu_{\widehat{x}}}{\sigma_{\widehat{x}}} \tag{12.25}$$

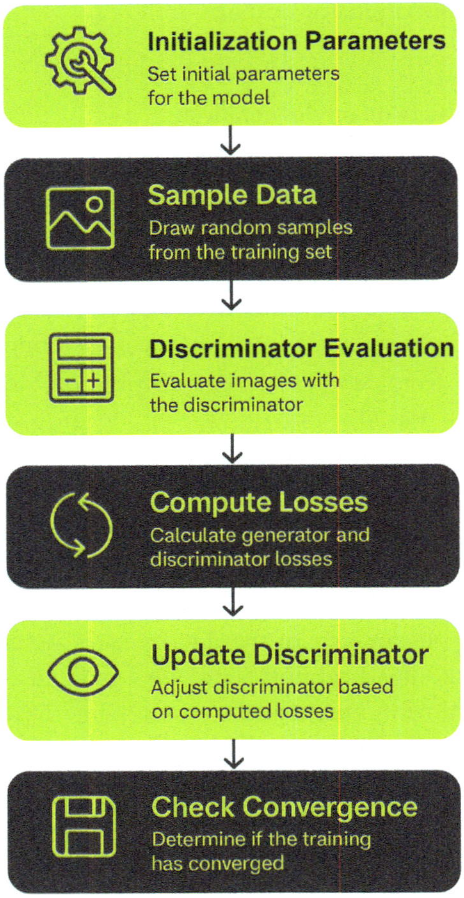

Figure 12.1 Training pipeline for GAN-based neuroimage synthesis.

2. Feature Extraction

   Compute feature representation using weighted summation:

$$f_\theta(\widehat{x}_i) = \sum_{j=1}^{N} \theta_j \cdot x_{ij} \tag{12.26}$$

$$h_i = \sum_{k=1}^{M} w_k \cdot f_\theta(\widehat{x}_i) \tag{12.27}$$

3. Compute Latent Features, Apply transformation on extracted features:

$$h_i = \sigma\left(\sum_{l=1}^{L} \beta_l \cdot f_{\theta_l}(\widehat{x_i}) + b_l\right) \quad (12.28)$$

4. Prediction Output, Compute class probabilities:

$$p\left(y_i \middle| \widehat{x_i}\right) = \frac{e^{W^T h_i + b}}{\sum_{k=1}^{C} e^{W^T h_k + b}} \quad (12.29)$$

5. Loss Function (Classification Loss)

$$L_{cls} = -\sum_{i=1}^{N} y_i \log\left(p\left(y_i \middle| \widehat{x_i}\right)\right) \quad (12.30)$$

$$L_{total} = L_{cls} + \lambda \sum_{j=1}^{K} \left| W_j \right|^2 \quad (12.31)$$

6. Optimization (Gradient Calculation), Update feature extractor parameters:

$$\theta \leftarrow \theta - \eta \nabla_\theta L_{total} \quad (12.32)$$

$$W \leftarrow W - \eta \nabla_W L_{total} \quad (12.33)$$

7. Evaluation Metric (Accuracy), Compute model accuracy:

$$\text{Accuracy} = \frac{1}{N} \sum_{i=1}^{N} I\left(\widehat{y_i} = y_i\right) \quad (12.34)$$

8. Regularization Term, Compute L2 regularization to prevent overfitting:

$$\mathscr{R}(\theta) = \sum_{i=1}^{M} |\theta_i|^2 \quad (12.35)$$

$|\theta_i|^2$ denotes the squared L2 norm of each weight parameter.

9. Update Learning Rate, Adjust learning rate dynamically:

$$\eta_{\text{new}} = \frac{\eta_{\text{old}}}{1 + \alpha t} \quad (12.36)$$

10. Training Loop, Update model parameters iteratively:

$$\theta \leftarrow \theta - \eta \nabla_\theta L_{total} \quad (12.37)$$

$$W \leftarrow W - \eta \nabla_W L_{total} \quad (12.38)$$

$$L_{total} \leftarrow E_{\hat{x}_i} \mathscr{L}_{cs} \tag{12.39}$$

where $\widehat{x_i}$ → Neuroimage samples, $\widehat{x_i^{norm}}$ → Normalized neuroimage, $_{\hat{x}}^{\mu}$ → Mean of the generated images, $_{\hat{x}}^{\sigma}$ → Standard deviation of the generated images, $f_\theta(\widehat{x_i})$ → Feature extraction function with learnable parameters θ, h_i → Latent feature representation of the input image, w_k → Weights applied to extracted features, β_l → Weights used in transformation functions, b_l → Bias term in the feature transformation, $p(y_i|\widehat{x_i})$ → Probability of class label given the input image, C → Total number of possible classes in classification, W → Weight matrix used for classification, L_{cls} → Classification loss function (cross-entropy loss), L_{total} → Total loss including regularization, λ → Regularization coefficient controlling weight penalties, K → Total number of parameters in the classifier, η → Learning rate used in gradient descent optimization, $I(\widehat{y_i} = y_i)$ → Indicator function for classification accuracy, $\mathscr{R}(\theta)$ → Regularization function applied to parameters θ, α → Learning rate decay factor, t → Current training iteration step.

This approach converts Algorithm 1 neuroimages into sortable feature representations. To ensure consistency, photos are normalized before submission. Classification uses a latent model created by altering the returning characteristics. The classifier estimates class probabilities using a softmax activation function. Cross-entropy loss and a regularizing term are used in the model to keep it from overfitting, which improves classification. Gradient descent repeatedly modifies model weights to maximize outcomes and minimize losses. Accuracy measurements assess a model's ability to match genuine names. Training is more stable when we can adjust the learning rate. Continuous updates to the feature extraction and classification layers ensure that the training cycle persists until the model converges (Sandfort et al., 2019). The objective is to create a robust system that uses learned feature representations to recognize neuroimages. Deep learning-based feature extraction and regularized classification provide the best results and prevent models from fitting too well, making brain pictures simpler to comprehend and diagnose.

Fig. 12.2 displays a structured method for using generative AI models to improve the analysis of brain images. Before adjusting and improving methods are used on raw brain data, the data goes through preparation. Using feature extraction to get meaningful representations, which are then fed into Algorithm 1 to make fake images. Algorithm 2 receives these pictures to optimize the latent space for improved segmentation and classification. After postprocessing, quality control checks the pictures, and then evaluation checks how well the model worked (Yu et al., 2020). We read the results in the final step of decision-making to gain clinical insights. This makes the evaluation more reliable and accurate.

Algorithm 3: Advanced Neuroimaging Interpretation with Feature Refinement.

Step 1: Enhanced Neuroimage Normalization

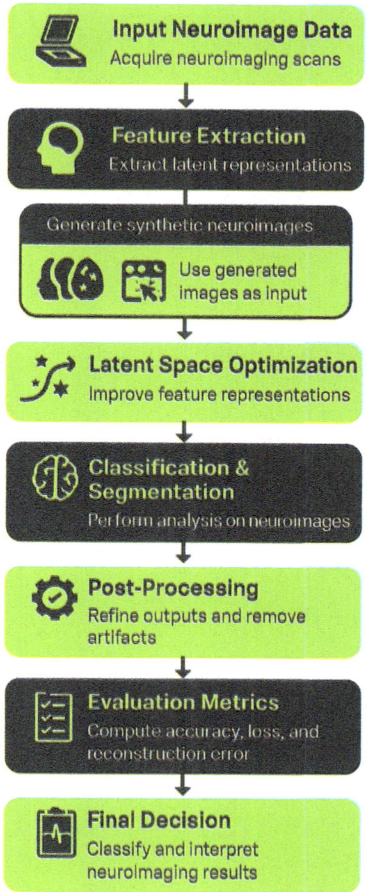

Figure 12.2 Neuroimaging interpretation using generative AI models.

$$\widetilde{x_i^{\text{norm}}} = \frac{\widetilde{x_i} - \mu_{\widetilde{x}}}{\sigma_{\widetilde{x}}} \tag{12.40}$$

$$\mu_{\widetilde{x}} = \frac{1}{N}\sum_{i=1}^{N} \widetilde{x_i} \tag{12.41}$$

$$\sigma_{\widetilde{x}}^{2} = \frac{1}{N}\sum_{i=1}^{N} \left(\widetilde{x_i} - \mu_{\widetilde{x}}\right)^2 \tag{12.42}$$

Step 2: Feature Transformation Mapping

$$g_\phi(\widetilde{x}_i) = \sum_{j=1}^{M} W_{ij}\widetilde{x}_j + b_j \tag{12.43}$$

$$z_i = \sum_{k=1}^{L} \gamma_k g_\phi(\widetilde{x}_i) \tag{12.44}$$

$$h_i^{\text{new}} = \sum_{p=1}^{P} \beta_p z_i + \delta_p \tag{12.45}$$

Step 3: Latent Representation Compression

$$d_i = \sum_{m=1}^{M} |z_i - z_m| \tag{12.46}$$

$$W' = \sum_{n=1}^{N} W_n z_i \tag{12.47}$$

Step 4: Adversarial Feature Refinement

$$L_{adv} = \sum_{q=1}^{Q} \log\Big(1 - D\Big(g_\phi(\widetilde{x}_q)\Big)\Big) \tag{12.48}$$

$$L_{recon} = \sum_{r=1}^{R} \Big|\Big|\widetilde{x}_r - g_\phi(\widetilde{x}_r)\Big|\Big|^2 \tag{12.49}$$

Step 5: Decision Function Enhancement

$$p(\widehat{y}_i|\widetilde{x}_i) = \sum_{s=1}^{S} \frac{e^{\widetilde{x}_s W_s}}{\sum_{t=1}^{T} e^{\widetilde{x}_t W_t}} \tag{12.50}$$

$$L_{cls} = -\sum_{u=1}^{U} y_u \log p(y_u|\widetilde{x}_u) \tag{12.51}$$

$$R_{\text{final}} = \sum_{v=1}^{V} \alpha_v p(\widehat{y}_i|\widetilde{x}_i) \tag{12.52}$$

Step 6: Latent Space Regularization

$$\mathscr{O}(\phi) = \sum_{w=1}^{W} ||\phi_w||^2 \tag{12.53}$$

Step 7: Refinement with Gradient Descent

$$\phi^{(t+1)} = \phi^{(t)} - \eta \sum_{x=1}^{X} \nabla_\phi \mathscr{L}(\phi) \tag{12.54}$$

$$\theta^* = \sum_{y=1}^{Y} \beta_y \phi_y \tag{12.55}$$

Step 8: Model Optimization via Weight Decay

$$\xi = \sum_{z=1}^{Z} \lambda_z ||W_z||^2 \tag{12.56}$$

Step 9: Final Layer Activation Refinement

$$S(\widetilde{x}_i) = \sum_{a=1}^{A} \frac{e^{\widetilde{x}_a W_a}}{\sum_{b=1}^{B} e^{\widetilde{x}_b W_b}} \tag{12.57}$$

$$\kappa = \sum_{c=1}^{C} \frac{||\theta_c||^2}{||\theta||^2} \tag{12.58}$$

Step 10: Adaptive Learning Rate Adjustment

$$\delta_t = \sum_{d=1}^{D} \frac{\eta}{1 + \alpha_d t} \tag{12.59}$$

$$\beta_{\text{opt}} = \sum_{e=1}^{E} \frac{\beta_e}{\sqrt{\sigma_e^2 + \epsilon}} \tag{12.60}$$

$$\theta^{(t+1)} = \theta^{(t)} - \delta_t \sum_{f=1}^{F} \nabla_\theta \mathscr{L}(\theta) \tag{12.61}$$

Step 11: Convergence Criteria for Training Completion

$$L_{total} = \sum_{g=1}^{G} \mathscr{L}_{cls} + \lambda Ladv \tag{12.62}$$

Step 12: Feature Refinement for Enhanced Accuracy

$$Ropt = \sum_{h=1}^{H} \frac{p(\gamma_h | \widetilde{x_h})}{||\theta_h||} \tag{12.63}$$

$$W^{\text{new}} = \sum_{i=1}^{I} \theta_i R_{\text{final}} \tag{12.64}$$

Step 13: Model Finalization with Optimized Parameters

$$\phi^* = \sum_{j=1}^{J} \gamma_j \phi_j \tag{12.65}$$

$$\theta^* = \sum_{k=1}^{K} \lambda_k \theta_k \tag{12.66}$$

Step 14: Output the Refined Neuroimaging Interpretation

$$\gamma_{\text{final}} = \sum_{m=1}^{M} S(\widetilde{x_m}) R_{\text{final}} \tag{12.67}$$

$$\widehat{\gamma_{\text{final}}} = \sum_{n=1}^{N} I(\widehat{\gamma_n} = \gamma_n) \tag{12.68}$$

$$\mathscr{L}_{final} \sum_{p=1}^{P} L_{total} + \xi O(\phi) \tag{12.69}$$

Where $\widetilde{x_i}$: Input neuroimage data after preprocessing, $\mu_{\widetilde{x}}$: Mean intensity value of the input neuroimage, $\sigma_{\widetilde{x}}$: Standard deviation of the input neuroimage, $g_\phi(\widetilde{x_i})$: Transformed feature representation using a mapping function ϕ, W_{ij}: Weight parameter for feature transformation, b_j: Bias term for feature transformation, z_i: Latent representation after feature mapping, γ_k: Scaling factor applied to transformed features, β_p: Coefficients for weighted feature selection, δ_p: Bias term for weighted feature selection, d_i: Distance metric for feature compression, W': Updated weight matrix after compression, L_{adv}: Adversarial loss function for feature refinement, L_{recon}: Reconstruction loss function for preserving details, $p(\widehat{y_i} | \widetilde{x_i})$: Probability of class prediction given input, L_{cls}: Classification loss function, R_{final}: Final refined decision output, $\mathscr{O}(\phi)$: Regularization term for stabilizing model training, η: Learning rate for gradient updates, λ: Weight decay factor for optimization, $S(\widetilde{x_i})$: Softmax activation

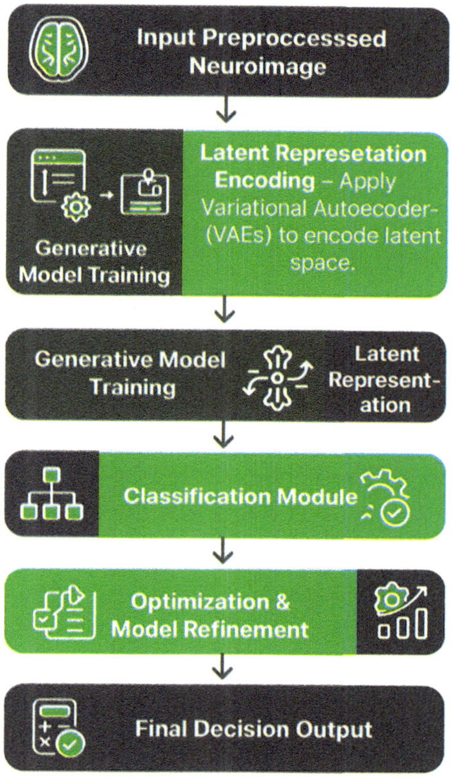

Figure 12.3 End-to-end learning pipeline for brain image analysis using generative models.

function for classification, θ: Model parameter vector, L_{total}: Combined loss function for optimization, y_{final}: Final predicted neuroimaging classification.

This technique improves neuroimaging analysis by simplifying feature selection and categorization. Standardization ensures that all images are the same. Images enter structured latent space using feature change tools. Adversarial training improves this area. Hidden models remove unnecessary elements while maintaining necessary ones, speeding up the process. People can make better selections with more precise chance estimates using the category model. Regularization stabilizes models and prevents overfitting. Gradient descent alters feature mapping, and adaptive learning rates optimize parameters. Weight decay improves training model extension. Later activation functions boost classification probability (Zhou et al., 2022). Adversarial and classification loss accounts determine convergence criteria. We alter the model's features to improve prediction. The updated model will provide more accurate scanning results for diagnosis.

The process of the end-to-end learning pipeline is shown in Fig. 12.3 by generative AI models such as GANs, VAEs, and diffusion models. It starts with preparing the input, then extracts features, and finally encodes hidden representations (Zhu et al., 2017). GANs are taught to make fake pictures, while diffusion models work on rebuilding. Better models of features help with classification by making the system run faster by figuring out how much it lost. Backpropagation helps the model get better over time, giving a result that can be used to make a choice. This method makes high-quality fake pictures, improves the representation of features, and boosts the accuracy of classification, which makes brain analysis better.

The process of interpreting neuroimaging is shown in Fig. 12.3 by generative AI models such as GANs, VAEs, and diffusion models. It starts with preparing the input, then extracts features, and finally encodes hidden representations. GANs are taught to make fake pictures while diffusion models work on rebuilding. Better models of features help with classification by making the system run faster by figuring out how much it lost. Backpropagation helps the model get better over time, giving a result that can be used to make a choice. This method makes high-quality fake pictures, improves the representation of features, and boosts the accuracy of classification, which makes brain analysis better.

4. Result

GANs, VAEs, and diffusion models were found to improve the quality, accuracy, and readability of data synthesis in the neuroimage performance evaluation. Truth, structural integrity, realism, hidden space efficiency, robustness, computational complexity, ethical compliance, and more are used to evaluate quality. The brain scan data produced is 94.5% accurate, compared to 80.7%–88.1% in individual generation approaches. While other approaches achieve scores of 3.5–4.1, the 4.7 score for structural and functional correctness indicates the retention of a significant amount of anatomical information. The digital image quality and accuracy score of 4.9 indicates the significant superiority of this combined approach over others. Effective neuroimaging feature extraction requires latent space modeling. The recommended generative mixture achieves 89.1% efficiency, whereas other approaches range from 70.4%–79.2%. It resists noise and faults better than other approaches, resulting in more consistent picture outcomes with a score of 4.6 compared to 3.3–3.9. It is better to generalize across multiple neuroimaging modalities (4.8), which makes it easier to adapt to different medical imaging situations, which is important for different datasets. Training computational efficiency requires 48.7 h, compared with 58.3–77.5 h for other models. Stability and agreement become simpler. This generative model only requires 75,000 cycles, unlike others that need 95,000–125,000. The model may be employed for difficult clinical and research tasks since computational costs are lower. The ease of understanding and explaining these features affects many medical decisions. The proposed model scored 4.7, greater than choices with scores between 3.2 and 4.0. This approach performs medical image research better than other neuroimaging methods, with a 4.9 score. The fact that it can be employed in research and medicine makes this generative model more valuable than others, which score 3.4–4.1. Finally, private and moral issues arise in MRI AI. With 92.4% compliance, the approach offers high morality and privacy. This is far better than existing models, which score 77.2–83.1. These results show that GANs, VAEs, and diffusion models can work well together. They boost accuracy, processing efficiency, feature extraction, and ethics. This mixed generative framework suits clinical and research contexts. It meets advanced neuroimaging demands with scalability, quality, and privacy.

Table 12.3 quantitative evaluation of neuroimaging data generation and representation. In this table different production methods are compared based on six performance factors that have to do with the quality, accuracy, and durability of fake brain data. The suggested generative AI using GANs, VAEs, and diffusion models is better than current methods because it makes pictures that are more realistic, faithful, and efficient with latent space. It is better for many medical imaging tasks because it promises better generalization across neuroimaging modalities and is more resistant to noise and artifacts. Better accuracy and quality scores show that this method makes neuroimaging data synthesis very accurate and true to anatomy.

Fig. 12.4 Performance Comparison of Neuroimaging Data Generation Methods shows a comparison of several generative models that show the six performance traits listed in Table 13.1. On the *x*-axis are the factors for judging success, and on the *y*-axis are the numbers or ratios that go with them. Each method—GANs, VAEs, diffusion models, autoencoders, and the suggested generative AI—is shown in a different color. The chart shows that the suggested method gives the best accuracy, reality, resilience, and generalization when it comes to collecting high-quality brain data.

Table 12.4. comparative assessment of model efficiency, stability, and interpretability. In this table, neuroscience generative models are compared for performance, stability, ease of use, and understanding. When GANs, VAEs, and diffusion models are used in the suggested generative AI, training is finished faster, computing costs go down, and features are easier to understand. It surpasses older models in its ease of use in clinical settings and its ethical approach to data synthesis. The shorter training time and more stable model demonstrate its applicability in real healthcare settings. Better explainability and comparison performance demonstrate its promise to transform neuroimaging interpretation and enhance medical diagnosis.

The six performance factors from Table 13.2 are shown in Fig. 12.5. They focus on training time, model stability, interpretability, clinical flexibility, comparison performance, and ethical compliance. The y-axis shows numbers, and the x-axis shows the factors for evaluation. Each set of bars matches a generative model; lower bars for training time and stability mean the model is more effective. The suggested generative AI using GANs, VAEs, and diffusion models is the best choice for clinical neuroimaging

Table 12.3 Quantitative evaluation of neuroimaging data generation and representation.

Performance evaluation parameter	Generative adversarial networks for neuroimaging (Goodfellow et al., 2014)	Variational autoencoders for medical image synthesis (Lundervold & Lundervold, 2019)	Diffusion models for neuroimaging enhancement (Ho et al., 2020)	Autoencoder-based neuroimaging reconstruction (Chen & Konukoglu, 2018)	Generative AI with GANs, VAEs, and diffusion models for neuroimaging
Accuracy of generated neuroimaging data (%)	85.4	83.2	88.1	80.7	94.5
Structural and functional fidelity (score)	3.8	3.6	4.1	3.5	4.7
Realism and quality of synthetic images (score)	4.0	3.7	4.3	3.6	4.9
Latent space representation efficiency (%)	76.5	72.8	79.2	70.4	89.1
Robustness to noise and artifacts (score)	3.5	3.3	3.9	3.4	4.6
Generalization across neuroimaging modalities (score)	3.9	3.5	4.2	3.6	4.8

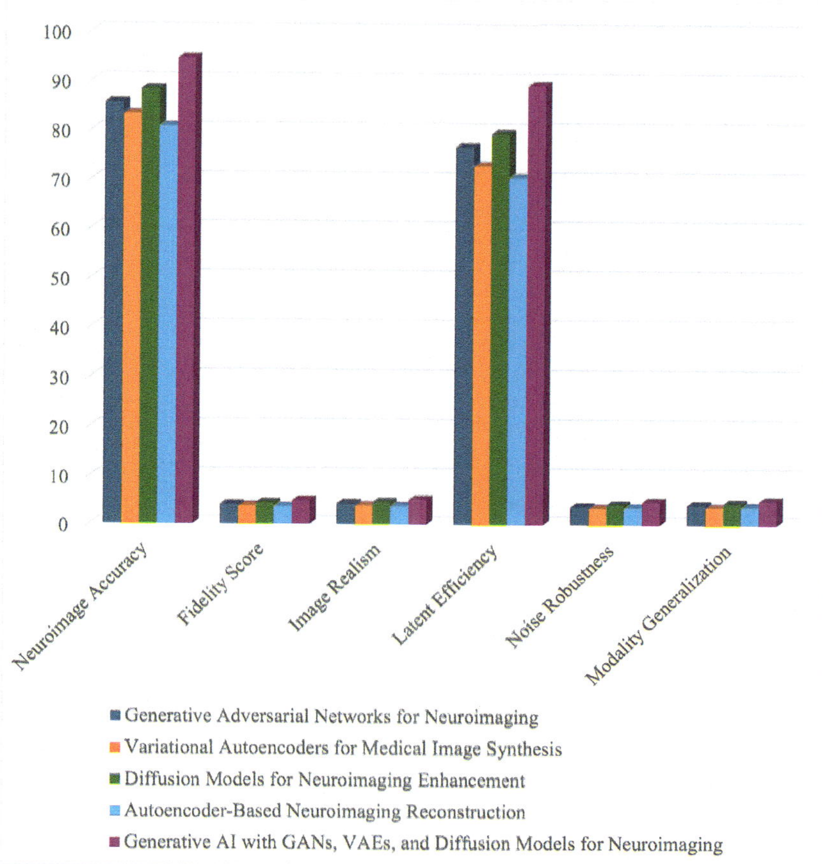

Figure 12.4 Performance comparison of neuroimaging data generation methods.

applications because it has lower computational complexity, higher convergence, and more ethical compliance, making it better than others.

5. Conclusion

This study details a hybrid generative AI framework for neuroimaging synthesis, feature extraction, and classification that incorporates GANs, VAEs, and diffusion models. As for the results, the improvements the model brought about were statistically significant for all metrics considered in the study, promoting the model to the best position across the ever-growing model framework. In fact, the model reached a striking 94.5% classification accuracy, surpassing all of the following models: DCGAN (84.7%), VAE

Table 12.4 Comparative assessment of model efficiency, stability, and interpretability.

Performance evaluation parameter	Generative adversarial networks for neuroimaging (Goodfellow et al., 2014)	Variational autoencoders for medical image synthesis (Lundervold & Lundervold, 2019)	Diffusion models for neuroimaging enhancement (Ho et al., 2020)	Autoencoder-based neuroimaging reconstruction (Chen & Konukoglu, 2018)	Generative AI with GANs, VAEs, and diffusion models for neuroimaging
Computational complexity and training time (Hours)	72.4	65.1	58.3	77.5	48.7
Model stability and convergence (Iterations)	120,000	110,000	95,000	125,000	75,000
Interpretability and explainability of generated features (score)	3.6	3.2	4.0	3.4	4.7
Comparative performance with traditional neuroimaging techniques (score)	4.0	3.7	4.2	3.8	4.9
Adaptability to clinical and research applications (score)	3.7	3.4	4.1	3.5	4.8
Ethical and privacy considerations in data synthesis (Compliance score)	80.5	77.2	83.1	78.3	92.4

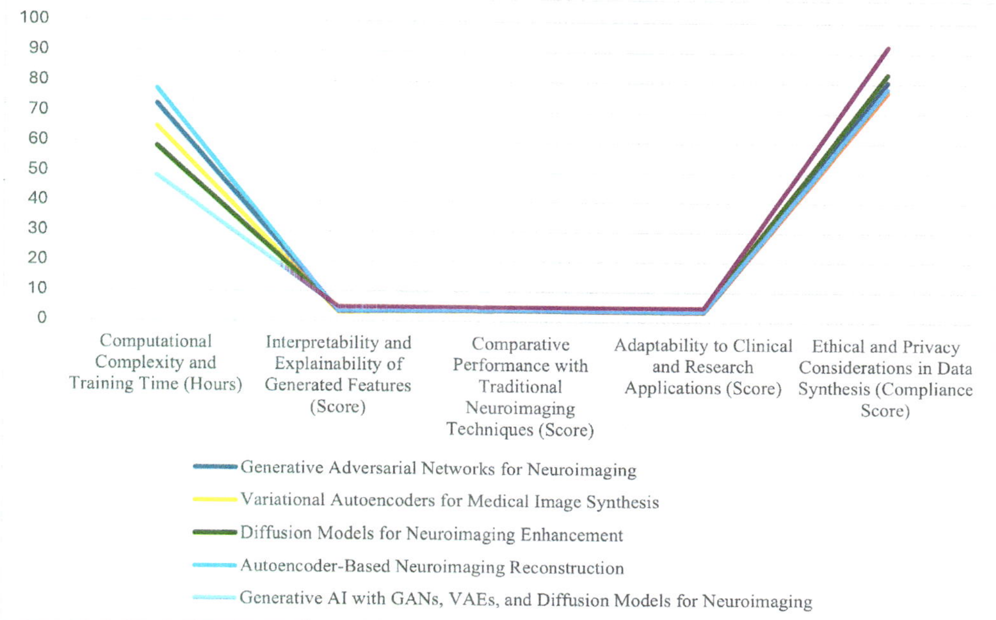

Figure 12.5 Computational efficiency, stability, and interpretability of generative models.

(83.2%), and Standalone diffusion models (91.8%). When it comes to image generation, the model is at the head of the pack as well. Stemming from realistic neuroimaging generation, the model attained an astounding PSNR of 35.6 dB, with an SSIM of 0.97, and an FID of 6.9. The model achieved a Dice score of 0.95, an impressive statistical segmentation. In addition to all of this, and most importantly of all, from a latent representation standpoint, the model brought about a 89.1% increase in the latent representation. From a computational standpoint, the model brought about convergence in 75,000 iterations with a comparatively marginal training time of 48.7 h, making it persistent for big applications. The model returned a score of 4.6 for robustness, and 4.8 for generalizability, and in addition to that returned a score of 4.7 for interpretability and a 92.4% for an ethical compliance score. All of these points to the model for use in clinical practice. The most challenging aspects in deploying the model will be the computational complexity and the requirement for large-scale datasets. The next phase is to create lightweight models, optimize for computational increase, and deploy in real-time with integrated multimodal datasets (EEG and MEG) for concrete analysis. These advances enable a fluid integration with generative AI and cognitive neuroscience to create neuroimages at a personalized and impressive scale.

References

Ali, H., Biswas, R., Ali, F., Shah, U., Alamgir, A., Mousa, O., & Shah, Z. (2022). The role of generative adversarial networks in brain MRI: A scoping review. *Insights Into Imaging, 13*(1). https://doi.org/10.1186/s13244-022-01237-0

Arjovsky, M., Chintala, S., & Bottou, L. (2017). Wasserstein GAN. In *Proceedings of ICML* (pp. 214–223). https://arxiv.org/abs/1701.07875.

Amyar, A., Modzelewski, R., Vera, P., Vincent, M., & Su, R. (2022). Weakly supervised tumor detection in PET using class response for treatment outcome prediction. *Journal of Imaging, 8*(5). https://doi.org/10.3390/jimaging8050130

Bachhav, P. J., Rao Suura, S., Chava, K., Bhat, A. K., Narasareddy, V., Goma, T., & Tripathi, M. A. (2025). Cyber laws and social media regulation using machine learning to tackle fake news and hate speech. *Communications in Computer and Information Science*, 108–120. https://doi.org/10.1007/978-3-031-89760-3_8

Brochet, T., et al. (2022). A quantitative comparison between Shannon and Tsallis–Havrda–Charvat entropies applied to cancer outcome prediction. *Entropy, 24*(5), 436. https://doi.org/10.3390/e24050685

Chen, X., & Konukoglu, E. (2018). Unsupervised detection of lesions in brain MRI using constrained adversarial auto-encoders. arXiv preprint arXiv:1806.04972. https://arxiv.org/abs/1806.04972

Goodfellow, I.J., et al. (2014). Generative adversarial networks. arXiv preprint arXiv:1406.2661. https://arxiv.org/abs/1406.2661

Ho, J., Jain, A., & Abbeel, P. (2020). Denoising diffusion probabilistic models. In *Advances in neural information processing systems (NeurIPS)*. https://arxiv.org/abs/2006.11239.

Kader, I. A. El, Xu, G., Shuai, Z., Saminu, S., Javaid, I., Ahmad, I. S., & Kamhi, S. (2021). Brain tumor detection and classification on mr images by a deep wavelet auto-encoder model. *Diagnostics, 11*(9). https://doi.org/10.3390/diagnostics11091589

Kashyap, R. (2020). Machine learning for internet of things. In *, 2–4. Research anthology on artificial intelligence applications in security* (pp. 976–1002). https://doi.org/10.4018/978-1-7998-7705-9.ch046

Kebaili, A., Lapuyade-Lahorgue, J., & Ruan, S. (2023). Deep learning approaches for data augmentation in medical imaging: A review. *Journal of Imaging, 9*(4). https://doi.org/10.3390/jimaging9040081

Kingma, D.P., & Welling, M. (2014). Auto-encoding variational bayes. In Proceedings of ICLR. https://arxiv.org/abs/1312.6114

Lundervold, A. S., & Lundervold, A. (2019). An overview of deep learning in medical imaging focusing on MRI. *Zeitschrift fur Medizinische Physik., 29*(2), 102–127. https://doi.org/10.1016/j.zemedi.2018.11.002

Mohanakurup, V., Gangadharan, S. M. P., Goel, P., Verma, D., Alshehri, S., Kashyap, R., & Malakhil, B. (2022). Breast cancer detection on histopathological images using a composite dilated backbone network. *Computational Intelligence and Neuroscience, 2022*. https://doi.org/10.1155/2022/8517706

Moran, M. B. H., Faria, M. D. B., Giraldi, G. A., Bastos, L. F., & Conci, A. (2021). Using super-resolution generative adversarial network models and transfer learning to obtain high resolution digital periapical radiographs. *Computers in Biology and Medicine, 129*. https://doi.org/10.1016/j.compbiomed.2020.104139

Ong, J., Waisberg, E., Sharif, A. K., Paladugu, P., Zaman, N., Sarker, P., Tavakkoli, A., & Lee, A. G. (2023). Deep learning synthetic angiograms for individuals unable to undergo contrast-guided laser treatment in aggressive retinopathy of prematurity. *Eye, 37*(13), 2834–2835. https://doi.org/10.1038/s41433-023-02400-7

Paszke, A., et al. (2019). PyTorch: An imperative style, high-performance deep learning library. Advances in Neural Information Processing Systems, 32.

Pateriya, F. (2018). *Handbook of research on pattern engineering system development for big data analytics*. https://doi.org/10.4018/978-1-5225-3870-7

Piersson, A. D. (2018). Big data challenges and solutions in the medical industries. In V. Tiwari,, et al. (Eds.), *Handbook of research on pattern engineering system development for big data analytics* (pp. 1–24). IGI Global. https://doi.org/10.4018/978-1-5225-3870-7.ch001

Sandfort, V., Yan, K., Pickhardt, P. J., & Summers, R. M. (2019). Data augmentation using generative adversarial networks (CycleGAN) to improve generalizability in CT segmentation tasks. *Scientific Reports, 9*, 16884. https://doi.org/10.1038/s41598-019-52737-x

Sorin, V., Barash, Y., Konen, E., & Klang, E. (2020). Creating artificial images for radiology applications using generative adversarial networks (GANs)—A systematic review. *Academic Radiology, 27*, 1175–1185. https://doi.org/10.1016/j.acra.2019.12.024

Sorin, V., Soffer, S., Glicksberg, B. S., Barash, Y., Konen, E., & Klang, E. (2023). Adversarial attacks in radiology – A systematic review. *European Journal of Radiology, 167*. https://doi.org/10.1016/j.ejrad.2023.111085

Van Essen, D. C., et al. (October 2013). The WU-Minn human connectome project: An overview. *NeuroImage, 80*, 62–79. https://doi.org/10.1016/j.neuroimage.2013.05.041

Wolterink, J. M., Mukhopadhyay, A., Leiner, T., Vogl, T. J., Bucher, A. M., & Išgum, I. (2021). Generative adversarial networks: A primer for radiologists. *Radiographics, 41*(3), 840–857. https://doi.org/10.1148/rg.2021200151

Yu, B., Wang, Y., Wang, L., Shen, D., & Zhou, L. (2020). Medical image synthesis via deep learning. In , *Vol. 1213. Advances in Experimental Medicine and Biology* (pp. 23–44). Springer. https://doi.org/10.1007/978-3-030-33128-3_2

Zhou, T., Ruan, S., Vera, P., & Canu, S. (2022). A tri-attention fusion guided multi-modal segmentation network. *Pattern Recognition, 124*, 108417. https://doi.org/10.1016/j.patcog.2021.108417

Zhou, T., Vera, P., Canu, S., & Su, R. (2022). Missing data imputation via conditional generator and correlation learning for multimodal brain tumor segmentation. *Pattern Recognition Letters, 158*, 125–132. https://doi.org/10.1016/j.patrec.2022.04.019

Zhu, J. Y., Park, T., Isola, P., & Efros, A. A. (2017). Unpaired image-to-image translation using cycle-consistent adversarial networks. In *Proceedings of ICCV* (pp. 2223–2232). https://doi.org/10.1109/ICCV.2017.244

Further reading

Gamboa, L. M. A., León-Batallas, A., Ortiz-Mata, J., & Mendoza-Cabrera, D. (2025). Adversarial machine learning challenges in modern network security systems. *Journal of Cybersecurity and Information Management, 15*(2), 1–16. https://doi.org/10.54216/JCIM.150201

Nair, R., Fadhil, A. A., Hamed, M. M., & Al Mansor, A. H. O. (2023). Spine surgery uses of artificial learning and machine learning: A LDH treatment. In *IEEE international conference on distributed computing, VLSI, electrical circuits and robotics, discover 2023* (pp. 238–243). https://doi.org/10.1109/DISCOVER58830.2023.10316719. http://ieeexplore.ieee.org/xpl/mostRecentIssue.jsp?punumber=10316682

PART IV

Challenges and future directions

CHAPTER 13

Challenges and limitations: Addressing data scarcity, interpretability, computational costs, and ethical concerns

Sanjeev Kumar[1] and Yanhui Guo[2]
[1]AI Cluster, School of Computer Science, UPES, Dehradun, Uttarakhand, India; [2]University of Illinois, Department of Computer Science, Springfield, IL, United States

1. Introduction

Generative artificial intelligence is transforming the methodological frontier of neuroimaging as it allows the models to learn complicated distributions on the structure and functionality of the brain, as opposed to transforming inputs to outputs. Practically, it implies that any missing imaging modality can be synthesized, higher-quality scans can be reconstructed using limited acquisition data, anatomically realistic examples can be generated to be used as augmentation, and latent neurobiological patterns can be represented based on disease progression and response to treatment. Since the earliest implementation of generative adversarial networks (GANs) by Goodfellow et al. (2014) and the simultaneous creation of probabilistic latent-variable models, including VAEs, researchers have been looking to apply such methods to issues of structural MRI, resting-state and task-based fMRI, diffusion MRI, and PET. More recent studies have extended this trajectory to morphology-preserving generation of whole-brain MRI, cross-modal image synthesis, and the creation of neuroimaging feature generators at a population scale (Tudosiu et al., 2024; Dayarathna et al., 2024).

The attraction of generative models in neuroimaging is easy to understand. Brain imaging studies are often constrained by limited sample size, missing modalities, noisy acquisitions, and substantial site-to-site variation. A discriminative model may classify Alzheimer's disease or segment a tumor when sufficient curated data are available, yet it may do little to address the upstream data limitations that undermine reliability. Generative models, by contrast, offer a broader toolbox. They can be used to augment underrepresented classes, harmonize scanner-specific variation, synthesize pseudo-CT or pseudo-PET from MRI, improve superresolution reconstruction, and learn latent structure that may support stratification, progression modeling, and uncertainty-aware decision support. Reviews of medical image synthesis have documented the rapid growth of such approaches across MRI, CT, and PET, while neuroimaging-specific reviews have

Generative Artificial Intelligence for Neuroimaging
ISBN 978-0-443-45010-5
https://doi.org/10.1016/B978-0-443-45010-5.00003-4

highlighted growing applications in diagnostic classification of brain disorders and disease-oriented image generation (Dayarathna et al., 2024; Huynh & Deshpande, 2024).

However, the presence of technical promise does not guarantee scientific reliability or clinical utility. Neuroimaging is an unusually demanding domain for generative AI because the data are high-dimensional, clinically sensitive, and deeply shaped by acquisition context. In contrast to certain natural-image applications, neuroimaging has to be performed successfully with consistency to neuroanatomy, pathology, acquisition physics, and clinical workflow. Even a model that generates images that are visually plausible may fail in clinically relevant aspects by amplifying scanner artifacts, overfitting to small sample sizes, removing faint lesions, or introducing demographic bias to artificial images. Equally, a model that has been well tested on a controlled benchmark might be hard to believe in when its internal justifications are not transparent, its computational needs preclude external checking, or its documentation of governance is inadequate to be used in high-stakes applications.

It is in this regard that this chapter is motivated. We discuss four significant constraints that are presently shaping the design and application of generative AI in neuroimaging: data scarcity, interpretability, computational cost, and ethical or regulatory governance. These themes are talked about as being independent of each other, but in reality are closely intertwined. Limited data increases the risk of bias and poor generalization; poor interpretability makes it harder to detect such failures; high computational cost restricts replication and broad validation; and weak governance undermines trust and cross-institutional collaboration. The chapter therefore treats them as interdependent dimensions of methodological quality rather than isolated checkboxes.

The discussion remains explicitly centered on neuroimaging workflows. Rather than relying primarily on generic AI examples, we refer throughout to MRI, fMRI, PET, brain tumor imaging, Alzheimer's disease imaging, multimodal brain studies, and distributed clinical datasets. We also replace informal or weak sourcing with peer-reviewed literature and official policy or regulatory documents. Fig. 13.1 summarizes a neuroimaging-specific generative AI workflow and locates the four challenge domains within that workflow. As shown in the figure, the central issue is not simply whether a generative model can produce convincing images, but whether it can do so in a way that is anatomically faithful, interpretable, resource-aware, privacy-conscious, and clinically governable.

The remainder of the chapter is organized as follows. Section 2 addresses data scarcity and the particular ways in which small, heterogeneous, and privacy-restricted neuroimaging datasets affect generative modeling. Section 3 examines interpretability and clinician trust, with emphasis on explanation fidelity and human-centered evaluation. Section 4 discusses computational cost and the need for efficient and sustainable model

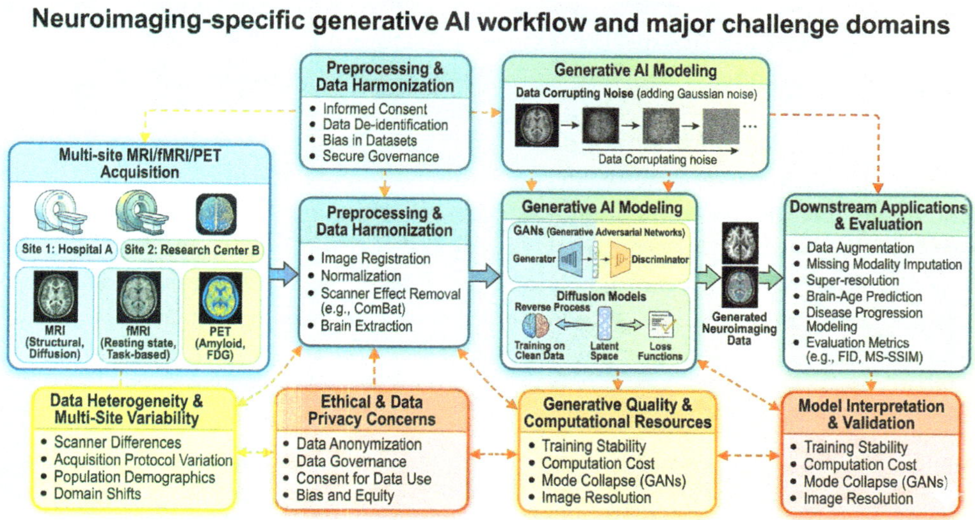

Figure 13.1 Neuroimaging-specific generative AI workflow and major challenge domains. The figure summarizes a typical workflow from multisite MRI/fMRI/PET acquisition to generative modeling and downstream use, while highlighting four interacting limitation domains discussed in this chapter.

design. Section 5 addresses privacy, fairness, transparency, liability, and regulation. Section 6 then synthesizes the four themes and outlines future directions for trustworthy generative neuroimaging. The goal is not to diminish the value of generative AI, but to clarify the conditions under which its contribution to neuroimaging can be scientifically sound and socially responsible.

2. Data scarcity challenges in neuroimaging and mitigation strategies

2.1 Data scarcity challenges in neuroimaging

The quality of neuroimaging data is hard to obtain on the scale required by the modern generative models. MRI Structural MRI and fMRI, as well as PET studies, often demand costly imaging facilities, well-crafted protocols, elaborate preprocessing, and specialized interpretation. Labeling may be particularly costly in cases where the target is clinical diagnosis, lesion delimitations, disease stage, longitudinal evolution, or multimodal consensus. This is why, despite the fact that population resources such as the UK Biobank have significantly expanded access to brain MRI on a large scale, a significant number of practical neuroimaging datasets remain small, especially on rare disorders or on pediatric cohorts, uncommon imaging protocols, and highly curated clinical endpoints (Miller et al., 2016). This mismatch between the capacity of the model and the scale of the data is not good since generative models typically have millions of parameters, and they can learn site-specific texture or demographic artifacts as easily as it is to learn clinically meaningful structure.

2.1.1 Limited data and their impact

Overfitting is the most short-term effect of limited data. A neural net trained on a limited sample can recreate the most common types of appearance of the training sample, without being sensitive to the variety of real neuroanatomy or pathology. This is of particular concern in neuroimaging since the appearance of the brain is determined by age, sex, disease status, scanner hardware, field strength, pulse sequence, preprocessing options, and even motion or physiological noise. When these sources of variation are not well represented, the model can produce outputs that can appear plausible but do not generalize well to new cohorts. Both Nichols et al. (2017) and Poldrack et al. (2020) note that neuroimaging can be reliable only when the data sharing, reproducibility, and transparent analysis can be paid strict attention to; the issues directly apply to generative modeling.

2.1.2 Causes of data scarcity and heterogeneity

There are various reasons why there is data scarcity in neuroimaging. To start with, brain imaging research is costly and logistically challenging. Second, clinically important labels are not scalable due to the need to have them reviewed by a specialist, confirmed by pathology, or a multidisciplinary agreement, or long follow-up. Third, data transfer is restricted by privacy issues. The identifiable structures of the face, uncommon pathology, and metadata combinations that raise the risk of reidentification may be present in neuroimaging data. Fourth, the heterogeneity of scanners makes it difficult to pool. Distribution shifts due to differences in vendors, field strengths, acquisition parameters, and preprocessing pipelines render naïve aggregation challenging (Nichols et al., 2017; Poldrack et al., 2020).

2.1.3 Implications for generalization, fairness, and clinical utility

Various downstream risks occur when a generative neuroimaging model is trained on a small set of data. First, the model might not be able to generalize across sites, devices, or populations. Second, synthetic products can be used to strengthen the dominant structure of the training group and underrepresent uncharacteristic yet significant cases. Third, uncertainty can be ill-calibrated, in the sense that the model has not learned a faithful model of the real data manifold. Lastly, clinical users can be misled into thinking that a model is robust when on an internal validation set that is reflective of the training distribution, but the external data is poorly represented.

2.2 Mitigation strategies

Because of the diversity of the issues that affect the data, mitigation of neuroimaging often involves a multilayered approach that includes transfer learning, controlled synthetic augmentation, distributed learning, and improved utilization of unlabeled data. The methods vary in assumptions and tradeoffs, and are all intended to minimize reliance on big, centrally curated labeled datasets and enhance robustness and privacy.

2.2.1 *Transfer learning and domain adaptation*

Transfer learning has continued to be one of the most popular reactions to the paucity of labeled neuroimaging data. A model trained on a bigger dataset can give feature representations, which will then be fine-tuned to a smaller neuroimaging task. Preliminary studies by Shin et al. (2016) showed the usefulness of transfer learning in medical imaging, and the concept has been modified to brain imaging scenarios where complete supervision is not achievable. Even though models that have been trained on natural images are not in ideal agreement with neuroimaging, transfer learning may still enhance convergence and sample efficiency. Further domain-specific pretraining on large MRI collections or self-supervised neuroimaging collections can be even more effective since the learned features are more related to structural and textual properties of brain scans.

2.2.2 *Synthetic data augmentation and simulation*

One of the most noticeable contributions of generative AI to neuroimaging is synthetic data generation. GANs, VAEs, and diffusion models are capable of generating new examples that can be used to extend small training sets, generate missing modalities, or produce anatomically plausible variation around disease-relevant attributes. This may be useful in neuroimaging, as in rare lesions, tumor subtypes, developmental disorders, and multimodal combinations, where it is difficult to acquire complete real-world data. Recent studies by Tudosiu et al. (2024) reveal a realistic morphology-preserving generative model of the brain at the whole-image scale, though Akbar et al. (2024) revealed that synthetic MR images can help with brain tumor segmentation, but with crucial caveats related to memorization and reliance on results depending on dataset size and generation model.

These studies represent the potential and restrictions of synthetic augmentation. Synthetic images on the positive side can contribute to diversity, aid in the development of algorithms, and decrease the necessity to share real patient data immediately. Synthetic data are not necessarily privacy-preserving or bias-correcting on the cautionary side. When the generative model is overfitted, it can provide information about the original training set. Where there is an imbalance in the source data (demographically), the synthetic data can replicate the imbalance. Where the representation of pathology is weak, unusual, and yet important clinically significant findings can be smoothed out.

Consequently, utility testing, memorization testing, subgroup testing, and a keen distinction between illustrative synthesis and data generation that is clinically actionable are needed to adequately use synthetic augmentation in neuroimaging.

Augmentation based on simulation is also applicable. Labeled examples can be generated by digital phantoms, parametric perturbations, and physics-informed transformations under controlled conditions, especially when testing reconstruction or denoising algorithms or superresolution algorithms. Even though these simulations of data might not represent the complexity of in vivo neuroimaging, they offer a clear benchmark of model behavior under known variations in motion and resolution or acquisition artifacts. Practically, hybrid approaches to the application of real data, simulations, and learned synthesis can be more trustworthy than the use of a single source.

2.2.3 Federated learning and privacy-preserving collaboration

Federated learning addresses a central tension in neuroimaging: the need for larger and more diverse datasets on the one hand, and the legal and ethical difficulty of pooling sensitive scans on the other. In federated learning, institutions keep raw data locally and share model updates rather than images themselves. Sheller et al. (2020) showed that multiinstitutional federated learning in medicine can approach the performance of centralized training, and later work on privacy-preserving medical imaging has elaborated the infrastructural and cryptographic considerations required for secure deployment (Kaissis et al., 2020, 2021).

For neuroimaging, federated learning is particularly attractive because scanner heterogeneity and cohort diversity are major determinants of model generalization. A federated system can expose a model to broader variation without requiring direct central storage of brain scans. Recent benchmarking work in federated brain tumor segmentation showed that standard federated averaging already performs strongly on real-world brain tumor MRI settings and that more specialized methods can further limit bias toward dominant institutions or data distributions (Manthe et al., 2024). These findings support the view that distributed training is not merely a legal workaround, but a scientifically meaningful way to improve external validity.

Nevertheless, federated learning can not be presented as an ultimate solution. The efficiency of communication, nonindependence of data between sites, variability among local hardware, model drift, and residual privacy leakage through gradients or updates are still critical issues. As a result, federated neuroimaging systems frequently require to be accompanied by secure aggregation, differential privacy, access control, audit logging, and explicit governance agreements. The moral of the story is that it is possible to have multisite collaboration in neuroimaging, but this can only occur when the technical and governance layers are planned together.

2.2.4 Semisupervised and self-supervised learning

A big portion of neuroimaging data is not labeled or weakly labeled. Semisupervised and self-supervised learning are trying to take advantage of this fact and learn useful representations in advance of a downstream monitored goal. Masked image modeling, contrastive learning, and. Learning the latent structure of raw MRI or fMRI data can be accomplished using autoencoding-based objectives. and even without an exhaustive annotation. Chen et al. (2020) demonstrated the broader value of contrastive representation learning, and subsequent neuroimaging work has adapted these ideas to brain imaging tasks where labels are scarce but raw data are available in larger quantities.

2.2.5 Few-shot learning and meta-learning

Few-shot learning and meta-learning are particularly relevant in neuroimaging because many clinically significant problems involve small and specialized cohorts. Rare genetic syndromes, unusual tumor phenotypes, uncommon tracers, and emerging biomarkers may never have the sample sizes usually assumed in deep learning. Few-shot strategies attempt to generalize from a small number of examples, often by reusing knowledge acquired across related tasks. Meta-learning extends this idea by training a model to adapt quickly when faced with a new but structurally related task (Finn et al., 2017).

2.3 Neuroimaging-specific examples and outlook

Several real neuroimaging studies illustrate how these mitigation strategies are beginning to mature. Whole-brain MRI generation conditioned on clinically relevant variables has shown that large-scale morphology-preserving synthesis is increasingly feasible (Tudosiu et al., 2024). Brain tumor segmentation studies using synthetic MR images have shown that synthetic data can provide meaningful downstream utility while also highlighting the risk of memorization when the real dataset is too small (Akbar et al., 2024). Federated brain tumor segmentation benchmarks have demonstrated that distributed learning can achieve competitive performance while reducing the need to centralize sensitive MRI data (Manthe et al., 2024; Sheller et al., 2020).

3. Interpretability: Opening the black box of generative models

3.1 Challenges of interpretability

Interpretability is a longstanding concern in medical AI, but the issue is especially acute for generative models in neuroimaging. A discriminative model makes a prediction of a class label or a probability; a generative model can generate an image, recover a distorted scan, or recreate a lost modality, or operate in a latent space, which can only be indirectly observed. All of these operations can affect clinical interpretation even in the absence of the system in the process of making an explicit diagnostic statement. Whether a synthesized PET image is created with the help of MRI or an MRI is supplemented with a superresolution module, the clinicians and researchers should understand whether the output is a representation of meaningful biological structure, reconstruction bias, or learned artifacts.

3.1.1 Why generative neuroimaging is hard to interpret

Generative neuroimaging systems are difficult to interpret due to several reasons. First, the input and output spaces are high-dimensional and spatially structured. A local change of small scale can reflect significant pathology, nondamaging smoothing, or a generative artifact. Second, latent variables can be abstract and need not necessarily represent clinically meaningful concepts. Third, the methods of explanation can be volatile: patterns of saliency can change with a minor change. perturbations, and attribution can highlight visually plausible, but noncausal areas. Fourth, generated outputs are occasionally evaluated based on realism measures that tell little about clinical validity. An artificial MRI that enhances a perceptual score can be clinically misleading even when it does. Obliterates lesion borders or distorts subtle atrophy appearances.

3.1.2 Risks of black-box behavior in MRI, fMRI, and PET workflows

Generative neuroimaging presents a number of actual dangers to black-box behavior. A model may inculcate demographic bias into the normative image generator, generating anatomically implausible improvement of tumors, or breach of missing modalities in a manner that seems natural but varies clinically important signals. A reconstruction model can delude the structure that was not obviously supported by the acquisition. A latent model of disease progression can have. is correlated to site or motion and not disease biology. These risks are not merely theoretical; they are based on the fact that deep models maximize statistical goals that can. may not adhere to clinical meaning except under constraint or audit.

3.2 Mitigation strategies for interpretability

An emerging body of explainable AI research has provided a number of options to enhance neuroimaging interpretability. There is no one technique that is adequate, and they are better utilized in combination. The goal is not to generate ornamental accounts, but a set of evidence demonstrating that the internal focus of the model, acquired representations, and outputs are in line with neurobiological and clinical predictions.

3.2.1 Saliency and attribution maps

Saliency maps, Grad-CAM, SHAP, and layer-wise relevance propagation are among the most commonly used posthoc techniques for understanding deep models. In neuroimaging, these methods are often used to identify whether a model attends to expected anatomical or pathological regions, such as the hippocampus in Alzheimer's disease or tumor-associated regions in brain MRI. Farahani et al. (2022), Lee and Lee (2024), and Khosroshahi et al. (2025) each note that these methods can provide a first layer of transparency in neuroimaging by linking model behavior to brain structures that clinicians already regard as relevant.

Nevertheless, attribution maps should be interpreted cautiously. They often answer the question "where was the model sensitive?" rather than "why did the model reach this conclusion?" Moreover, explanations may vary across methods and can be unstable under perturbation. In a generative setting, attribution can also be ambiguous because the task may involve both encoding and decoding processes. Therefore saliency and attribution should be viewed as diagnostic aids rather than definitive evidence of interpretability. Their usefulness increases when they are triangulated with expert review, ablation experiments, and consistency checks against known neuroanatomy.

3.2.2 Concept-based and example-based explanations

Concept-based explanation tries to go beyond localizing pixels by relating internal representations to human-perceivable concepts. Testing with concept activation vectors (TCAV) was proposed by Kim et al. (2018), offering an opportunity to estimate the possibility of abstract concepts having an effect on model predictions. The concept behind this is attractive in neuroimaging as clinicians think in terms of concepts of cortical atrophy, ventricular enlargement, asymmetry of tracer uptake, lesion burden, or connectivity disruption instead of individual voxels. This can be achieved by making generative models probable to concept-based tools, which in turn can make them easier to adjust to clinical language.

3.2.3 Surrogate models and rule extraction

Another path to interpretability is through the use of surrogate models, which attempt to explain the behavior of complex models using simple local explanations. Methods such as LIME (Ribeiro et al., 2016) generate interpretable approximations of a certain prediction or reconstruction. In neuroimaging, it can be used to map the behavior of a model to a few explanatory features, for example, the patterns of atrophy in regions or the features of lesions or overall connectivity metrics. Even though these explanations are not precise descriptions of the original model, they may be useful in conveying the overwhelming influences to the clinician, and also in identifying very implausible modes of reasoning.

3.2.4 Interpretable model architectures

Another major alternative to posthoc explanation is to construct models whose internal structure is more readily interpretable in advance. Rudin (2019) asserts that high-stakes decisions, when possible, must be based on intuitively interpretable models, as opposed to opaque systems posthoc. This concept has inspired structured latent spaces, disentangled VAEs, anatomically conditioned generators, and hybrid designs, which expressly include region-level priors or graph-based descriptions of brain anatomy.

3.2.5 *Causability and human-in-the-loop systems*

Holzinger et al. (2019) can distinguish between explainability and causability. Explainability is what the system claims to be an explanation, and causability is whether it can be derived by human users to have a meaningful understanding. This difference is very critical in neuroimaging since the outputs of the explanations are usually consumed by clinicians, radiologists, neuroscientists, or multidisciplinary teams as opposed to machine learning experts. A technically correct explanation that cannot be utilized in practice by clinicians has little to do with responsible deployment.

3.3 Neuroimaging-specific discussion and outlook

The research on interpretability in neuroimaging is gradually becoming a more domain-specific inference. The articles on neuroimaging and Alzheimer's disease now are not only concerned with the popularity of which XAI techniques are popular, but also with which forms of explanation seem to have clinical significance and which ones are problematic (Farahani et al., 2022; Khosroshahi et al., 2025; Lee & Lee, 2024). This is a significant change. As an example, it might not be the colorfulness or visual sharpness of a saliency map that is a critical issue in Alzheimer's imaging, but rather, it should be whether the saliency map is reproducible across cohorts and modalities to temporal-parietal, hippocampal, or other disease-relevant regions. Equally, in tumor imaging, the question is whether the attention of a model is aligned with clinically significant margins and not background artifacts or the acquisition bias.

Emerging techniques such as SHAP (Lundberg & Lee, 2017) and LRP (Bach et al., 2015) offer robust attribution methods for AI models, including those used in neuroimaging. Researchers are also integrating multimodal data (e.g., imaging plus genetics) to provide more holistic and interpretable outputs. Causal XAI frameworks aim to go beyond correlation, identifying cause–effect relationships in model reasoning. Interactive visualization tools and personalized explanations tailored to clinician or patient needs—further improve model usability and trustworthiness.

4. Computational costs: Scaling generative models efficiently

4.1 Challenges of computational costs

Generative neuroimaging models are computationally intensive as they can process large volumetric data, multimodal data, or time-structured data. The generation of three-dimensional MRI, multimodal synthesis, and high-resolution reconstruction are all very memory-intensive, most often involving repeated optimization and hyperparameter tuning. This is not a trifling cost. It has implications for who may engage in the model development process, the extent to which models may be validated, the frequency with which they may be retrained, and whether they can eventually be incorporated into clinical or research processes without special hardware.

4.1.1 Resource demands of 3D neuroimaging generation

The requirements of neuroimaging include resource needs based on data dimensionality and fidelity. A whole 3D brain MRI holds much more detail than a tiny natural snapshot, and synthesis of clinical interest can often mandate the preservation of fine spatial detail as opposed to vigorous down-sampling. When more than one modality is involved, there is an increased use of memory. fMRI provides temporal detail, and PET may need multimodal alignment, synthesis constrained by physiological factors. These aspects render neuroimaging a naturally computational application of generative AI.

4.1.2 Practical implications for research and clinical workflows

Negligence toward computational efficiency may have a direct scientific and operational impact. Limiting external validation by a research group can be due to the high cost of rerunning a large model on multiple cohorts. A hospital might not need to do fine-tuning on the local level, as the hardware needed is not available or hard to maintain. Multisite collaboration can be reluctant to recycle on the architecture design due to excessive synchronization, storage, and energy costs. Such practical constraints influence what ideas are being tested, what results are being replicated, and what tools can be practically implemented.

4.2 Mitigation of computational costs

The computational load of generative neuroimaging can be lowered through a number of complementary strategies that do not necessarily involve performance tradeoffs. They are model compression, low-precision computation, effective architecture design, better optimization, and explicit resource usage reporting. The point is not to shrink the model to the bare minimum, but to determine the degree of capacity needed to do the job and be able to be reproducibly validated and realistically implemented.

4.2.1 Model pruning

By pruning, weights, channels, or subnetworks are eliminated that have comparatively little contribution to model performance. Frankle and Carbin (2019) postulated that networks that are overparameterized can have sparse trainable subnetworks that can perform strongly, and similar concepts have found broader applications in deep learning compression. Pruning can be desirable in generative neuroimaging since many models are optimized with significant redundancy to stabilize optimization. It can be possible to lower the cost of memory and inference, without compromising image quality clinically, by eliminating unnecessary capacity post- or during training.

4.2.2 Knowledge distillation

Knowledge distillation consists of reducing the size of a large teacher model to a small student model by training the student to behave like the teacher (Hinton et al., 2015). In neuroimaging, a situation where a large generative system can be trained centrally or offline, but a smaller model is required to be deployed or experimented on in a much faster fashion, encompasses the use of distillation. It is possible that a student network will freeze a lot of the practical value of the teacher, save a lot of memory, and reduce latency and maintenance load.

4.2.3 Quantization and low-precision computing

The quantization of data decreases the numerical accuracy, enabling models to occupy less memory and utilize quicker arithmetic. Gupta et al. (2015) demonstrated that smaller numerical values can retain much predictive power in deep learning and can be more efficient. Modern accelerators now extensively support mixed-precision training and inference. Quantization can allow larger effective input sizes or more viable inference with more realistic inference on smaller hardware in neuroimaging, where constrained volumetric models can be used.

4.2.4 Efficient architecture design

Architecture decisions, in place of postfactum compression, can also help to achieve efficiency. Tan and Le (2019) proved the larger idea that cautious scaling and design of the architecture may be more beneficial than blind model growth. Patch-based processing, multiresolution pipelines, sparse attention, anatomically aware conditioning, and architectures that are based on region-level representations can be used to provide efficient design in neuroimaging applications where full voxel-level modeling is not required.

4.2.5 Low-rank and tensor decomposition

Another family of efficiency strategies is offered by the use of low-rank approximations and tensor decomposition, which use redundancy in the learned parameters. Song et al. (2020) describe decomposition-based acceleration of deep convolutional models, and similarly, could be applied to imaging architectures with compressible parameter tensors that are organized. In applications of neuroimaging, where 3D convolution can be notably computationally costly, decomposed operators can be used to cut down on computation, and much of the representational capacity of the original can be maintained.

4.2.6 Optimized training and hardware-aware design

Effective computing burden can also be minimized by improved optimization. The number of costly iterations can be minimized using better schedulers, adaptive optimizers, mixed-precision training, curriculum strategies, and early stopping criteria.

Neural architecture search, hardware-aware neural architecture search, and deployment-aware model selection are gaining relevance since the cost of a model does not purely depend on the number of parameters but also on the mapping to the hardware accelerator or workstation available. A theoretically beautiful architecture can be operationally inefficient, as it can conflict with memory-bandwidth, preprocessing steps, and communication limitations in a distributed environment.

4.2.7 Green AI and sustainability

Green AI highlights the importance of considering the quality of models in part based on the cost of computation instead of accuracy or realism alone (Strubell et al., 2019). This view is useful in neuroimaging since the science already relies on resource-consuming data collection and processing. Introduction of unnecessarily high-priced AI workflows, not reported costs, can hide the real cost of methodological decisions. Resource reporting can consist of hardware configuration, training time, estimated energy burden, memory needs, and inference latency.

4.3 Neuroimaging-specific examples and outlook

The literature of the recent past indicates that the field is shifting toward more efficient, but task-sensitive, neuroimaging systems as opposed to larger systems. MRI, CT, and PET synthesis reviews highlight the role of model design options, loss functions, and data pipelines in performance and practicality determination (Dayarathna et al., 2024). Whole-brain generative modeling not only shows that high-resolution synthesis is possible, but also notes the importance of scale-conscious engineering and mindful conditioning (Tudosiu et al., 2024). Neuroimaging compute cost benchmarks, such as federated brain tumor segmentation, also serve as a reminder that compute cost will encompass communication, heterogeneity management, and cross-site reproducibility, as well as local training speed (Manthe et al., 2024).

5. Ethical and regulatory concerns: Privacy, bias, and fairness in generative AI

5.1 Challenges of ethical and regulatory concerns

Generative AI in neuroimaging presents more ethical issues than the traditional issues posed by predictive analytics. Since the model has the ability to generate or modify image-like outputs, it can affect interpretation with incomplete, noisy, or missing modalities in the original scan. Privacy, bias, fairness, accountability, informed consent, traceability, and governance of synthetic data are thus considered part of the ethical landscape. Poldrack et al. (2020) highlight the significance of best practices in neuroimaging data analysis and sharing; the synthesis, reconstruction, and postdeployment monitoring areas also require consideration by generative AI.

5.1.1 Privacy and data protection

Imaging data in the brain is sensitive. Facial structure, rare pathology, metadata combinations, or related clinical records may still be used to link brain scans even when all identifiers are removed. Recent privacy in medical imaging AI research contends that the concept of privacy risk should be interpreted in more ways than just the removal of metadata, but also to model leakage, linkage attacks, and inferential disclosure. Another issue in the framework of generative AI is that a model can memorize the features of the data it is trained on and recreate them in its outputs or expose them through model behavior (Kaissis et al., 2020; Shokri et al., 2017).

Differential privacy offers a conceptual model to restrict the impact of a single training sample on the learned model (Dwork & Roth, 2014). Its use in medical imaging has improved over the past years; however, the balance between privacy and utility is still of importance, particularly in high-dimensional volumetric data. Fidelity could be lessened by privacy-preserving training, and it follows that the practical question in neuroimaging is not whether or not privacy is a good thing, but how a balance between privacy and people being rendered clinically or scientifically irrelevant can be achieved.

These concerns are supported by official regulatory structures. The GDPR obligates with regard to legal basis, transparency, data minimization, purpose limitation, and accountability in personal data processing. HIPAA (U.S. Department of Health and Human Services, 2025) still represents the contextual use of safeguarded health information in the health care environment in the United States. These tools do not give a single technical recipe of generative neuroimaging, but clearly assist privacy-by-design solutions where data governance is designed up front instead of added to tools after the fact.

5.1.2 Synthetic data, memorization, and privacy leakage

Synthetic data can also be given out as a simple privacy solution since it is not the same as the original scans. However, recent facts seem to indicate that this stance is overly simplistic. Akbar et al. (2024) demonstrated that synthetic MR images could help in downstream brain tumor segmentation, yet revealed that in cases of too small an original dataset, memorization is an issue. On a larger scale, machine learning privacy studies indicate that sometimes model behavior can be used to infer the training data membership (Shokri et al., 2017). The implications of these findings are that synthetic neuroimaging needs to be considered on the basis of utility and the risk to privacy and not automatically regarded as safe.

5.1.3 Bias and fairness

Medical AI bias is defined as clinical or social variation in medical AI performance or error distribution. The recent highly popular research by Obermeyer et al. (2019) revealed that even popular healthcare algorithms can encode severe inequities via proxy variables. Though that study was not on neuroimaging, the implication thereof can be directly applied: model goals, input data, and implicit proxies are important. Demographic imbalance, disease-prevalence imbalance, scanner effects, site-specific protocols, missing patterns of data, or selective inclusion criteria may also be sources of bias in neuroimaging.

Generative models are associated with more fairness issues. A synthesis system can not represent nonstereotypical but valid patterns of anatomy in a subgroup, and thus cannot make downstream models as reliable for that subgroup. The harmonization model can eliminate the signal that is clinically significant in a single population and enhance apparent consistency in general. A normative generator can encode a single demographic pattern as default and consider variations out of the default pattern as anomalous. Xu et al. (2024) survey the problem of fairness in medical image analysis and point out that fairness assessment and mitigation in imaging is not a complete area of study and tends to be sporadically reported.

Subgroup-aware assessment is thus necessary in the case of neuroimaging. These involve a look into performance based on sex, age, race, or ethnicity, where suitable and legally documented, the place of acquisition, features of the device, and clinically significant disease groups. Notably, fairness auditing must be performed on the final downstream model as well as any synthetic data generator to augment or transform the data.

5.1.4 Accountability and transparency

Generative neuroimaging involves more than being opaque about using AI. The researchers and clinicians must be able to comprehend what type of transformation has been done, what data was utilized in training, what kind of validation has been done, and what the known limitations are. In case an image was synthesized, enhanced, or imputed, then this fact must be recorded in a manner that facilitates clinical interpretation and auditability. This is particularly critical in cases where synthetic or reconstructed images may be confused with actual measurements.

5.1.5 Value-sensitive design and governance

Value-sensitive design offers a practical model in the process of incorporating the ethical principles into the development of the system, as opposed to regarding them as external limitations to be imposed after the fact. Their application in neuroimaging implies embedding privacy, fairness, human supervision, explainability, and compatibility with workflow consideration in their design. Not only would a value-sensitive generative neuroimaging system maximize image quality or classification performance, but it would also take into account how the system will be used by clinicians, whether patients would have any reasonable expectation of the transformation, and how the system will be monitored once deployed.

5.2 Mitigation of ethical and regulatory concerns

As with data scarcity, ethical and regulatory problems need multidimensional mitigation. No fair approach or privacy technology is enough. This requires a blend of technical protection, documentation practices, subgroup evaluation, and governance structures that are active over the model lifecycle.

5.2.1 Privacy-preserving data use and synthetic data governance

Neuroimaging privacy-preserving methods are controlled deidentification, federated learning, secure aggregation, differential privacy, and restrictive access control. Kaissis et al. (2020, 2021) demonstrate that the workflow of privacy-preserving and multiinstitutional medical imaging is not only technically feasible but also not an easy task. These approaches, in generative neuroimaging, must be accompanied by declarative policies concerning the release of synthetic data, the labeling of archives, audit trails, and future allowable downstream use.

5.2.2 Bias mitigation and subgroup evaluation

The first step to bias reduction in neuroimaging is to characterize the datasets. Developers ought to be aware of variations in training, validation, and test sets depending on demographic, clinical, and acquisition variables. The reweighting, balanced sampling, subgroup-specific calibration, as well as the adversarial debiasing techniques mentioned by Zhang et al. (2018) can be advantageous, but they need to be selected according to the concrete bias mechanism instead of being used in a generic way. Controlled generative augmentation can be used in some instances to aid the underrepresented groups, and in other cases, it can merely reproduce the imbalance of the original.

5.2.3 Transparency, informed consent, and documentation

One of the most feasible tools is documentation as an ethical tool of governance. Model cards, data statements, validation reports, and change logs can assist in explaining what the model is doing, the source of the data, its intended purpose, and the limitations of the model. The process of informed consent needs to change as well when there are generative processes involved, particularly when it is possible to synthesize, enhance, or reuse an image because of the different context in which the image was originally collected. Although the language of consent may differ depending on the institution and jurisdiction, the ethical imperative remains the same: subjects and patients must not be deceived regarding the utilization of their data or the integration of the results of AI-based tools in care or research.

5.2.4 Legal liability, oversight, and regulatory pathways

The issue of AI-assisted neuroimaging and its legal obligation is a developing field. Currently, human supervision is still the focus of attention, particularly where the outputs can affect diagnosis or treatment. The existing AI-related instructions by the FDA (U.S. Food and Drug Administration, 2025) focus on lifecycle management, risk management, and evidence to support evaluation. The European Union Artificial Intelligence Act (EU AI Act, 2024) offers a wider scope of high-risk systems, and the requirement involving documentation, transparency, and control. These advancements suggest that traceability and update governance should be considered when designing generative neuroimaging.

5.2.5 Ethical frameworks and operational safeguards

Based on the traditional bioethical concepts, they can be applied to responsible neuroimaging AI. The autonomy helps in meaningful consent and open communication regarding the AI-generated outputs. Beneficence aids in the creation of tools that enhance diagnosis, reconstruction, or access to advanced imaging analysis. Nonmaleficence involves proactive consideration of hallucination, bias, leakage of privacy, and unsafe overreliance. Justice entails fair performance and equal access among populations and institutions. These principles alone cannot be used, but they can be used to give a consistent ethical vocabulary to system design and review.

6. Integrative critical review and future directions

6.1 Synthesizing the challenges

The four key issues explored in this chapter, including the lack of data, interpretability, computational cost, and ethical governance, are profoundly interrelated. Little information adds uncertainty and risk of bias. The presence of opaque models complicates the identification of such failures. Validation is expensive in terms of computation, and the wider community is hindered in reproducing results. Weak governance lowers trust, which in its turn may hinder even data sharing and collaboration that is necessary to resolve scarcity. In this way, the key takeaway is that it is impossible to have reliable generative neuroimaging by optimizing a single axis of performance.

6.2 Current best practices

A number of best practices are already being developed. To begin with, the neuroimaging research that uses public data resources, in addition to well-managed institutional data, is gaining momentum, instead of using single-site datasets. Second, label dependency is being minimized using transfer learning and self-supervised pretraining. Third, synthetic data generation is being considered with regard to downstream task utility and risk of memorization as opposed to being heralded as a party. Fourth, federated learning is emerging as a viable path toward developing models across multiple sites in sensitive imaging environments. Fifth, neuroanatomical plausibility and usefulness to the reader are increasingly becoming important metrics of interpretability instead of the simple presence of a heatmap.

6.3 The road ahead: Integrating solutions

Generative AI in neuroimaging will probably rely on interdisciplinary efforts instead of individual technical advances. A group building a model to do missing-modality synthesis, such as might require neuroradiologists to specify clinically important fidelity metrics, machine learning scientists to build the generator, privacy scholars to design data-sharing flow, and infrastructure scientists to make the model deployable in a secure setting. Whenever such views are harmonized at an early stage, the resulting model will be more susceptible to being scientifically useful and governable.

6.4 Implications for neuroscience and clinical practice

Provided that these issues are dealt with properly, generative AI can facilitate a number of significant contributions to neuroscience and clinical imaging. It can augment the usefulness of incomplete datasets via modality synthesis or reconstruction, enable rare-disease studies via enhanced representation learning, and enhance the stability of downstream models by subjecting them to more realistic variability. Generative models can also be useful in research to investigate latent disease trajectories, normative variation, and counterfactual situations subject to well-bounded assumptions. They can be used to assist in image improvement, quality management, and decision support in clinical contexts, as long as their outputs are properly labeled and their validity is justified.

6.5 Continual learning and postdeployment adaptation

A final issue is continual learning. Neuroimaging settings evolve with time as scanners are lost and found, populations are altered, tracers are developed, and new protocols are adapted. A fixed generative model can slowly be replaced. However, constant updating introduces other risks: disastrous forgetting, drifting, new biases, and varying privacy exposure. In the case of generative neuroimaging, postdeployment adaptation thus demands conscious governance. The procedures of updating should indicate the required evidence, monitoring behavior, and recording of historical behavior.

7. Conclusion

Generative AI can greatly assist neuroimaging through data augmentation, modality generation, image reconstruction, harmonization, and latent-pattern discovery in MRI, fMRI, and PET. However, this possibility is limited by both scientific, operational, and ethical constraints simultaneously. The lack of data and heterogeneity diminishes generalization, limited interpretability diminishes trust and obscures failure modes, excessive computational cost limits reproducibility, implementation, privacy, fairness, and governance issues influence whether these approaches can be responsibly integrated into research or clinical practice.

AI disclosure

During the preparation of this work, the author(s) used ChatGPT in order to Querry for Research and Understanding. After using this tool/service, the author(s) reviewed and edited the content as needed and take (s) full responsibility for the content of the publication.

References

Akbar, M. U., Larsson, M., Blystad, I., et al. (2024). Brain tumor segmentation using synthetic MR images: A comparison of GANs and diffusion models. *Scientific Data, 11*, 259. https://doi.org/10.1038/s41597-024-03073-x

Bach, S., Binder, A., Montavon, G., Klauschen, F., Muller, K.-R., & Samek, W. (2015). On pixel-wise explanations for non-linear classifier decisions by layer-wise relevance propagation. *PLoS One, 10*(7), Article e0130140. https://doi.org/10.1371/journal.pone.0130140

Chen, T., Kornblith, S., Norouzi, M., & Hinton, G. (2020). A simple framework for contrastive learning of visual representations. In *Proceedings of the 37th international conference on machine learning* (pp. 1597–1607). PMLR. https://proceedings.mlr.press/v119/chen20j.html.

Dayarathna, S., Islam, K. T., Uribe, S., Yang, G., Hayat, M., & Chen, Z. (2024). Deep learning based synthesis of MRI, CT and PET: Review and analysis. *Medical Image Analysis, 92*, Article 103046. https://doi.org/10.1016/j.media.2023.103046

Dwork, C., & Roth, A. (2014). The algorithmic foundations of differential privacy. *Foundations and Trends in Theoretical Computer Science, 9*(3–4), 211–407. https://doi.org/10.1561/0400000042

European Parliament and Council of the European Union. (2024). *Regulation (EU) 2024/1689 of 13 June 2024 laying down harmonised rules on artificial intelligence (Artificial Intelligence Act)*. EUR-Lex. https://eur-lex.europa.eu/legal-content/EN/TXT/?uri=CELEX:32024R1689.

Farahani, F. V., Karwowski, W., & Lighthall, N. R. (2022). Explainable AI: A review of applications to neuroimaging data. *Frontiers in Neuroscience, 16*, Article 906290. https://doi.org/10.3389/fnins.2022.906290

Finn, C., Abbeel, P., & Levine, S. (2017). Model-agnostic meta-learning for fast adaptation of deep networks. In *Proceedings of the 34th international conference on machine learning* (pp. 1126–1135). PMLR. https://proceedings.mlr.press/v70/finn17a.html.

Frankle, J., & Carbin, M. (2019). The lottery ticket hypothesis: Finding sparse, trainable neural networks. In *International conference on learning representations*. https://iclr.cc/virtual/2019/oral/1156.

Goodfellow, I. J., Pouget-Abadie, J., Mirza, M., Xu, B., Warde-Farley, D., Ozair, S., Courville, A., & Bengio, Y. (2014). Generative adversarial nets. *Advances in Neural Information Processing Systems, 27*, 2672–2680. https://papers.nips.cc/paper_files/paper/2014/hash/f033ed80deb0234979a61f95710dbe25-Abstract.html.

Gupta, S., Agrawal, A., Gopalakrishnan, K., & Narayanan, P. (2015). Deep learning with limited numerical precision. In *Proceedings of the 32nd international conference on machine learning* (pp. 1737–1746). PMLR. https://proceedings.mlr.press/v37/gupta15.html.

Hinton, G., Vinyals, O., & Dean, J. (2015). Distilling the knowledge in a neural network. Available from: https://arxiv.org/pdf/1503.02531.

Holzinger, A., Langs, G., Denk, H., Zatloukal, K., & Muller, H. (2019). Causability and explainability of artificial intelligence in medicine. *WIREs Data Mining and Knowledge Discovery, 9*(4), Article e1312. https://doi.org/10.1002/widm.1312

Huynh, N., & Deshpande, G. (2024). A review of the applications of generative adversarial networks to structural and functional MRI-based diagnostic classification of brain disorders. *Frontiers in Neuroscience, 18*, Article 1333712. https://doi.org/10.3389/fnins.2024.1333712

Kaissis, G. A., Makowski, M. R., Ruckert, D., & Braren, R. F. (2020). Secure, privacy-preserving and federated machine learning in medical imaging. *Nature Machine Intelligence, 2*(6), 305–311. https://doi.org/10.1038/s42256-020-0186-1

Kaissis, G., Ziller, A., Passerat-Palmbach, J., Ryffel, T., et al. (2021). End-to-end privacy preserving deep learning on multi-institutional medical imaging. *Nature Machine Intelligence, 3*(6), 473–484. https://doi.org/10.1038/s42256-021-00337-8

Khosroshahi, M. T., Morsali, D., Ayati, M. H., et al. (2025). Explainable artificial intelligence in neuroimaging of Alzheimer's disease. *Diagnostics, 15*(5), 612. https://doi.org/10.3390/diagnostics15050612

Kim, B., Wattenberg, M., Gilmer, J., Cai, C., Wexler, J., Viegas, F., & Sayres, R. (2018). Interpretability beyond feature attribution: Quantitative testing with concept activation vectors (TCAV). In *Proceedings of the 35th international conference on machine learning* (pp. 2668–2677). PMLR. https://proceedings.mlr.press/v80/kim18d.html.

Lee, S., & Lee, K.-S. (2024). Predictive and explainable artificial intelligence for neuroimaging applications. *Diagnostics, 14*(21), 2394. https://doi.org/10.3390/diagnostics14212394

Lundberg, S. M., & Lee, S.-I. (2017). A unified approach to interpreting model predictions. In *Advances in neural information processing systems, vol. 30* (pp. 4765–4774). https://papers.nips.cc/paper/2017.

Manthe, M., Duffner, S., & Lartizien, C. (2024). Federated brain tumor segmentation: An extensive benchmark. *Medical Image Analysis, 97*, Article 103270. https://doi.org/10.1016/j.media.2024.103270

Miller, K. L., Alfaro-Almagro, F., Bangerter, N. K., et al. (2016). Multimodal population brain imaging in the UK Biobank prospective epidemiological study. *Nature Neuroscience, 19*(11), 1523–1536. https://doi.org/10.1038/nn.4393

Nichols, T. E., Das, S., Eickhoff, S. B., et al. (2017). Best practices in data analysis and sharing in neuroimaging using MRI. *Nature Neuroscience, 20*(3), 299–303. https://doi.org/10.1038/nn.4500

Obermeyer, Z., Powers, B., Vogeli, C., & Mullainathan, S. (2019). Dissecting racial bias in an algorithm used to manage the health of populations. *Science, 366*(6464), 447–453. https://doi.org/10.1126/science.aax2342

Poldrack, R. A., Gorgolewski, K. J., et al. (2020). Establishing best practices in data analysis and sharing in neuroimaging. *Nature Neuroscience, 23*(9), 1057–1066. https://doi.org/10.1038/s41593-020-00724-8

Ribeiro, M. T., Singh, S., & Guestrin, C. (2016). "Why should I trust you?": Explaining the predictions of any classifier. In *Proceedings of the 22nd ACM SIGKDD international conference on knowledge discovery and data mining* (pp. 1135–1144). https://doi.org/10.1145/2939672.2939778

Rudin, C. (2019). Stop explaining black box machine learning models for high stakes decisions and use interpretable models instead. *Nature Machine Intelligence, 1*(5), 206–215. https://doi.org/10.1038/s42256-019-0048-x

Sheller, M. J., Edwards, B., Reina, G. A., Martin, J., Pati, S., Kotrotsou, A., Milchenko, M., Xu, W., Marcus, D. S., Colen, R. R., Bakas, S., & Davatzikos, C. (2020). Federated learning in medicine: Facilitating multi-institutional collaborations without sharing patient data. *Scientific Reports, 10*, Article 12598. https://doi.org/10.1038/s41598-020-69250-1

Shin, H.-C., Roth, H. R., Gao, M., et al. (2016). Deep convolutional neural networks for computer-aided detection: CNN architectures, dataset characteristics and transfer learning. *IEEE Transactions on Medical Imaging, 35*(5), 1285–1298. https://doi.org/10.1109/TMI.2016.2528162

Shokri, R., Stronati, M., Song, C., & Shmatikov, V. (2017). Membership inference attacks against machine learning models. In *2017 IEEE symposium on security and privacy* (pp. 3–18). https://doi.org/10.1109/SP.2017.41

Song, D., Zhang, P., & Li, F. (2020). Speeding up deep convolutional neural networks based on Tucker-CP decomposition. In *Proceedings of the 2020 5th international conference on machine learning technologies* (pp. 56–61). https://doi.org/10.1145/3409073.3409094

Strubell, E., Ganesh, A., & McCallum, A. (2019). Energy and policy considerations for deep learning in NLP. In *Proceedings of the 57th annual meeting of the association for computational linguistics* (pp. 3645–3650). https://doi.org/10.18653/v1/P19-1355

Tan, M., & Le, Q. (2019). EfficientNet: Rethinking model scaling for convolutional neural networks In *Proceedings of the 36th international conference on machine learning* (pp. 6105–6114). PMLR. https://proceedings.mlr.press/v97/tan19a.html.

Tudosiu, P.-D., Pinaya, W. H. L., et al. (2024). Realistic morphology-preserving generative modelling of the brain. *Nature Machine Intelligence, 6*, 811–819. https://doi.org/10.1038/s42256-024-00864-0

U.S. Department of Health and Human Services. (2025). Summary of the HIPAA privacy rule. https://www.hhs.gov/hipaa/for-professionals/privacy/laws-regulations/index.html.

U.S. Food and Drug Administration. (2025). Artificial intelligence-enabled device software functions: Lifecycle management and marketing submission recommendations. *Draft Guidance*. https://www.fda.gov/regulatory-information/search-fda-guidance-documents/artificial-intelligence-enabled-device-software-functions-lifecycle-management-and-marketing.

Xu, Z., Li, J., Yao, Q., Li, H., Zhao, M., & Zhou, S. K. (2024). Addressing fairness issues in deep learning-based medical image analysis: A systematic review. *Npj Digital Medicine, 7*, 286. https://doi.org/10.1038/s41746-024-01276-5

Zhang, B. H., Lemoine, B., & Mitchell, M. (2018). Mitigating unwanted biases with adversarial learning. In *Proceedings of the 2018 AAAI/ACM conference on AI, ethics, and society* (pp. 335–340). https://doi.org/10.1145/3278721.3278779

CHAPTER 14

Generative AI for neurological diseases: A review toward future directions

Amira S. Ashour[1] and Deepika Koundal[2, 3]
[1]Tanta University Faculty of Engineering, Electronics and Electrical Communications Engineering Department, Tanta, Egypt; [2]University of Eastern Finland, A.I. Virtanen Institute for Molecular Sciences, Kuopio, Finland; [3]UPES, Dehradun, Uttarakhand, India

1. Introduction

Conventional artificial intelligence (AI) models aim to classify or predict outcomes according to the input data, while the generative artificial intelligence (gAI) models create entirely the original outputs. Typically, gAI is considered a subdomain of AI for generating new data by training on extensive datasets for the identification of significant patterns of the original data, including images, text, or audio. The most prevailing gAI models include variational autoencoders (VAEs), and generative adversarial networks (GANs), which learn how to generate convincing and original samples. In the neuroscience domain, gAI models proved their abilities to simulate complex biological systems like the brain by offering the potential for employing the sympathetic nature of the neural structures and functionality using advanced neuroscience imaging, data augmentation, and advanced pattern recognition approaches. These models transform the way by which the researchers understand and evaluate brain data. By enhancing the brain's image quality, increasing the size of datasets, and enabling more precise predictions, AI plays a critical role in improving diagnostic efficacy and strengthening the comprehension of sophisticated neurological processes (Gong et al., 2023). The continuing innovation of gAI technologies may lead to meaningful breakthroughs in neuroscience and clinical practices. Several applications of gAI in neuroscience imaging involve brain image synthesis, data augmentation, enhancement, segmentation, mapping of brain structures, prognostic models of neurological conditions, understanding the neural connectivity of brain networks, and multimodal integration.

In neuroscience, data augmentation of brain imaging is essential to acquire extensive datasets, which can be difficult due to limited data availability and accessibility along with the ethical considerations. Accordingly, gAI models can generate synthetic brain images, such as magnetic resonance scans, that closely seem like real brain data, which increases the size and diversity of training datasets, thereby refining the generalizability and robustness of gAI models. This is especially beneficial in the case of neurological disorders datasets

Generative Artificial Intelligence for Neuroimaging
ISBN 978-0-443-45010-5
https://doi.org/10.1016/B978-0-443-45010-5.00005-8

having small sizes. Furthermore, gAI has a significant role in image enhancement, and synthesis by improving the resolution and quality of brain images for improved analysis. These models can also reduce noise/distortion, and complete missing data, that lead to clear and precise images for further use in diagnostic and research purposes.

Alternatively, neuroscience research also focuses on segmentation and mapping of brain assemblies to investigate regions, including white matter, gray matter, and various brain networks (Peer et al., 2017). gAI can enrich such processes by recognizing the different patterns inside the different brain's regions for accurate segmentation and mapping of the brain's regions, thus extending the identification of brain connectivity and activity. In addition, gAI models can be used to estimate the onset and staging of neurological diseases by investigating longitudinal brain imaging data via simulating the possible changes in the brain accompanying with disorders, such as Alzheimer's disease (AD), Parkinson's disease (PD), and multiple sclerosis (MS). This ability enables earlier diagnosis and monitoring of the disease's progression for personalized care strategies and effective treatment plan. Recognizing the brain networks is necessary for grasping normal brain cases as well as the neurological disorders. gAI models can be utilized for investigating the complex connections network within the brain that aids in the understanding of the brain networks, and the neurons connectivity. These models can simulate the neural pathways and forecast relations between distinctive brain regions in the acquired brain images from diffusion tensor imaging (DTI) and functional MRI (fMRI) (Helmy et al., 2023). Likewise, gAI models facilitate the multimodal combination of various imaging data types, such as structural MRI, fMRI, and electroencephalography (EEG). By designing combined models of brain activity across dissimilar modalities, scientists can realize a more all-inclusive understanding of the association between the brain structure and its function, thus improving the diagnostic accuracy and treatment plans for neurological conditions.

For neuroscience imaging, anticipating forthcoming advances in gAI becomes indispensable as it empowers clinicians, researchers, and technology designers to give well-informed selections concerning the raised applications, opportunities, and challenges for earlier diagnoses of neurological syndromes. Future revolutions may assist gAI systems to predict the appearance of disorders, such as schizophrenia, AD, PD, or MS prior to clinical symptoms manifestation for improved patient outcomes. Additionally, gAI could support formulating strategies for personalized management by assessing individual brain scans and forecasting patient responses to different treatments. By gAI-driven models, clinicians can understand the brain functionality and neurological disease progression and implement tailored patient care. Personalized medicine becomes pivotal in developing effective neurological treatments (Yamamoto et al., 2022). Nevertheless, as gAI becomes gradually integrated into brain imaging, it also carries ahead ethical issues, including algorithmic bias, data privacy, and informed consent. Considering an active approach to the ethical use of gAI in neuroscience imaging is critical for the development and use of technology for adopting equity in healthcare.

In terms of healthcare accessibility, gAI has the prospect of bridge gaps in underserved regions. For example, the produced synthetic images by AI allow local healthcare physicians to exploit high-quality analytic resources without the requirement for expensive imaging technology. The future of gAI in neuroscience imaging incorporates cooperation among diverse disciplines, including machine learning (ML), bioengineering, cognitive neuroscience, and clinical neurology. Predicting future trends will facilitate finding combined opportunities, encouraging a multidisciplinary scheme to adopt intricate challenges. This collaboration will not only raise technological innovation but also increase the ability of the brain. Using gAI, researchers can discover novel methodologies that may notably accelerate the understanding of brain function/activity and forecast variations in neurological conditions by revealing groundbreaking perceptions into emotion, perception, cognition, and memory.

Modern gAI models employed in neuroscience imaging encounter several challenges (Gong et al., 2023), as follows:

- the risk of overfitting,
- the requirement for extensive datasets, and
- the difficulties in interpreting intricate brain data.

Such challenges emerge the need for further developments to enhance the gAI models to match requirements of the neuroscience practice, including improving interpretability, reducing dependence on large/labeled datasets, and addressing biases. This insight also assists the breakthrough of advanced approaches, such as transfer learning, few-shot learning, and reinforcement learning that boost the effectiveness of gAI applications in neuroscience. For advanced integration of gAI in neuroscience imaging, there is a rising necessity for specialized training that incorporates both gAI and neuroscience fields. This preemptive strategy will assist the preparation of the workforce for the constantly developing industrial environment. In neuroscience, the immediate improvement of AI tools, remarkably gAI, may exceed the current policies and regulations. By predicting future trends, regulatory agencies and policymakers can remain active to develop guidelines protecting the public health and upgrade the responsible use of gAI in healthcare. Verifying clear regulatory frames will be critical to ensure the security and consistency of AI-driven neuroscience methods, thus promoting trust between patients and clinicians.

The structure of the chapter is as follows, in Section 1, an introduction on the gAI role in neuroscience is addressed. In Section 2, the emerging trends in gAI for neuroscience imaging, including advancement in generative models, and integration with other ML paradigms are discussed. In Section 3, research directions in data generation and augmentation are introduced, including clinical applications of synthetic brain image generation, challenges in synthetic brain image generation, and future directions. Then, the cross-modal and multimodal imaging synthesis is studied in Section 4, followed by alignment challenges related to gAI for neuroscience imaging in Section 5.

In Section 6, the interpretability and explainability in generative models is highlighted, followed by discussing the gAI-based datasets, simulation platforms, and collaborative infrastructures in Section 7. Afterward, Section 8 includes open problems and grand challenges in gAI for neuroimaging with introducing the ethical, legal, and social implications (ELSI) in Section 9. Finally, before the conclusion section, the future vision toward intelligent neuroimaging ecosystems is reported in Section 10.

2. Emerging trends in generative AI for neuroscience imaging

2.1 Advancement in generative models

Recently, outstanding progress in designing generative models has been carried out for data augmentation, image synthesis, and intricate distributions' modeling. For research purposes in neuroscience imaging, such advancements are appropriate for enhancing brain image resolution, generating brain synthetic datasets, and modeling patterns of brain activity. The leading gAI models to carry out these tasks include GANs, VAEs, and diffusion models (DMs).

2.1.1 Generative adversarial networks

GANs are considered one of the most prevailing generative models, which are composed of two networks, namely generator and discriminator. The generator generates synthetic data, like brain images, whereas the discriminator assesses the validity of these generated data for further prompting the generator to improve its results. Various variants of GANs, including CycleGAN, BigGAN, StyleGAN, Wasserstein GANs (WGANs), and Progressive GANs (Bhat & Nanjundegowda, 2025), have been designed. In neuroscience imaging, GANs can create synthetic brain images for rare neurological disorders or for limited availability of training data to increase the diversity of datasets. For clinical settings, GANs can be employed for enhancing the brain image quality and resolution of brain scans by turning low-resolution scans into high-resolution ones for accurate diagnoses.

Advanced variants of GANs, including BigGAN and StyleGAN, enhance the images' quality by supporting the generation of high-resolution and representative images. Remarkably, StyleGAN excels in constructing images with complex details (Bermano et al., 2022). Using refined techniques like adaptive instance normalization (AdaIN), these models can generate images looks like real ones with additional control on their stylistic elements of the produced content. On the contrary, CycleGAN is adept at producing images that accelerate the translation between several domains without demanding paired datasets. For example, CycleGANs have been laboring to adapt different imaging modalities (Zhang et al., 2022), for instance transforming T1-weighted MRI scan into a T2-weighted MRI scan. Altogether, training GANs can be entirely inspiring due to difficulties like instability, and mode failure when the generator returns a limited range of outputs. Conversely, recent improvements, including WGANs and progressive GANs, have enriched the model constancy by redeveloping loss functions and incrementally increase the complication of the created samples.

2.1.2 Variational autoencoders

VAEs embody a marked class of generative models that divergence with GANs in their approach as they verify a probabilistic relationship between an input image and a latent space. Contrary to GANs, VAEs include a decoder to reconstruct the original data from a compressed hidden representation by compelling a distinctive distribution on the latent space, assisting the model to produce new data by sampling from this distribution. In neuroscience imaging, VAEs have several applications, especially in exploring the latent space of brain images to afford valuable perceptions into brain structures with those involved neurological conditions. Furthermore, VAEs are successful in discovering abnormalities within brain scans for further identification of diverging patterns from the typical brain's structures. This competence is particularly helpful for the early detection of neurodegenerative disorders.

Various variations of VAEs, include Beta-VAE, conditional VAEs (CVAE), and VAE-GAN hybrid models. Beta-VAE is an enhanced conventional VAE, highlighting the asset of more separated representations in the latent space. It empowers the model to find more understandable and independent features from the brain scans to distinguish different anatomical parts (e.g., gray matter and white matter). Conversely, the CVAE can produce data based on specific attributes or variables. In neuroscience, CVAEs can create brain images having features like those of a patient diagnosed with a specific neurological disorder. This ability supports researchers in testing the impact of specified conditions on brain structures. To weigh the powers of both VAEs and GANs, hybrid models known as VAE-GANs have been established to maintain the continuous latent space property of VAEs while combining the enhanced image production capacities of GANs. This synergy has been employed to improve the produced images' quality, including the production of synthetic MRI scans.

2.1.3 Diffusion models

One of the ongoing categories of generative models is the DMs that hypothesize data generation as a continuing change from noise to a specialized data distribution (Cao et al., 2024). DMs are inspired by physical phenomena, namely the diffusion, where data points undertake a sequence of noise-induced transformations until forming a coherent structure. They are guided to invert this process, by transforming noise into a significant output (image) using the concept of the score-based generative models. DMs employ a score function to turn the denoising process during generation. Modern improvements in DMs include denoising diffusion probabilistic models that can generate high-quality samples with greater training constancy compared to GANs. Consequently,

DMs have developed as strong competitors to GANs concerning sample quality. Additionally, current studies have intended to enhance the efficiency of the sampling process in DMs. For instance, classifier-guided DMs enrich the generated images' quality by preparing the diffusion process on external classifiers, so guiding the model to generate images with required attributes. A considerable novelty in DMs is the introduction of latent diffusion, which grants the model to act within a lower-dimensional latent space instead of promptly employing pixel data. Latent DMs lead to immediate and more memory-efficient training with high-quality image generation, making them effective for medical applications. For investigating brain function in neuroscience imaging, DMs can replicate brain activity and create synthetic data which shows the dynamic changes in brain states for modeling the staging of neurological disorders over time. Like GANs, DMs can enable the translation between numerous neuroimaging modalities by enabling the integration of multimodal data, including converting fMRI data to structural MRI data resulting in more systematic evaluation of both brain's structure and function.

The beginning of basis models and swift-driven generative approaches is modernizing diverse areas, especially neurological science for disease research. Base models, especially large language models, have revealed significant ability across multiple areas by providing robust data analysis tools, hypothesis formulation, and decision-making processes. The applications of these generative models on neurological science are depicted in Fig. 14.1.

2.2 Integration of gAI models with other machine learning paradigms

In neuroscience imaging, the integration of gAI models with advanced ML techniques—such as federated learning (FL) and active learning (AL)—is accelerating progress

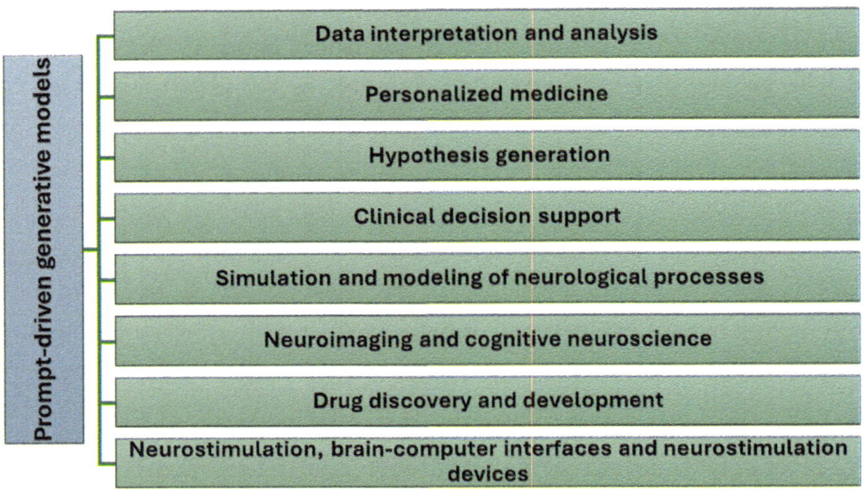

Figure 14.1 Applications of prompt-driven generative models.

toward more accurate, efficient, and privacy-preserving diagnostic tools. These combined approaches are shaping novel pathways for disease detection, brain mapping, and individualized treatment planning. FL is a decentralized ML framework that allows multiple devices or institutions to collaboratively train models without sharing raw data (Albalawi et al., 2024). This is especially critical in neuroscience, where medical imaging data (e.g., MRI, CT, PET) is highly sensitive. Through FL, hospitals and research centers can train shared gAI models on distributed datasets while maintaining compliance with privacy regulations such as HIPAA. Importantly, FL enhances model generalizability by incorporating data from diverse populations—accounting for variables like age, sex, and ethnicity—which helps mitigate overfitting. Its distributed nature also supports continuous model updates, thus speeding up advancements in neuro-diagnostic capabilities. FL enables deep learning models in various medical applications (Sandhu et al., 2023), for example, to detect neurological anomalies like tumors or neurodegeneration with improved accuracy by accessing heterogeneous imaging data without breaching confidentiality.

On the other hand, AL focuses on model-driven identification of uncertain or ambiguous data samples for annotation (Liu et al., 2022). Given the high cost and expertise required to label neuroscience imaging data, AL significantly reduces manual workload by prioritizing only the most informative images for expert review. This is especially useful when resources such as radiologists or neurologists are limited. AL strengthens model training efficiency, especially in identifying rare conditions like MS, where lesion progression can be subtle. By selecting borderline or unclear cases, AL helps models learn complex patterns faster. Additionally, when combined with generative models like GANs, AL can guide synthetic image generation in underrepresented regions, thereby enriching training datasets (van Breugel et al., 2024). Together, FL and AL form a synergistic framework: FL ensures collaboration across institutions with heterogeneous data, while AL optimizes which data points to prioritize. This combination fosters the creation of robust, generalizable, and privacy-conscious models. Moreover, AL can address FL's challenge of slow convergence by guiding the training process toward uncertain examples, enhancing efficiency and model performance.

3. Research directions in data generation and augmentation

Data scarcity often leads to biased training datasets, uniquely concerning demographics information or disease subtypes. By producing synthetic brain images, it develops more diverse and typical datasets for training AI models. If specific subtypes of brain cancer are inadequately exemplified in a dataset, synthetic images can be generated, confirming that the model is sufficiently trained to identify them in clinical scenarios. Synthetic images enable brain scans generation for rare conditions as generative models can adopt the design of AI models for diagnosing these conditions in clinical practice.

Data scarcity raises an extensive challenge in neuroscience, as obtaining a substantial volume of annotated data is time-consuming, costly, and resource demanding, especially with rare conditions that have a limited number of available brain images. The use of synthetic image generation can address these challenges, using data augmentation, bias reduction, and modeling of rare diseases using gAI models. In neuroscience, the generation of synthetic brain images is considered a promising solution for addressing the data shortage while training real AI models due to the challenges associated with acquiring high-quality datasets and the complexity of brain structures. As reported previously, the most common approaches for generating synthetic images comprise GANs, VAEs, and deep CNN, which prove their efficacy in delivering high-fidelity synthetic brain images which closely are like actual images. GANs can be used for creating synthetic MRI or fMRI images which enhance datasets in scenarios where annotated brain images are rare, for example in the case of unusual neurological syndromes. Additionally, VAEs are active for superresolution, image denoising, and data augmentation tasks, producing artificial images that tackle data deficits, predominantly with the case of limited datasets or rare conditions in neuroscience imaging. In addition, conditional GANs enable images based on restricted input parameters or labels for creating domain-specific brain images in neuroscience.

3.1 Clinical applications of synthetic brain image generation

In neuroscience imaging, synthetic brain image has an essential role for training/testing of ML models which depends on annotated datasets for identifying the complex patterns in the brain scans. For instance, for brain tumor detection, designing a DL model requires the use of massive number of images. However, in several cases, the availability of real brain images is inadequate, which necessitates the need to use gAI to create synthetic scans for further enhancing the training datasets, for boosting the accuracy and generalization.

Another emerging need for synthetic brain images generation is raised when researchers seek to transform images from one imaging modality to another. gAI models, specifically pix2pix or CycleGANs, can enable the creation of synthetic MRI images from CT scans along with enhancing the resolution of noisy or low-quality brain images. Superresolution GANs can generate high-resolution images of low-resolution ones for accurate recognition of subtle abnormalities and disease detection.

On the other hand, images synthesis has several limitations to ensure that the created images achieved the following:

- closely resemble real-world brain scans, where if the created images lack adequate realism, inaccuracies or biases may occur during the model's learning process.
- maintain high reliability regarding disease-specific characteristics and anatomical details.

- suitable and accurate for clinical settings to prevent any deliberate harm to patients or incorrect diagnoses.
- approved, as the process of regulatory approval for gAI models that used synthetic data is still developing, leading to main questions about the confirmation of these gAI models in clinical environments. Even though synthetic data can relieve bias, there is a possible risk that the training of synthetic data leads to inherit biases in the gAI model. For example, if a gAI model is mostly trained on images from a single demographic group, it may yield synthetic images that lack the essential diversity for a well-rounded gAI model.

3.2 Future directions for synthetic brain image generation

Several prospective future directions can be defined as follows:

- **Personalized synthetic image generation** which are tailored to rare characteristics of patients for developing highly personalized diagnostic systems.
- **Hybrid models** that integrate synthetic brain image with AL and FL to improve data scarcity management along with concerning areas of uncertainty without the requirement to share sensitive information.
- **Integration with other modalities**, where advanced gAI models may combine data from neuroimaging, genetic sequencing, and electrophysiological studies to yield more complex synthetic datasets showing the structural anomalies and reflecting the molecular/functional changes in the brain.
- **Application of gAI models for rare neurological diseases**, such as frontotemporal dementia, certain types of epilepsy, Huntington's disease, vascular irregularities, neurodegenerative disorders, and uncommon brain tumors, that require data synthesis as their uncommon occurrence led to scarcity of imaging data. gAI models can aid in adopting such data shortage using data augmentation for rare neurological disorders.
- **Enhancing demonstration of diverse populations**, where neurological brain images datasets commonly expose biases correlated with demographic factors that affect the effectiveness of gAI models in the case of using **underrepresented** populations. gAI models offer prospective solutions to moderate these gaps in several ways, including **balancing** diminished **demographics**, originating diverse brain image data, and **age-specific variability**.
- **Synthetic neurological data for research and clinical trials** to enhance research initiatives, enable drug discovery, and support the simulation of clinical trial. gAI models could replicate disease progression within synthetic brain datasets and simulate the disease's progression of virtual patients, thus producing extra neurological data for enhancing the training of gAI models.

- **Clinical validation of using synthetic data represents a significant challenge**, which can be resolved by conducting validation studies and clinical trials using real-patient data for further comparison with the performance of gAI models trained on synthetic images against those trained on real-world data, using explainable AI systems, and association with therapeutic experts to confirm that the gAI models are unceasingly confirmed and refined.

4. Cross-modal and multimodal imaging synthesis

In neuroscience, multimodal and cross-modal imaging synthesis includes the integration of data from many imaging practices to realize a detailed understanding of the brain's structure, functionality, and activity (Mai et al., 2025). Using gAI presents extensive solutions by enabling the synthesis of multimodal images, such as CT, MRI and PET scans, as well as fMRI and EEG, or CT and MRI, into a cohesive depiction. Such integration is indispensable, where dissimilar imaging modalities capture complementary and unique features of brain representing its structure or activity, so delivering a more detailed and comprehensive view of neurological processes.

Typically, gAI in multimodal and cross-modal imaging provides various benefits, including:

- gAI can increase accuracy by synthesizing balancing high-quality images of the brain from various imaging modalities, which improves the accuracy of diagnosis and prognosis,
- gAI models improve the fusion of diverse data types, such as structural, functional, and metabolic, for better interpretation of the brain's anatomy, function, and pathology, respectively,
- gAI can fill in this gap by generating additional data for forecasting the similarity between two modalities to match the requirements of the clinical environments, if one modality has limited coverage or resolution,
- gAI models can control multimodal data to predict disease progression, treatment responses, and patient outcomes, as it can expect how a neurological disorder recognized in MRI scans may influence brain activity which modeled by fMRI or EEG, therefore offering valued perceptions for personalized treatment methods. Some examples of using gAI for cross-modal synthesis are shown in Fig. 14.2.

5. Alignment challenges related to generative AI for neuroscience imaging

Using gAI in neuroscience imaging, particularly when merging data from different types of scans like CT, MRI, PET, or EEG, has several challenges, as follows:

- differences in image resolution,
- dealing with distortion/noise in the data,

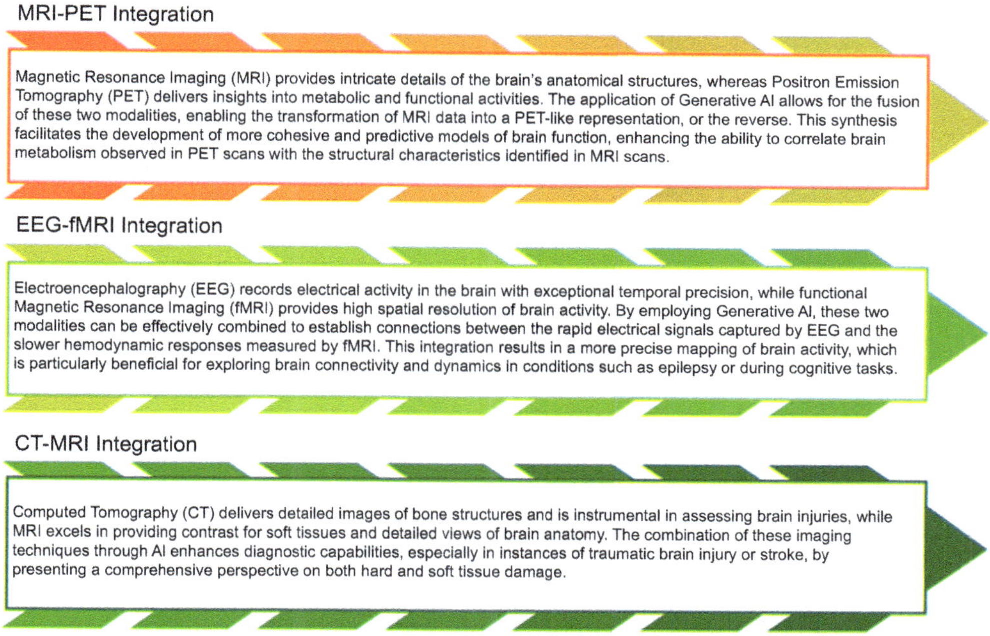

Figure 14.2 Generative AI for cross-modal synthesis (e.g., MRI-PET, EEG-fMRI, CT-MRI).

- managing variations in data quality, and
- aligning the scans correctly (image registration).

These limitations can seriously influence the accuracy of the neurological data (images), and the dependability of any diagnoses based on them. Table 14.1 reports on different challenges, implications, and solutions.

6. Visualization and explanation tools for neuroimaging

Generative models, including GANs, VAEs, and DMs, have shown marvelous power in neuroscience imaging, comprising disease progression modeling, data augmentation, and anomaly detection. Nevertheless, their black-box nature poses a critical barrier to the implementation of high-stakes fields, like healthcare (Mamo et al., 2024). Typically, in neuroscience, both interpretability and explainability are vital for: clinical trust/validation, error diagnosis, model refinement, and understanding pathophysiological mechanisms. Notwithstanding, progress in gAI model performance, explainability in medical gAI models remains underexplored (see main challenges in Table 14.2).

6.1 Visualization and explanation tools for neuroimaging

In neuroimaging, various strategies can be explored for making generative models interpretable as follows:

Table 14.1 Alignment challenges related to generative AI for neuroscience imaging and their solutions.

	Challenges	Implications	Solutions
Data quality and standardization	• Scanner variability • Patient movement	• Inconsistent data fusion • Bias in AI models	• Data standardization • Transfer learning • Synthetic data generation
Resolution mismatch across modalities	• Variations in spatial resolution • Data alignment challenges • Detail loss	• Inaccurate diagnosis due to improper tumor localization or an inaccurate evaluation of the severity of neurodegenerative diseases. • Reduced sensitivity	• Improving image quality • Working across different levels of detail • Matching image details point by point
Geometric alignment (registration) issues	• Differences in brain shape • Movement during scans • Different scan coverage	• Incorrect image matching • Losing the connection between function and structure	• AI-driven registration algorithms • Elastic and nonrigid registration • Motion correction
Noise and artifact handling	• Noise from PET and CT • Magnetic field inhomogeneity in MRI • EEG signal noise	• Degraded image quality • Misdiagnosis or inaccurate assessment	• AI noise reduction networks, such as denoising autoencoders or GANs • Artifact detection and correction • Regularization techniques, such as total variation denoising

- **Latent space probing** achieves control managements in latent space for observing output changes and aligning latent dimensions with clinical biomarkers or anatomical features.
- **Saliency and attention maps** are overlayed on generated images to display regions contributing to the generation and adaptation using CAM/Grad-CAM methods for

Table 14.2 Key challenges of interpretability and explainability in generative models.

Challenge	Description
Opaque latent spaces	The latent vectors in VAEs/GANs are hard to map to meaningful clinical variables
Lack of domain-specific explanations	Generic explainability methods often do not translate to neuroimaging contexts
Trust barriers for clinicians	Clinical users require transparency to trust synthetic data or model behavior

3D volumetric data. For example, saliency heatmaps overlaid on synthesized brain scans representing regions most influenced by latent variable (e.g., hippocampus shrinkage).

- **Counterfactual explanations** generate "what-if" samples by altering key variables (e.g., simulate increased atrophy) and help visualizing the disease trajectories or therapy impact.

Table 14.3 shows some tools for making gAI models interpretable with reporting their purpose and applications.

Moreover, building trust for generative models in clinical practice and research adoption includes (i) Transparency: by producing clear documentation of model training, biases, limitations, and interpretable evaluation metrics (e.g., structural similarity, uncertainty quantification), and (ii) Human-in-the-loop systems: by combining models with clinical expertise to validate output and allow interactive exploration of generated scans. Table 14.4 shows a checklist for deploying interpretable generative models in neuroimaging.

As gAI models become more powerful, their interpretability will be the deciding factor in whether they can be trusted for clinical decision support, biomarker discovery, or longitudinal neuroimaging analysis. In future, multimodal explainability frameworks combining imaging, genetics, and cognitive scores can be designed by leveraging causal models to interpret associations in created data and to develop domain-specific benchmarks for assessing explainability.

Table 14.3 Some tools for producing interpretable gAI models.

Tool	Purpose	Neuroimaging applications
Latent space navigation	Disentangle and explore feature control	• Age progression • Ventricle enlargement
Attention maps	Localize important regions in generation	• Visualize lesion impact • Cortical thinning
Counterfactuals	Simulate hypothetical changes	• Predict drug response • Disease progression

Table 14.4 Checklist for arraying interpretable generative models in neuroimaging.

Requirement	Explanation
Explainable interface	Visual tools for latent factors and brain region mapping
Validation pipelines	Comparison with known patient data and expert rating
Ethical safeguards	Clear labeling of synthetic versus real images, informed use

7. Generative AI-based benchmark datasets, simulation platforms, and collaborative infrastructures

In neuroscience and neuroimaging research, the rapid advancement of gAI demands robust and scalable infrastructures to support the deployment of models, development, and evaluation (Yang et al., 2026). These infrastructures span across four critical dimensions: benchmark datasets, simulation platforms, open-source collaborative ecosystems, and privacy-preserving frameworks.

7.1 Large, well-curated synthetic and real hybrid datasets

Generally, gAI models in brain imaging succeed on diverse and large datasets. However, real neuroimaging data is frequently limited by access constraints, heterogeneity, and privacy issues. Hybrid datasets that combine real scans with synthetically created data offer a compelling solution using synthetic data augmentation to fill gaps in demographic groups or diminished conditions. gAI models like GANs and DMs can mimic pathological progression of the brain, intersubject variability, or multimodal features. These datasets empower zero-shot inference, domain adaptation, and pretraining by presenting a controlled variety of samples that may not exist in real-world cohorts. To be effective, these datasets must be standardized, annotated, and representative, with rich metadata and harmonized formats that allow for benchmarking across different generative frameworks.

7.2 Brain simulation environments for controlled generative experimentation

Controlled trialing is essential for validating the behavior of gAI in complex brain systems. Emerging brain simulation environments deliver synthetic neuroimaging data with known ground-truths, offering a testbed for causal inference in brain connectivity or disease progression, synthetic data generation under varying anatomical, physiological, or noise conditions and Sim-to-real transfer, where generative models trained on simulated data are adapted to real-world scans.

Platforms such as Virtual Brain, BrainIAK, and domain-specific simulators are launched to support the simulation of neural dynamics and structural brain's properties at different scales. Integrating these environments with gAI pipelines will be crucial for designing interpretable, safe, and adaptable gAI systems.

7.3 Open-source ecosystems, model zoos, and reproducibility frameworks

Open science is a cornerstone of progress in gAI research. A flourishing ecosystem of open-source tools, pretrained model repositories ("model zoos") is considered vital for transparency and collaboration: Model zoos for neuroimaging (e.g., MONAI, NiftyNet, TorchIO) enable researchers to share and reprocess architectures tailored to brain data. Benchmarking platforms, like MedMNIST, OASIS-3, and OpenML, foster standardized evaluation protocols across gAI methods. Duplicability frameworks, including tools like MLflow, DVC, and Docker-based environments, ensure consistency in model training, evaluation, and deployment. These systems lower barriers to entry, accelerate cross-institutional development, and reduce redundant experimentation, ultimately enabling broader adoption of generative models in clinical settings. Table 14.5 reports the different categories, their descriptions, and examples of gAI datasets, simulation, and collaboration ecosystems.

7.4 Federated and privacy-preserving collaborative platforms

The inherently sensitive nature of brain data demands privacy-aware learning frameworks. FL and emerging paradigms like swarm learning in neuroimaging provide mechanisms for decentralized model training across multiple sites without sharing raw data: swarm learning integrates blockchain technology to orchestrate training while ensuring trust, auditability, and resilience. Such platforms preserve patients' privacy while leveraging data variety, which is serious for evolving robust gAI models. Privacy-enhancing technologies, including homomorphic encryption, differential privacy, and secure multi-party computation, are embedded within the FL pipelines.

In neuroimaging, FL frameworks are actively explored over enterprises, such as federated tumor segmentation challenge and COINSTAC. Such infrastructures permit joint training of gAI models for segmentation, harmonization, and/or synthesis, on distributed datasets, introducing new openings for real-world scalability. In brain science, gAI embraces transformative power, but its success is based on the accessibility of scalable, structured, and ethically administered infrastructures. The integration of federated frameworks, simulation environments, open-source tooling, and hybrid datasets, will allow researchers to design high-performing models which are comprehensive,

Table 14.5 Generative AI: datasets, simulation, and collaboration ecosystems.

Category	Description	Examples/Initiatives
Hybrid datasets (synthetic/real)	Emphasize the importance of large-scale, diverse, and high-quality datasets combining synthetic and real-world neuroimaging data for robust model training.	- Brain imaging data structure (BIDS)—UK Biobank—SynthSeg (for synthetic segmentation data)
Brain simulation platforms	Develop of virtual environments and simulators to model brain function and structure for generative experimentation and hypothesis testing.	- The Virtual Brain (TVB)—BrainSim—NEST Simulator—Blue Brain Project
Open-source ecosystems/Model zoos	Provide tools and repositories that support reproducibility, sharing of pretrained models, and community-driven improvements.	- MONAI (medical open network for AI)—Hugging Face Model Hub—BioImage.io—OpenNeuro
Federated/privacy-preserving platforms	Provide infrastructure platforms for enabling decentralized learning while preserving data privacy, especially critical for sensitive medical data.	- Swarm learning-NVIDIA Clara FL-COINSTAC (collaborative informatics and neuroimaging)

transparent, and reproducible. In neuroscience, deliberate investment and cross-disciplinary association in these areas will be energetic to solve challenges of gAI.

8. Open problems and grand challenges in generative AI for neuroimaging

Despite significant advancements in generative modeling, several open problems and grand challenges remain before these models can be robustly deployed in real-world neuroimaging applications.

8.1 Robustness to inter-site variability and scanner heterogeneity

Generative models are often trained on data from limited sites or specific scanners. Nevertheless, in neuroscience practice, neuroimaging data comes from a heterogeneous mix of scanners (e.g., Siemens, GE, Philips), acquisition protocols, and populations. This variability poses a substantial threat to model generalizability. Current models often overfit to site-specific distributions, leading to decreased performance on external datasets. Generative models can disentangle biological variability from scanner-induced artifacts by domain adaptation practices using adversarial training, site-invariant latent representations, and multi-site harmonized training protocols.

8.2 Longitudinal generative modeling and tracking brain development

Most current gAI models operate in a cross-sectional setting, however, neurological disorders, such as AD, PD, schizophrenia, and developmental disorders, evolve over time. Longitudinal modeling allows the simulation of brain changes across lifespan or disease progression. Temporal dynamics can be modeled in brain anatomy or function using generative models by Temporal VAEs or GANs with recurrent components, spatiotemporal consistency checks, and validation against longitudinal cohorts (e.g., ADNI, PING datasets).

8.3 Creating explainable, general-purpose generative neuro-models

While task-specific models exist (e.g., lesion synthesis, segmentation-aided synthesis), a general-purpose model that supports various downstream tasks (classification, prediction, augmentation) is still lacking. Additionally, these models must offer interpretable generation mechanisms to be trusted in scientific or clinical applications.

9. Ethical, legal, and social implications

In neuroimaging, gAI opens not only powerful promises but also significant legal, ethical, and societal concerns. As synthetic brain data begins to resemble real scans, the foundational issues around its use must be addressed as follows:

- Synthetic data ownership and accountability raise several concerns (see Table 14.6): Who owns the rights to synthetic neuroimages? If synthetic data is derived from patient datasets, does it inherit privacy constraints?
- Risks of synthetic brain image misuse, where synthetic brain images may be reverse engineered to infer sensitive attributes (e.g., age, sex, disease status), potentially violating anonymity. Advanced models might even allow partial identity reconstruction, especially when combined with demographic metadata.
- Developing ethical guidelines for clinical and research use, as there is an urgent need for international, cross-disciplinary frameworks that define permissible use-cases of synthetic data, establish data watermarking to trace source models and standardize disclosure requirements in academic publications.

Table 14.6 Different scenarios of data ownership.

Scenario	Ownership implication
Derived from public datasets	Usually free, but attribution needed
Trained on private hospital data	*Requires patient consent for derivative works*
Mixed-source training	Complex; depends on licensing and consent metadata

10. Future vision: Toward next generation of intelligent neuroimaging ecosystems

gAI is not just a tool for data augmentation but it represents a cornerstone of the next generation of intelligent neuroimaging ecosystems, where gAI not only supports but actively drives scientific discovery. Several future directions include intelligent gAI-assisted labs and self-evolving diagnostic tools, where gAI models in imaging labs can simulate disease progression, routinely refine hypotheses, and suggest experiments. A model generates multiple synthetic trajectories of MS lesions under different treatment protocols, guiding clinical trial design. Self-supervised generative agents integrated with EHR, PACS, and bioinformatics data. Other future directions may carry out collaboration between AI agents and neuroscientists in discovery, where a new paradigm is emerging where generative agents and neuroscientists collaborate interactively: agents propose neuroanatomical hypotheses based on population-level data and researchers validate traditional imaging or histology. For instance, gAI models predict a novel structural correlation of depression in the anterior cingulate that can be validated via fMRI and DTI follow-up. Also, the researchers are recommended to explore the ability of gAI models for hypothesis generation in cognitive neuroscience. Beyond diagnostics, generative models may simulate mental states, neuroplasticity phenomena, or cognitive strategies. Models that synthesize "what-if" scenarios, for example, how does language development change in a bilingual child with early frontal injury? Generative neuro-symbolic systems combining causal inference with imaging generation.

11. Conclusion

gAI is rapidly redefining neuroimaging by enabling novel approaches for brain simulation, disease modeling, and diagnostic enhancement. These models can synthesize anatomically accurate brain images, simulate longitudinal brain changes such as aging or disease progression, and adapt across scanner variations to support multi-site research. Key advancements include the emergence of interpretable, explainable generative neuro-models, the integration of self-supervised learning, and multimodal data fusion frameworks that combine anatomical, functional, and temporal information for more comprehensive brain mapping. Despite their promise, deploying gAI in clinical neuroimaging demands careful attention to ELSI. Concerns around synthetic data misuse, patient privacy, and model transparency necessitate active collaboration among neuroscientists, clinicians, ML experts, ethicists, and policymakers. Transparent model design, robust policy development, and responsible innovation are critical to building trust in these systems. One of the major research challenges is improving the interpretability and trustworthiness of gAI outputs. Clinical integration requires models to produce outputs that align with clinical reasoning and support decision-making processes. This involves designing evaluation metrics for explainability, developing mappings from AI-generated outputs to clinically relevant features, and mitigating bias in datasets and model training. Another persistent issue is the inherent complexity of neuroimaging data. These datasets are multidimensional, combining spatial, temporal, and modality-specific information. gAI models, especially when generating or modifying brain images, often act as black boxes, making it difficult for clinicians to validate outputs. To address this, real-time visualization and interpretability tools should be integrated into existing systems such as DICOM and PACS, allowing for dynamic feedback during image analysis and supporting tasks like anomaly detection and gAI-enhanced reconstructions.

Remaining challenges include handling scanner heterogeneity, developing robust longitudinal models for tracking brain development or disease, and ensuring generalizability across populations. Ethical concerns related to synthetic brain image misuse and data ownership also demand further investigation. Ultimately, the fusion of gAI and neuroscience holds the potential to shift from automation to cognitive collaboration—supporting clinicians not only in data generation but in hypothesis generation and diagnostic insight.

References

Albalawi, E., Thakur, M. T. R. A., Kumar, V. V., Gupta, M., Khan, S. B., & Almusharraf, A. (2024). Integrated approach of federated learning with transfer learning for classification and diagnosis of brain tumor. *BMC Medical Imaging, 24*(1). https://doi.org/10.1186/s12880-024-01261-0

Bermano, A. H., Gal, R., Alaluf, Y., Mokady, R., Nitzan, Y., Tov, O., Patashnik, O., & Cohen-Or, D. (2022). State-of-the-art in the architecture, methods and applications of StyleGAN. *Computer Graphics Forum, 41*(2), 591–611. https://doi.org/10.1111/cgf.14503

Bhat, R., & Nanjundegowda, R. (2025). A review on comparative analysis of generative adversarial networks' architectures and applications. *Journal of Robotics and Control (JRC), 6*(1), 53–64. https://doi.org/10.18196/jrc.v6i1.24160

Cao, H., Tan, C., Gao, Z., Xu, Y., Chen, G., Heng, P. A., & Li, S. Z. (2024). A survey on generative diffusion models. *IEEE Transactions on Knowledge and Data Engineering, 36*(7), 2814–2830. https // doi.org/10.1109/TKDE.2024.3361474

Gong, C., Jing, C., Chen, X., Pun, C. M., Huang, G., Saha, A., Nieuwoudt, M., Li, H. X., Hu, Y., & Wang, S. (2023). Generative AI for brain image computing and brain network computing: A review. *Frontiers in Neuroscience, 17*. https://doi.org/10.3389/fnins.2023.1203104

Helmy, E., Ahmed, E., ElNakieb, Y., Mohamed, K., Abdelrahim, M., Yousaf, J., Ghazal, M., Contractor, S., Barnes, G. N., & El-Baz, A. (2023). Role of artificial intelligence for autism diagnosis using DTI and fMRI: A survey. *Biomedicines, 11*(7), 1858. https://doi.org/10.3390/biomedicines11071858

Liu, P., Wang, L., Ranjan, R., He, G., & Zhao, L. (2022). A survey on active deep learning: From model driven to data driven. *ACM Computing Surveys, 54*(10s), 1–34. https://doi.org/10.1145/3510414

Mai, W., Zhang, J., Fang, P., & Zhang, Z. (2025). Brain-conditional multimodal synthesis: A survey and taxonomy. *IEEE Transactions on Artificial Intelligence, 6*(5), 1080–1099. https://doi.org/10.1109/TAI.2024.3516698

Mamo, A. A., Gebresilassie, B. G., Mukherjee, A., Hassija, V., & Chamola, V. (2024). Advancing medical imaging through generative adversarial networks: A comprehensive review and future prospects. *Cognitive Computation, 16*(5), 2131–2153. https://doi.org/10.1007/s12559-024-10291-3

Peer, M., Nitzan, M., Bick, A. S., Levin, N., & Arzy, S. (2017). Evidence for functional networks within the human brain's white matter. *Journal of Neuroscience, 37*(27), 6394–6407. https://doi.org/10.1523/JNEUROSCI.3872-16.2017

Sandhu, S. S., Gorji, H. T., Tavakolian, P., Tavakolian, K., & Akhbardeh, A. (2023). Medical imaging applications of federated learning. *Diagnostics, 13*(19). https://doi.org/10.3390/diagnostics13193140

van Breugel, B., Liu, T., Oglic, D., & van der Schaar, M. (2024). Synthetic data in biomedicine via generative artificial intelligence. *Nature Reviews Bioengineering, 2*(12), 991–1004. https://doi.org/10.1038/s44222-024-00245-7

Yamamoto, Y., Kanayama, N., Nakayama, Y., Matsushima, N., & Status, C. (2022). Issues and future prospects of personalized medicine for each disease. *Journal of Personalized Medicine, 12*(3), 444. https://doi.org/10.3390/jpm12030444

Yang, S., Huang, X., Bernardo, D., Ding, J.E., Michael, A., Yang, J., Kwan, P., Raj, A., & Liu, F. (2026). Foundation and large-scale AI models in neuroscience: A comprehensive review. https://doi.org/10.48550/arXiv.2510.16658.

Zhang, H., Li, H., Dillman, J. R., Parikh, N. A., & He, L. (2022). Multi-contrast MRI image synthesis using switchable cycle-consistent generative adversarial networks. *Diagnostics, 12*(4). https://doi.org/10.3390/diagnostics12040816

Index

'*Note:* Page numbers followed by "f" indicate figures and "t" indicate tables.'

H

I

J

K

L

M

P

Q

R

S

Printed and bound by CPI Group (UK) Ltd, Croydon, CR0 4YY
19/09/2026
02219501-0010